T0329814

MASS SPECTROMETRY ANALYSIS FOR PROTEIN–PROTEIN INTERACTIONS AND DYNAMICS

MASS SPECTROMETRY ANALYSIS FOR PROTEIN–PROTEIN INTERACTIONS AND DYNAMICS

Edited by

Mark Chance
Case Western Reserve University

WILEY

A John Wiley & Sons, Inc., Publication

Published by John Wiley & Sons, Inc., Hoboken, New Jersey
Published simultaneously in Canada

For general information on our other products and services or for technical support, please contact our Customer Care Department within the United States at (800) 762-2974, outside the United States at (317) 572-3993 or fax (317) 572-4002.

Wiley also publishes its books in a variety of electronic formats. Some content that appears in print may not be available in electronic formats. For more information about Wiley products, visit our web site at www.wiley.com.

Library of Congress Cataloging-in-Publication Data:

ISBN 978-0-470-25886-6

■ CONTENTS

Foreword   xiii

Contributors   xv

**1   Overview of Mass Spectrometry Technologies for Examining
     Protein Structure: Current and Future Directions**   1
     Shannon M. Swiatkowski and Mark R. Chance

     1.1   Introduction   1
     1.2   Hydrogen/Deuterium Exchange Mass Spectrometry   3
     1.3   Hydroxyl-Radical-Mediated Protein Footprinting   5
     1.4   Chemical Cross-linking   8
     References   8

**2   Hydrogen Exchange Mass Spectrometry:
     Principles and Capabilities**   11
     Sébastien Brier and John R. Engen

     2.1   The Chemistry of Hydrogen Exchange   11
           2.1.1   Principles of Proton Transfer   11
           2.1.2   Mechanisms of Backbone Amide Hydrogen Exchange   14
           2.1.3   Factors Affecting Hydrogen Exchange   15
     2.2   HX Mechanisms in Proteins   19
     2.3   Deuterium Incorporation into Proteins   23
           2.3.1   Continuous Labeling   23
           2.3.2   Pulse Labeling   25
           2.3.3   Other Labeling Strategies   26
     2.4   Measuring HX with Mass Spectrometry   26
           2.4.1   Global Versus Local Exchange   27
           2.4.2   Back Exchange   27
           2.4.3   Proteolysis Before MS   28
           2.4.4   Mass Measurements and Data Processing   29
     2.5   Capabilities of HX MS in Structural Biology   31
           2.5.1   Protein Folding Studies   32
           2.5.2   Quality Control   32
           2.5.3   Aid in Structure Elucidation   34
           2.5.4   Interactions and Dynamics   35

Acknowledgment   36
References   36

**3 Covalent Labeling Methods for Examining Protein Structure
   and Protein Interactions**   **45**
Keiji Takamoto and Janna Kiselar

3.1 Introduction   45
3.2 Chemistry of Hydroxyl Radical Footprinting   46
   3.2.1 Generation of Hydroxyl Radicals   47
   3.2.2 Reactions of Hydroxyl Radical Products: Nature of
          Amino Acid Modifications   48
   3.2.3 Relative Reactivity of Amino Acid Side Chains   49
   3.2.4 Principles of Hydroxyl Footprinting and Protein
          Integrity During Radiolysis   50
3.3 Mass Spectrometry Approaches for Quantitative
   Protein Footprinting   52
   3.3.1 Quantification of Peptide Oxidation Using LC–MS   52
   3.3.2 Confirmation of Peptide Identity and Determination
          of Modification Sites by MS/MS   54
3.4 Examples of Various Methods that Generate Hydroxyl Radicals
   in Solution to Examine Protein Structure   55
   3.4.1 Radiolytic Footprinting of Cytochrome *c*   55
   3.4.2 Fenton Hydroxyl Radical Footprinting   56
   3.4.3 Laser Photolysis of H_2O_2 57
   3.4.4 Radiolysis by High-Voltage Electric Discharge
          Within ESI Ion Source   57
   3.4.5 Synchrotron X-Ray Footprinting for Protein Complexes
          and Assembly Studies: Probing Arp2/3 Complex Activation
          by ATP and WASp Binding Proteins   58
3.5 The Future: Hybrid Approaches that Combine Experimental
   and Computational Data   62
References   63

**4 Complementary Methods for Structure Determination:
   Hydroxyl-Radical-Mediated Footprinting and Deuterium
   Exchange Mass Spectrometry as Applied to Serpin Structure**   **69**
Xiaojing Zheng and Patrick L. Wintrode

4.1 Introduction   69
4.2 Technical Comparison of Hydroxyl-Radical-Mediated
   Footprinting and H/D Exchange Methodologies   73
4.3 Structural Mass Spectrometry Data   76
   4.3.1 H/D Exchange Data   76
   4.3.2 Synchrotron Footprinting Data   76

4.4 Solvent Accessibility 79
4.5 Dynamics 84
4.6 Significance for Serpin Structure and Function 87
4.7 Summary 87
Acknowledgment 88
References 88

5 Deuterium Exchange Approaches for Examining Protein Interactions: Case Studies of Complex Formation 91
Elizabeth A. Komives

5.1 Interactions of Regulatory and Catalytic Subunits of Protein Kinase A 91
   5.1.1 Interaction of the Catalytic Subunit with a Pseudosubstrate 92
   5.1.2 Interaction of the Catalytic Domain with the RIα Regulatory Domain 92
   5.1.3 Combination Of H/D Exchange Data and Computational Docking 93
5.2 Allostery in Protein–Protein Interactions Revealed by H/D Exchange 95
   5.2.1 Allostery within the Regulatory Subunit Revealed by H/D Exchange 95
   5.2.2 Allostery in the Thrombin-Thrombomodulin Interaction 97
5.3 Interactions of the Inhibitor IκBα with the Transcription Factor NF-κB 97
   5.3.1 H/D Exchange of IκBα Reveals Partially Unfolded Regions 98
   5.3.2 H/D Exchange Reveals IκBα Folds upon Binding to NF-κB 98
References 101

6 Hydrogen/Deuterium Exchange Studies of Viruses 105
Sebyung Kang and Peter E. Prevelige Jr.

6.1 Overview of Virus Lifecycles 105
6.2 Structural Investigations of Viral Capsids 105
6.3 Dynamics of Viral Capsids 106
6.4 Hydrogen/Deuterium Exchange Studies of Virus Capsid Structure 107
   6.4.1 Bacteriophage P22 107
   6.4.2 HIV 109
   6.4.3 Brome Mosaic Virus 113
6.5 Hydrogen/Deuterium Exchange Studies of Viral Protein Dynamics 114
   6.5.1 Bacteriophage Phi-29 Scaffolding Protein 114
   6.5.2 Packaging Motor P4 from dsRNA Bacteriophages Phi-8 and Phi-12 117

6.6 Technical Aspects of Performing Hydrogen/Deuterium Exchange
Experiments on Viruses   118
   6.6.1 Dissociation of Structures   118
   6.6.2 Presence of Nucleic Acid   118
   6.6.3 Potential for Strain Variation   119
   6.6.4 Presence of BSA   119
   6.6.5 Complexity and Size   119
References   119

**7 Use of Enhanced Peptide Amide Hydrogen/Deuterium
Exchange-Mass Spectrometry (DXMS) in the Examination
of Protein–Protein Interactions   123**
*Yoshitomo Hamuro, Stephen J. Coales, Lora L. Hamuro,
and Virgil L. Woods Jr.*

7.1 Introduction   123
7.2 Theory of H/D Exchange   124
   7.2.1 Amide H/D Exchange   124
   7.2.2 Protection Factor   125
   7.2.3 Backbone Amide Hydrogens as Thermodynamic Sensors 125
   7.2.4 H/D Exchange for Protein–Protein Interactions   126
7.3 Overview of DXMS Technology for Protein–Protein Interactions   126
   7.3.1 On-Exchange Reaction   126
   7.3.2 Quench of Exchange Reaction   127
   7.3.3 Protein Fragmentation by Proteolysis   127
   7.3.4 HPLC Separation   128
   7.3.5 Mass Analysis   128
   7.3.6 DXMS of a Protein with or without Protein Binding Partner 128
7.4 DXMS of Human Growth Hormone and Its Binding Protein   129
   7.4.1 Human Growth Hormone High Affinity Variant   129
   7.4.2 DXMS Experiments of Human Growth Hormone
       and Its Binding Protein   130
   7.4.3 DXMS of hGHwt and hGHv without hGHbp   131
   7.4.4 DXMS of hGHwt and hGHv with hGHbp   131
   7.4.5 DXMS of hGHbp with or without hGH   132
   7.4.6 Enhanced Affinity by Increasing the Free Energy
       of the Unbound State   133
7.5 DXMS of PKA Regulatory Subunits   133
   7.5.1 Protein Kinase A (PKA) Regulatory Subunits   133
   7.5.2 DXMS Experiments of PKA Regulatory Subunits   134
   7.5.3 DXMS of cAMP-Bound PKA R-Subunits   134
   7.5.4 Interaction between R-Subunits and C-Subunit   136
   7.5.5 Interaction between R-Subunits and cAMP   138
   7.5.6 Lack of Significant Effects on the D/D Domain
       upon Binding to cAMP or C-Subunit   139

7.6 DXMS of PKA R-Subunit D/D Domains and
D-AKAP2 AKB Domain   139
  7.6.1 PKA R-Subunit D/D Domains and D-AKAP2
AKB Domain   139
  7.6.2 DXMS Experiments of PKA R-Subunit D/D Domains
and D-AKAP2 AKB Domain   140
  7.6.3 DXMS of D-AKAP2 AKB Domain with or
without PKA R-Subunit D/D Domains   141
  7.6.4 DXMS of PKA R-Subunit D/D Domains with or
without D-AKAP2 AKB Domain   143
7.7 Epitope Mapping by DXMS   146
  7.7.1 Epitope Mapping   146
  7.7.2 DXMS Experiments of Cytochrome c in the Presence
and Absence of Antibody   147
  7.7.3 Antibody Binding Site on Cytochrome c by DXMS   147
  7.7.4 Comparison with X-ray Crystallographic Structure   147
7.8 Conclusions   148
Abbreviations   149
Acknowledgments   150
References   150

**8 Cross-linking as a Tool to Examine Protein Complexes:
Examples of Cross-linking Strategies
and Computational Modeling   157**
Evgeniy V. Petrotchenko and Christoph H. Borchers

8.1 Introduction   157
8.2 Cross-linking Strategies   157
8.3 Cross-linking Methodology   158
8.4 Challenges Associated with Combining Cross-linking
with Mass Spectrometry   159
8.5 Advances in Mass Spectrometry Instrumentation and Capabilities   160
8.6 Novel Cross-linking Reagents for Mass Spectrometry
Applications   162
8.7 Analytical Software   165
8.8 Using Cross-linking Distance Constraints to Build
Experimental Models of Protein Complexes   167
References   167

**9 Complex Formation in the Actin Cytoskeleton: Cross-linking
Tools to Define Actin Protein Structure and Interactions   169**
Sabrina Benchaar and Emil Reisler

9.1 Introduction   169
9.2 Mapping Cross-linking with Methods Other than
Mass Spectrometry   171

9.3 Actin-Actin Cross-linking 171

  9.3.1 Intermolecular Cross-linking in F-Actin by N, N'-p-Phenylene-Dimaleimide between Lysine 191 and Cysteine 374 171

  9.3.2 Intermolecular Cross-linking in F-Actin by N-(4-Azidobenzoyl)-Putrescine between Glutamine 41 and Lysine 113 172

9.4 Intrastrand Cross-linked Actin between GLN41 and CYS374 174

9.5 Regulation of Cytoskeleton by ABPs and Mapping their Interfaces with Actin by Cross-linking 175

  9.5.1 Actin-Depolymerizing Factor/Cofilin 175

  9.5.2 Mapping the Interaction of Cofilin with Subdomain 2 on G-Actin 176

  9.5.3 Cofilin-Induced Switch from Intramolecular to Intermolecular Cross-linking in Skeletal F-Actin 178

  9.5.4 The Main Cofilin Binding Site on G-Actin 178

9.6 Cross-linking of Actin and Muscle Proteins—Examples of Experimental Approaches 181

9.7 Concluding Remarks 182

Acknowledgment 183

References 183

10 Computational Approaches to Examining Protein–Protein Interactions: Combining Experimental and Computational Data in the Era of Structural Genomics **189**

J.K. Amisha Kamal

10.1 Interactome in Structural Genomics 189

10.2 Importance of Computational Methods in Structural Genomics 190

10.3 Combining Computational Method with Experimental Data in Modeling the Structure of Protein Binary Complex 190

  10.3.1 General Strategy of the Method 191

  10.3.2 Docking Complexes of Known Crystal Structures without Using Footprinting Constraints 192

  10.3.3 Radiolytic Footprinting: G-Actin/GS1 and G-Actin/Cofilin 198

  10.3.4 Docking Complex of Known Crystal Structure Using Footprinting Constraints 201

  10.3.5 Docking Complex of Unknown Crystal Structure Using Footprinting Constraints 205

10.4 Method Summary 208

10.5 Experimental Methods 210

  10.5.1 Radiolysis 210

  10.5.2 Mass Spectrometry 210

  10.5.3 Solvent-Accessible Surface Area Calculation 210

10.5.4 Homology Modeling 211
10.5.5 Protein–Protein Docking 211
10.5.6 Docking with Experimental Constraints 211
10.5.7 Electrostatic Potential Surface Mapping and Calculation
of Interface Parameters 212
10.5.8 Footprinting Interface Consistency Score 212
Acknowledgment 212
References 213

**11 Studies of Intact Proteins and Protein Complexes:
ESI MS Approaches 217**
*Igor A. Kaltashov, Rinat R. Abzalimov, Agya K. Frimpong,
and Stephen J. Eyles*

11.1 Introduction 217
11.2 Tertiary Structure Integrity and Conformational Heterogeneity
(Charge State Distributions) 220
11.3 Quaternary Structure Integrity and Composition
of Non-Covalent Complexes 224
11.4 Functional Competence 226
11.5 Flexibility Maps and Binding Interfaces 229
11.6 Gas Phase Ion Chemistry and Its Influence on the Measurement
of Protein Properties in Solution 231
11.7 Challenges and Future Outlook 234
Acknowledgments 237
References 237

**12 Two Approaches to Mass Spectrometric Protein Footprinting:
PLIMSTEX and FPOP 243**
Michael L. Gross, Mei M. Zhu, and David M. Hambly

12.1 Introduction: Protein–Ligand Interactions by Mass Spectrometry,
Titration, and Hydrogen/Deuterium Amide Exchange
and Fast Photochemical Oxidation of Proteins 243
12.2 Protein–Ligand Interactions by Mass Spectrometry, Titration,
and H/D Amide Exchange (PLIMSTEX) 245
12.2.1 General Protocol for PLIMSTEX 245
12.2.2 Titration Curves 245
12.3 Applications of PLIMSTEX 247
12.3.1 Determination of Association Constant (K_a),
Stoichiometry (n), and Protection (ΔD_i) 247
12.3.2 Ras–GDP Interacting with Mg^{2+}: A 1:1 Protein:
Metal Ion Interaction 247
12.3.3 The Interactions of Apo-Calmodulin
with Ca^{2+}: A 1:4 Protein: Metal Ion Interaction 249

12.3.4 Applications in Biologically Relevant Media 250

12.3.5 The Interaction of Holo-CaM and Peptides 251

12.4 Self-Association of Insulin: A Protein/Protein Interaction 253

12.5 Features of PLIMSTEX 254

12.6 Fast Photochemical Oxidation of Proteins:
An Example of Fast Protein Footprinting 256

12.6.1 Hydroxyl Radicals as a Probe 256

12.6.2 Fast Hydroxyl-Radical Footprinting 258

12.6.3 Locating the Sites of Radical Reaction 258

12.6.4 Application of FPOP to Apomyoglobin 259

12.7 Features of FPOP 263

12.8 Future 264

Abbreviations 265

Acknowledgments 265

References 265

Index **271**

■ FOREWORD

This book would not have been possible without the efforts of scientists in the field who have labored to advance the field of structural mass spectrometry over the last several years. These efforts, and the germ of an idea that a book such as this was timely and possible, came together in January 2006 in the form of a mass spectrometry meeting, the 18th Sanibel Conference on Mass Spectrometry, titled "Focus on Biomolecular Structure, Dynamics and Function: Hydrogen Exchange and Covalent Labeling Techniques", organized by Igor Kaltashov and John Engen. At this three-day meeting, experts in mass spectrometry, many of which are authors of chapters in this book, exchanged the latest ideas related to understanding protein structure and dynamics, and found that mass spectrometry based approaches were converging on a common goal: to fill gaps in our understanding of protein structure and conformational dynamics, built on a firm foundation of high resolution structure data.

This group has since that time strengthened their interactions, forming a Hydrogen Exchange and Covalent Labeling interest group within the American Society for Mass Spectrometry; this group has grown in two short years to over 750 members. This book provides a milestone in the efforts of this group to present the state-of-the art in their field and disseminate that art as widely as possible. The future of this field looks very bright indeed.

On a personal note, I wish to thank all of my co-authors, I earnestly hope they find their efforts to be rewarded in this volume. I also wish to acknowledge the outstanding editing assistance from Shannon Swiatkowski and the support of all the faculty and staff of the Case Center for Proteomics, whose single minded pursuit of excellence makes being Center director a very rewarding occupation.

Mark Chance
Cleveland, Ohio
May 2008

■ CONTRIBUTORS

Rinat R. Abzalimov
Department of Chemistry
University of Massachusetts at Amherst
710 North Pleasant Street
Lederle Graduate Research Tower 701
Amherst, MA 01003, USA
E-mail: abzalimov@nsm.umass.edu

J. K. Amisha Kamal
Center for Proteomics
Case Western Reserve University
10900 Euclid Avenue
Cleveland, OH 44106-4988, USA
E-mail:

Sabrina Benchaar
Department of Chemistry
and Biochemistry
Molecular Biology Institute
University of California
Los Angeles, CA 90095, USA
E-mail: sabbenchaar@yahoo.fr

Christoph H. Borchers
University of Victoria-Genome BC
Proteomics Center
3101-4464 Markham Street
Vancouver Island Technology Park
Victoria V8Z 7X8, BC, Canada
E-mail: christoph@proteincentre.com

Sébastien Brier
Department of Chemistry
and Chemical Biology
The Barnett Institute of Chemical
& Biological Analysis

Northeastern University
Boston, MA 02115, USA
E-mail: s.brier@hotmail.fr

Mark R. Chance
Center for Proteomics
Case Western Reserve University
10900 Euclid Avenue, BRB 930
Cleveland, OH 44106-4988, USA
E-mail: Mark.chance@case.edu

Stephen J. Coales
ExSAR Corporation
11 Deer Park Drive, Suite 103
Monmouth Junction, NJ 08852, USA
E-mail: scoales@exsar.com

John R. Engen
341 Mugar Life Sciences
The Barnett Institute
Northeastern University
360 Huntington Avenue
Boston, MA 02115-5000, USA
E-mail: j.engen@neu.edu

Stephen J. Eyles
Department of Polymer Science
& Engineering,
University of Massachusetts at
Amherst
710 North Pleasant Street
Lederle Graduate Research
Tower 701
Amherst, MA 01003, USA
E-mail: eyles@polysci.umass.edu

Agya K. Frimpong
Department of Chemistry
University of Massachusetts at Amherst
710 North Pleasant Street
Lederle Graduate Research Tower 701
Amherst, MA 01003, USA
E-mail: afrimpong@chem.umass.edu

Michael L. Gross
Department of Chemistry
Washington University in St. Louis
One Brookings Drive, Box 1134
St. Louis, MO 63130, USA
E-mail: mgross@wustl.edu

David Hambly
Amgen Inc.
1201 Amgen Court West
Seattle, WA 98119-3105, USA
E-mail: dhambly@amgen.com

Lora L. Hamuro
Provid Pharmaceutical
671 US Route 1
North Brunswick, NJ 08902, USA
E-mail:
lora.hamuro@providpharma.com

Yoshitomo Hamuro
ExSAR Corporation
11 Deer Park Drive, Suite 103
Monmouth Junction, NJ 08852, USA
E-mail: yhamuro@exsar.com

Igor A. Kaltashov
Department of Chemistry
University of Massachusetts at
Amherst
710 North Pleasant Street
Lederle Graduate Research Tower 701
Amherst, MA 01003, USA
E-mail: kaltashov@chem.umass.edu

Sebyung Kang
Chemistry and Biochemistry
Building RM 116

Department of Chemistry and
Biochemistry
Montana State University
Bozeman, MT, 59717, USA
E-mail:
Sabsab7@chemistry.montana.edu

Janna Kiselar
Center for Proteomics
Case Western Reserve University
10900 Euclid Avenue, BRB 934
Cleveland, OH 44106-4988, USA
E-mail: Janna.kiselar@case.edu

Elizabeth A. Komives
Department of Chemistry
and Biochemistry
University of California San Diego
9500 Gilman Drive
La Jolla, CA 92093-0378, USA
E-mail: ekomives@ucsd.edu

Evgeniy V. Petrotchenko
University of Victoria-Genome BC
Proteomics Center
Department of Biochemistry
and Microbiology
University of Victoria
Victoria V8Z 7X8, BC, Canada
E-mail: jenya@proteincentre.com

Peter E. Prevelige Jr.
Department of Microbiology
University of Alabama
at Birmingham
Birmingham, AL 35294, USA
E-mail: prevelig@uab.edu

Emil Reisler
Department of Chemistry
and Biochemistry
Molecular Biology Institute
University of California
Los Angeles, CA 90095, USA
E-mail: reisler@mbi.ucla.edu

Shannon M. Swiatkowski
Center for Proteomics
Case Western Reserve University
10900 Euclid Avenue,
BRB 9th Floor
Cleveland, OH 44106-4988, USA
E-mail:
shannon.swiatkowski@case.edu

Keiji Takamoto
Center for Proteomics
Case Western Reserve University
10900 Euclid Avenue, BRB 934
Cleveland, OH 44106-4988, USA
E-mail: keiji.takamoto@case.edu

Patrick L. Wintrode
Department of Physiology
and Biophysics
Case Western Reserve University
10900 Euclid Avenue
Cleveland, OH 44106-4988, USA
E-mail: patrick.wintrode@case.edu

Virgil L. Woods Jr.
Department of Medicine
and Biomedical Sciences
University of California San Diego
Basic Science Building, Room 4011
9500 Gilman Drive, Dept 0656
La Jolla, CA 92093-0656, USA
E-mail: vwoods@ucsd.edu

Xiaojing Zheng
Case Center for Proteomics
Case Western Reserve University
10900 Euclid Avenue
BRB 9th Floor
Cleveland, OH 44106-4988, USA
E-mail:
Xiaojing.zheng@case.edu

Mei M. Zhu
Millennium Pharmaceuticals, Inc.
40 Landsdowne Street
Cambridge, MA 02139, USA
E-mail: May.Zhu@mpi.com

Overview of Mass Spectrometry Technologies for Examining Protein Structure: Current and Future Directions

SHANNON M. SWIATKOWSKI and MARK R. CHANCE

Center for Proteomics, Case Western Reserve University, Cleveland, OH, USA

1.1 INTRODUCTION

Understanding the molecular structure and dynamics of macromolecules at high resolution and with high throughput is a topic of great importance in biology. Nuclear magnetic resonance (NMR) and crystallographic approaches are the foundation of rapid progress in this area. Access to genome sequences and cloning resources from an ever-increasing number of organisms and allied high-throughput structure and modeling studies are likely to enable resolution of the structure of most protein domains in the near future (Chance et al., 2004). However, the machinery of eukaryotic cell biology involves multidomain proteins that interact in large complexes as molecular machines (Sali et al., 2003; Russell et al., 2004). Understanding how these domains interact is crucial in understanding their function. As this "database" of structural information evolves and develops, examination of the structure–function relationships of a wide range of proteins becomes possible. In addition, many biological questions of interest invoke questions of protein dynamics, ligand binding, complex formation, or the structural effects of posttranslational modifications. Many of these experiments are beyond the range of classical structural biology approaches (see below) and structural mass spectrometry (MS) methods have been very successful in filling this technological gap. The fundamental contributions of mass spectrometry to structural biology studies have grown dramatically due to increases in instrument sensitivity and resolution that have accrued over the past 10 years. This has advanced our ability to reliably sequence and identify protein

Mass Spectrometry Analysis for Protein–Protein Interactions and Dynamics, Edited by Mark Chance
Copyright © 2008 John Wiley & Sons, Inc.

fragments and their modified products, a feature on which structural mass spectrometry fundamentally relies. This book catalogs the state of the art in these approaches and provides a perspective on the future prospects for the field. The three main technologies of structural mass spectrometry that have rapidly evolved and grown, include covalent labeling strategies, hydrogen–deuterium (H/D) exchange, and chemical cross-linking.

Although the technologies have a great many differences in their sample preparation, instrumentation requirements, and other details of the approaches, their similarities must not be overlooked. First, they all rely on detailed identification and sequencing of peptide fragments generated by specific or nonspecific cleavage of intact and (generally) purified protein species (or complexes). Second, they infer structural information based on a mass shift of these peptide species after exposure to the labeling reagents of choice. The target atoms that are labeled must be solvent accessible, at least transiently. Third, the value of the structural information is greatly enhanced by having a structural model of the protein or proteins. It is, in fact, very clear that the advancement of these approaches will be significantly accelerated by a union of these experimental technologies with computational modeling approaches in the context of the rapidly expanding structure databases (Chance et al., 1997; Guan et al., 2004; Kamal and Chance, 2008; Takamoto and Chance, 2006).

Structural models for most protein domains, providing a foundation for structural mass spectrometry, are accumulating rapidly (Eswar et al., 2007). Advances in protein structure determination and computational modeling mediated by structural genomic initiatives throughout the world promise to correlate sequence and structure for most protein domains within the next 5 years (Burley et al., 1999; Chance et al., 2002, 2004). Coincident with progress toward this milestone is the realization of the importance of macromolecular interactions and even the fundamental significance of large macromolecular complexes mediating most normal and aberrant biological functions (Gavin, 2005). Solving the structure and connecting it to function for these large complexes are two of the most important challenges in structural biology today. Unlike solving the structure of protein domains or short nucleic acids that contain tertiary structure, this effort is far from high throughput and likely involves a combination of computational and experimental approaches, tailored specifically to the problem at hand.

The barriers to determining the structure and dynamics of proteins and their complexes include known limitations in crystallography and NMR technologies. Issues such as complex size, crystallizability, solubility, and amounts of materials are well known. In recent years, electron microscopy (EM) and tomography techniques, particularly at low temperatures, have substantially improved and are making important contributions to determining the structure of complexes (Sali et al., 2003). These approaches have resolution limitations for many samples and are better for larger complexes or cells due to sample dose tissues. This leaves a gap in technological progress for the "medium" size complexes, particularly medium-sized binary complexes (50–200 kDa). This has spurred the development of a host of computational methods that can fill in the gap and contribute to understanding the relationship between protein structure and function.

1.2   HYDROGEN/DEUTERIUM EXCHANGE MASS SPECTROMETRY

Deuterium exchange is a very powerful technique in the repertoire of structural mass spectrometry. H/D exchange MS methods were developed in the early 1990s, inspired by related NMR methods. The practicality of the method lies in the fact that amide hydrogens are sensitive probes for solvent accessibility, protein lability, and protein secondary structure. The H/D exchange method is shown in Fig. 1.1 (and compared with hydroxyl radical footprinting, a covalent labeling approach). The protein backbone amide hydrogens are exchangeable with deuterium atoms from the solvent surrounding the protein at different measurable exchange rates. The amide hydrogens at the surface of proteins exchange very rapidly, while amide hydrogens that are buried or are participating in stable hydrogen bonds have much slower exchange rates (Busenlehner and Armstrong, 2004). Thus, H/D exchange rates can be measured along the entire length of the protein backbone, providing a comprehensive measure of protein structure and solvent accessibility. Since backbone amide hydrogens are also involved in the formation of hydrogen bonds in protein secondary structures, their exchange rates are also a reflection of secondary structure and structural stability.

The protein of interest is subjected to a pulse of deuterium intended to label structural regions that are solvent accessible and to monitor changes in accessibility in response to the binding of a ligand (Katta and Chait, 1993; Zhang and Smith, 1993). After solvent labeling, the reaction is quenched, the protein is fragmented by proteolysis, the peptide fragments are separated by high-pressure liquid chromatography (HPLC), and mass spectrometry analysis is performed. Peptide fragments with increased mass relative to control experiments without addition of deuterium indicate *specific segments* that were solvent accessible and exchange competent during the deuterium pulse. The method used to minimize back exchange during the analytical steps is lowering the pH to ~2.5; thus, only proteases with activity at acid pH (e.g., pepsin) can be used to fragment the protein.

Chapter 2 describes the fundamental concepts that govern the hydrogen exchange (HX) reaction beginning with the chemistry of hydrogen exchange and continuing through to discuss HX mechanisms in proteins and how they can be assessed with mass spectrometry. These concepts build a foundation for a discussion of basic HX MS methodology and how it can be applied to various biological problems in subsequent chapters of this book. Similarly, Chapter 12 outlines PLIMSTEX (protein–ligand interactions in solution by mass spectrometry, titration, and H/D exchange). This strategy can be used to determine the conformational change, binding stoichiometry, and affinity for a variety of protein–ligand interactions. Chapters 5–7 provide a more comprehensive look at protein complex structure through several case studies of complex formation. In Chapter 5, HX MS data are used in combination with docking, biochemical, and genetic data to better understand the biophysics of protein–protein interactions in protein kinase A (PKA) and nuclear factor kappa B (NF-κB). Hydrogen/deuterium exchange (HDX) studies of protein complex formation can also be extended to viruses as explained in Chapter 6, where HX MS data of viral capsid structure and protein dynamics studies are presented

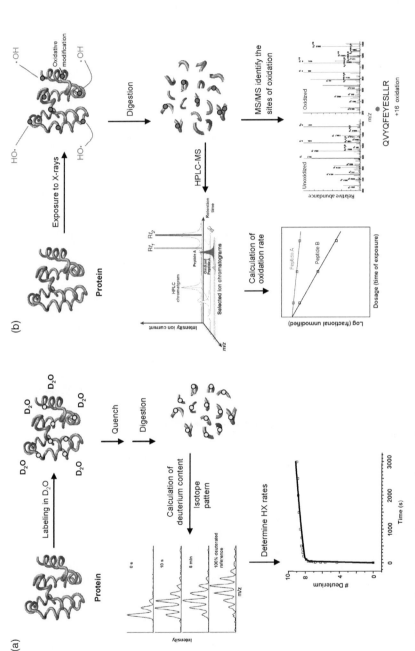

FIGURE 1.1 Schematic representations of synchrotron footprinting and H/D exchange mass spectrometry techniques. (a) By changing the solvent from H_2O to D_2O, amide protons in the backbone of protein exchange with the protons from the solvent. The exchange reaction is quenched by quickly changing solution pH to approximately 2.4 and usually combining with rapid freezing. Typically, the protein is digested with pepsin. The kinetics of amide H/D exchange is measured by mass spectrometry to provide essential dynamic information for protein. (b) When a protein is exposed to synchrotron X-rays, the hydroxyl radicals generated from water will modify side chains of the protein. After the X-ray exposure, the protein samples are digested by proteases and solvent accessibility information is provided by MS. The particular modification sites are determined by tandem MS and the side chain reactivity is accurately measured by quantitative liquid chromatography-coupled mass spectrometry. Figure reproduced, with permission from Elsevier Ltd., from Zhang et al., 2008.

4

along with several technical aspects of performing HX MS experiments on viruses. Chapter 7 highlights four examples of the use of enhanced peptide amide hydrogen/deuterium exchange coupled with proteolysis, liquid chromatography, and mass spectrometry (DXMS) to probe protein–protein interactions. The DXMS method can be used to complement thermodynamic information of a protein–protein interaction, monitor conformational changes associated with kinase activation, compare isoform-specific differences in binding of a common ligand, and map the epitopes of monoclonal antibodies. Chapter 4 provides a specific comparison of hydrogen/deuterium exchange and covalent labeling on the same protein, the first such direct comparison. It highlights the complementarity of the methods. Finally, Chapter 11 concludes the discussion of this method by explaining the many advantages of electrospray ionization (ESI) and matrix-assisted laser desorption/ionization (MALDI) MS compared to other means of monitoring the progress of hydrogen/deuterium exchange reactions.

1.3 HYDROXYL-RADICAL-MEDIATED PROTEIN FOOTPRINTING

The development of the hydroxyl radical as a modification reagent for footprinting and its application in conjunction with mass spectrometry were directly inspired by the development of deuterium exchange mass spectrometry methods. Hydroxyl-radical-mediated protein footprinting is similar. The overall method is outlined in Fig. 1.2. In this case, the protein solution is exposed to ionizing radiation and the hydroxyl radicals covalently react with surface-accessible residues, primarily side chain groups (Maleknia et al., 2001). As in the deuterium exchange methods, the protein is subjected to proteolysis. However, in contrast to deuterium exchange, the production of stable modifications through hydroxyl radical exposure allows a wide range of samples as well as proteases to be used to fragment the protein under a wide range of solution conditions and pH values. Also, the stable modification of side chains allows a specific probe site to be identified using tandem mass spectrometry methods, while, for deuterium exchange, typically the conformational change can only be localized to the specific peptide fragment. The drawback is that if a reactive side chain is not present in a particular peptide segment, there are no probes. However, the examination of side chains is complementary to the deuterium exchange method that examines backbone structure.

To generate the limited dose required for footprinting and to quantitatively examine the reactivity of the specific peptides in question, a series of samples are exposed to variable doses, the samples are digested, and the individual peptides are analyzed by HPLC and mass spectrometry. Thus, a dose–response curve is generated for each peptide of interest; this generates a quantitative biophysical measure (based on the observed rate of modification) of the relative reactivity of the sites in the different peptides. Consistent and reliable quantitation, which is essential to footprinting, is provided by measuring the relative amounts of the modified and unmodified peptide products in the same experiment. Since the modifications are stable, it is relatively straightforward to use tandem mass

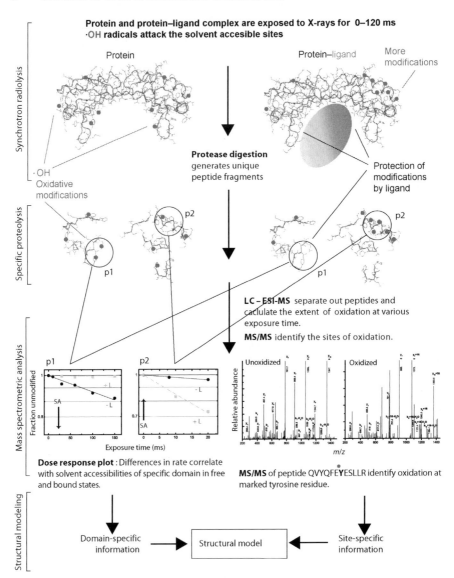

FIGURE 1.2   Schematic representation of protein footprinting using synchrotron radiolysis and mass spectrometry. The examples emphasize the protection formed in the interface of a protein–ligand complex as well as allosteric conformation changes that can result in increases in reactivity upon ligand binding; however, the comparison could be for any two (or multiple) functional states of the protein of interest. Two sets of samples, one free protein and the other a protein–ligand complex, are exposed to X-rays for different time intervals. The exposed samples are digested with specific digestion enzymes. The digested fragments are analyzed by ESI-MS to quantitate the extent of modification products and determine the fraction "unmodified" for a specific exposure time. A plot of fraction unmodified versus exposure time, known as the dose response plot, fit to a first-order function providing the rate of modification for the specific peptide. Comparisons of the dose response of the same peptide under different conditions provide structural information about ligand binding. MS/MS is used to determine the specific

spectrometric methods to specifically identify the amino acid positions of the protein that have been modified; these represent the probe sites for the analysis (Guan et al., 2002; Kiselar et al., 2002).

On the right-hand side of Fig. 1.2, where the reaction is carried out in the presence of a protein–protein complex, modification is suppressed at the site corresponding to the interacting surface and unchanged at sites distal to the contact. The decrease of reactivity of peptides in the interface is quantitatively measured using the dose–response curves; a sample dose–response for a peptide whose reactivity is suppressed in the presence of the ligand is shown at the bottom left of the figure. It is critical for the method that the dose–response curves indicate a linear regime extrapolated to zero fraction modified; this ensures the reactivities of particular sites are not changing due to the oxidation process itself. Also, the method of examining the loss of the unmodified fraction emphasizes the interrogation of intact material. Although these particular refinements are unique to the hydroxyl-radical-mediated protein footprinting approach, they are derived from a detailed knowledge and respect for safeguards that have evolved throughout the long history of development of "footprinting" research (Brenowitz et al., 2002; Takamoto and Chance, 2004); such safeguards ensure that the structural and biochemical information provided by footprinting methods is reliable. It must be emphasized, however, that footprinting (along with H/D exchange) provides only "local" information about the reactivity of the side chain probes. Allosteric changes in conformation induced by ligand binding can also give rise to either protections (decreases in side chain reactivity) or enhancements (increases in reactivity) depending on the induced conformational changes. This must be carefully borne in mind when interpreting data from these experiments, and various examples are mentioned in this review.

Chapter 3 reviews the chemistry of covalent labeling approaches for structural mass spectrometry and provides details of the various methods used to generate reactive species and define their reactions with proteins. Specifically discussed is the hydroxyl radical footprinting methodology and mass spectrometry approaches for quantitative footprinting. The complementarity of the hydroxyl-radical-mediated footprinting and hydrogen/deuterium exchange mass spectrometry techniques is discussed in Chapter 4. This chapter compares the structural results of the two methods on the trypsin inhibitor and also explores the ability of these methods to probe protein conformational dynamics. Chapter 10 provides a thorough discussion on the subject of using computational approaches to examine protein–protein interactions through the combination of computational and experimental data. This chapter specifically describes how a combination of radiolytic footprinting coupled to mass spectrometry analysis and docking with the ClusPro server can be used to derive a structure for the actin/cofilin binary complex. To obtain a "snapshot" of a protein that is uninterrupted by oxidation-induced protein unfolding, Chapter 12

---

modification site within the peptide and provide side-chain-specific structural resolution. Synchrotron protein footprinting data are often used as one of the constraints in model building for complexes. Figure reproduced, with permission from the International Union of Crystallography, from Gupta et al., 2007.

describes a fast radical footprinting technique termed fast photochemical oxidation of proteins (FPOP). This approach ensures that there is sufficient protein to be oxidized and then analyzed by employing a flow system coupled with a pulsed laser to produce OH radicals from hydrogen peroxide.

1.4   CHEMICAL CROSS-LINKING

Chemical cross-linking is also a valuable tool for examining higher-order structure of proteins and protein complexes. Many types of reagents are used in cross-linking experiments and can be used to cross-link multiple kinds of residues. Strategies to insert specific susceptible residues in sites of interest are also valuable. Cross-linking methodology is able to provide or confirm low-resolution structures when some model structure is available (Novak and Giannakopulos, 2007). Variations in the method (top-down versus bottom-up) are outlined in the book, and further potential applications of chemical cross-linking of proteins, as well as combinations with other techniques such as hydrogen/deuterium exchange and molecular modeling, are suggested.

Chapter 8 introduces the cross-linking methodology along with several strategies of the technique and how it can be used in conjunction with mass spectrometry to obtain structural information regarding the organization and function of protein complexes that may be otherwise impossible to obtain. Chapter 9 provides a specific example of how cross-linking can be used as a tool to probe actin structure and its interactions with actin binding proteins. It focuses on several examples of actin complexes with cytoskeletal proteins and considers some approaches that facilitate the mapping of cross-linked peptides by mass spectrometry. Finally, the book outlines HDX of intact proteins, an excellent tool for examining protein dynamics (Chapter 11) and methods for deriving biophysical data on ligand association (Chapter 12).

REFERENCES

Busenlehner, L. S. and Armstrong, R. N., 2004. Insights into enzyme structure and dynamics elucidated by amide H/D exchange mass spectrometry. *Arch Biochem Biophys* 433, 34–46.

Burley, S. K., Almo, S. C., Bonanno, J. B., Capel, M., Chance, M. R., Gasterland, T., Lin, D., Sali, A., Studier, F. W., and Swaminathan, S., 1999. Structural genomics: beyond the human genome project. *Nat Genet* 23, 151–157.

Brenowitz, M., Chance, M. R., Dhavan, G., and Takamoto, K., 2002. Probing the structural dynamics of nucleic acids by quantitative time-resolved and equilibrium hydroxyl radical 'footprinting'. *Curr Opin Struct Biol* 12, 648.

Chance, M. R., Sclavi, B., Woodson, S. A., and Brenowitz, M., 1997. Examining the conformational dynamics of macromolecules with time-resolved synchrotron X-ray 'footprinting'. *Structure* 5, 865–869.

Chance, M. R., Bresnick, A. R., Burley, S. K., Jiang, J. S., Lima, C. D., Sali, A., Almo, S. C., Bonanno, J. B., Buglino, J. A., Boulton, S., Chen, H., Eswar, N., He, G., Huang, R., Ilyin, V.,

McMahan, L., Pieper, U., Ray, S., Vidal, M., and Wang, L. K., 2002. Structural genomics: a pipeline for providing structures for the biologist. *Protein Sci* 11, 723–738.

Chance, M. R., Fiser, A., Sali, A., Pieper, U., Eswar, N., Xu, G., Fajardo, J. E., Radhakannan, T., and Marinkovic, N., 2004. High-throughput computational and experimental techniques in structural genomics. *Genome Res* 14, 2145–2154.

Eswar, N., Webb, B., Marti-Renom, M.A., Madhusudhan, M.S., Eramian, D., Shen, M.Y., Pieper, U., and Sali, A., 2007. Comparative protein structure modeling using MODELLER. *Curr Protoc Protein Sci* Chapter 2, Unit 2.9.

Gavin, A. C., 2005. Keystone symposia: proteomics and bioinformatics and systems and biology. *Expert Rev Proteomics* 2, 291–293.

Guan, J. Q., Vorobiev, S., Almo, S. C., and Chance, M. R., 2002. Mapping the G-actin binding surface of cofilin using synchrotron protein footprinting. *Biochemistry* 41, 5765.

Guan, J. Q., Almo, S. C., and Chance, M. R., 2004. Synchrotron radiolysis and mass spectrometry: a new approach to research on the actin cytoskeleton. *Acc Chem Res* 37, 221–229.

Gupta, S., Sullivan, M., Toomey, J., Kiselar, J., and Chance, M.R., 2007. The Beamline X28C of the Center for Synchrotron Biosciences: a national resource for biomolecular structure and dynamics experiments using synchrotron footprinting. *J Synchrotron Radiat* 14, 233–243.

Kamal, J. K. and Chance, M. R., 2008. Modeling of protein binary complexes using structural mass spectrometry data. *Protein Sci* 17, 79–94.

Katta, V. and Chait, B. T., 1993. Hydrogen/deuterium exchange electrospray ionization mass spectrometry: a method for probing protein conformational changes in solution. *J Am Chem Soc* 115, 6317.

Kiselar, J. G., Maleknia, S. D., Sullivan, M., Downard, K. M., and Chance, M. R., 2002. Hydroxyl radical probe of protein surfaces using synchrotron X-ray radiolysis and mass spectrometry. *Int J Radiat Biol* 78, 101.

Maleknia, S. D., Ralston, C. Y., Brenowitz, M. D., Downard, K. M., and Chance, M. R., 2001. Determination of macromolecular folding and structure by synchrotron X-ray radiolysis techniques. *Anal Biochem* 289, 103.

Novak, P. and Giannakopulos, A., 2007. Chemical cross-linking and mass spectrometry as structure determination tools. *Eur J Mass Spectrom* 13 (2), 105–113.

Russell, R. B., Alber, F., Aloy, P., Davis, F. P., Korkin, D., Pichaud, M., Topf, M., and Sali, A., 2004. A structural perspective on protein–protein interactions. *Curr Opin Struct Biol* 14, 313–324.

Sali, A., Glaeser, R., Earnest, T., and Baumeister, W., 2003. From words to literature in structural proteomics. *Nature* 422, 216–225.

Takamoto, K. and Chance, M. R. In: Myers, R. A., editor. *Encyclopedia of Molecular Cell Biology and Molecular Medicine*. 2nd ed. Weinheim: Wiley VCH; 2004, 521–548.

Takamoto, K. and Chance, M. R., 2006. Radiolytic protein footprinting with mass spectrometry to probe the structure of macromolecular complexes. *Annu Rev Biophys Biomol Struct* 35, 251–276.

Zhang, Z. and Smith, D. L., 1993. Determination of amide hydrogen exchange by mass spectrometry: a new tool for protein structure elucidation. *Protein Sci* 2, 522.

Zheng, X., Wintrode, P. L., and Chance, M. R., 2008. Complementary structural mass spectrometry techniques reveal local dynamics in functionally important regions of a metastable serpin. *Structure* 16, 38–51.

CHAPTER 2

Hydrogen Exchange Mass Spectrometry: Principles and Capabilities

SÉBASTIEN BRIER and JOHN R. ENGEN

Department of Chemistry and Chemical Biology, The Barnett Institute of Chemical & Biological Analysis, Northeastern University, Boston, MA, USA

2.1 THE CHEMISTRY OF HYDROGEN EXCHANGE

Hydrogen exchange (HX) detected by mass spectrometry (MS) is an extremely valuable method for understanding proteins. The hydrogen exchange reaction itself, which has been understood by examining the exchange behavior of small amide models and peptide analogues, imposes specific limits on the overall HX MS method. In this chapter, the fundamental concepts that govern the hydrogen exchange reaction will be described. These concepts build a foundation for a discussion of basic HX MS methodology and its application to various biological problems. Examples of application of the method to specific problems will be provided in subsequent chapters of this book.

2.1.1 Principles of Proton Transfer

Hydrogen exchange between dissolved macromolecules and water is driven by the strong base OH^- and the strong acid H_3O^+. The proton transfer reaction can be described in three steps (Fig. 2.1a): (i) formation of a hydrogen-bonded complex via diffusional collisions between a proton donor (A–H) and an acceptor (B); (ii) rapid equilibrium redistribution of the proton between the donor and the acceptor within the complex; (iii) dissociation of the hydrogen-bonded complex (Eigen, 1964). The transfer is productive if the proton is carried away by the acceptor and unproductive if

Mass Spectrometry Analysis for Protein–Protein Interactions and Dynamics, Edited by Mark Chance
Copyright © 2008 John Wiley & Sons, Inc.

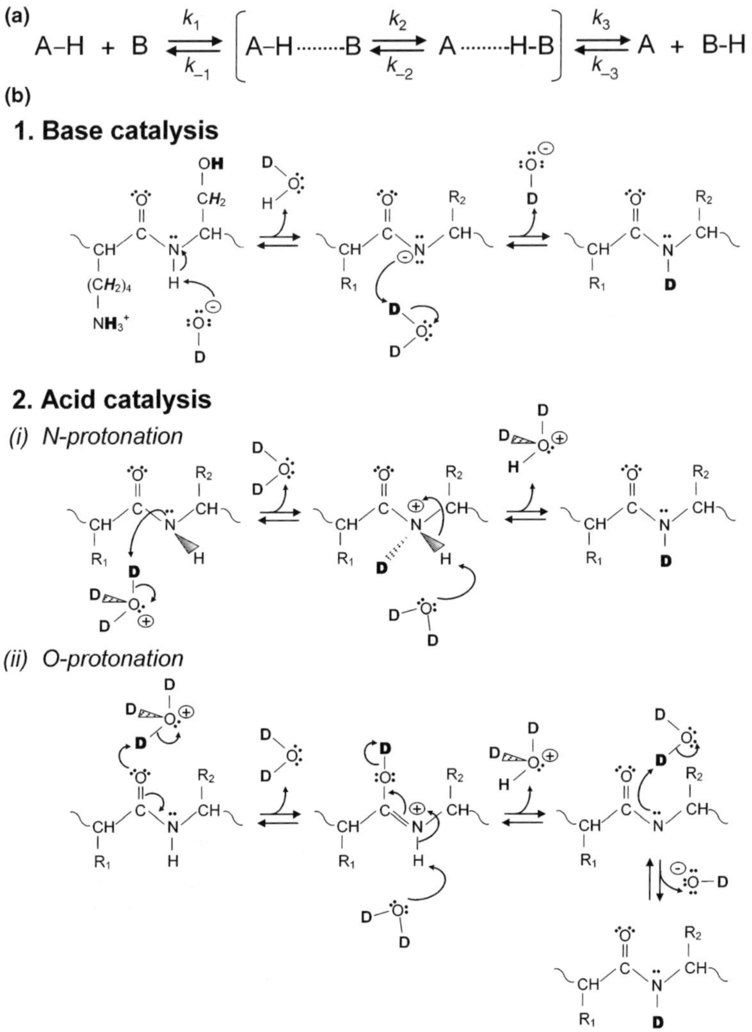

FIGURE 2.1   Chemistry of hydrogen exchange. (a) Proton transfer between a donor (A–H) and a potential acceptor (B). The exchange occurs via the formation of a collision complex and rapid equilibration of the proton across the hydrogen bridge at a rate constant k_2 (in brackets). (b) Hydrogen/deuterium exchange mechanisms in proteins and peptides. The exchangeable hydrogens are located on amino acid side chains bound to heteroatoms (bold, b1 left) or at peptide amide positions. Hydrogens covalently attached to carbon atoms (bold italic, b1 left) essentially do not exchange. The exchange process is both acid and base catalyzed. Base-catalyzed exchange (b1) occurs through the formation of an imidate anion, which is subsequently reprotonated. Acid-catalyzed exchange (b2) may proceed by two distinct pathways: (i) direct protonation of the nitrogen or (ii) protonation of the carbonyl oxygen and formation of an imidic acid intermediate. References for these mechanisms are found in the text.

the proton remains with the donor. The overall rate constant of the proton transfer reaction is approximated by Equation 2.1

$$k = k_1 \left(\frac{10^{\Delta pK}}{10^{\Delta pK} + 1} \right), \tag{2.1}$$

where k_1 represents the diffusion-limited collision rate constant and ΔpK is the difference between the pK of the proton acceptor and the donor (Englander et al., 1972; Englander and Kallenbach, 1984). When proton transfer occurs from a stronger to a weaker acid ($pK_{acceptor} \gg pK_{donor}$), the rate constant of exchange (k) is equivalent to the diffusion limit (k_1) (which was calculated to be $10^{10} \, M^{-1} \, s^{-1}$ (Englander et al., 1972)). In other words, for transfer from a stronger to a weaker acid, every collision between A–H and B leads to a successful reaction. On the contrary, when the transfer proceeds from a weaker to a stronger acid, the exchange rate constant k is reduced to $k_1(10^{\Delta pK})$. Proton transfer proceeds more slowly than the diffusion limit, and only a small fraction of collisions lead to productive transfer.

In proteins and peptides, hydrogens bonded to carbon (e.g., $-CH_3$ groups) essentially do not exchange with hydrogens in the surrounding solvent. The hydrogens located on polar side chains or the N/C termini and bonded to heteroatoms such as $-N$, $-O$, or $-S$ exchange quite easily. Finally, hydrogens located at backbone amide linkages are also able to undergo exchange (Fig. 2.1b). The capacity of hydrogens to exchange comes from their ability to form hydrogen-bonded complexes upon collisions and from the high rate of proton equilibration across the hydrogen bridge (Eigen, 1964; Englander et al., 1972). As reported above, the ΔpK term represents a key parameter that drives the rate of proton exchange. Most of the exchangeable protons on amino acid side chains display low pK values for deprotonation (pK < 13) and are therefore easily removed by OH^- at pH 7, 25°C (the pK for OH^- protonation is equal to 15.7 (Englander et al., 1972; Englander, 2006)) because their exchange rates are close to the diffusion-limited collision rate constant k_1. In contrast, the exchange of hydrogens at backbone amide linkages proceeds more slowly under the same experimental conditions due to their extreme pK value for deprotonation (pK ~ 18.0) (Molday and Kallen, 1972). The ΔpK term in Equation 2.1 is negative for backbone amide hydrogens (hereafter referred to as NHs) since the transfer proceeds from a weaker to a stronger acid (this principle explains why the rate of exchange of nonprotected hydrogens in polar side chains and N- and C-termini is much faster than the exchange of hydrogens at peptide amide positions, even when the parameters are adjusted to minimize the exchange (see Section 2.1.3). As a result, the hydrogen exchange that is most often measured in proteins and peptides is that of the backbone amide hydrogen (NH). NHs make excellent structural probes because they are found on all amino acids expect proline and are involved in the formation of secondary structure (Englander et al., 1996).

2.1.2 Mechanisms of Backbone Amide Hydrogen Exchange

Backbone amide hydrogen exchange can proceed via one of the three distinct mechanisms: base-catalyzed exchange and two different acid-catalyzed exchange reactions (Perrin, 1989). Although base-catalyzed exchange is far more important for the physiological conditions in which proteins and peptides exist, the acid-catalyzed mechanisms will also be reviewed and explained in this section.

The base-catalyzed mechanism was proposed in 1959 by Berger et al. based on proteolysis and ionization results of N-methylacetamide (Berger et al., 1959). Exchange begins by abstraction of the amide proton by a hydroxide ion and the formation of an imidate anion (Fig. 2.1b, 1). The imidate intermediate is subsequently reprotonated to complete the successful transfer. Acid-catalyzed exchange, in contrast, may occur by two distinct mechanisms. The first mechanism is analogous to the base-catalyzed reaction except that the sequence of protonation and deprotonation is reversed (Fig. 2.1b, 2i). When nitrogen is protonated to form an N-protonated intermediate, there is the loss of one hydrogen bound to the positively charged nitrogen to restore the neutrality. This mechanism has one significant problem: the most basic site of an amide is not the nitrogen but the oxygen (the experimental pK values for protonation of the nitrogen and the carboxyl oxygen are around -7 and 0, respectively (Fersht, 1971; Molday and Kallen, 1972)). As a result, it was postulated that acid-catalyzed amide exchange might occur via the imidic acid mechanism (Fig. 2.1b, 2ii). In this reaction, the protonation of the carbonyl oxygen lowers the pK for deprotonation of the amide group from 18.0 to about 7.5 (Pletcher et al., 1968; Molday and Kallen, 1972; Eriksson et al., 1995). The acidified amide proton is then removed by a water molecule to produce the imidic acid intermediate that returns to an amide by reversing the steps. One particular feature of this later acid-catalyzed exchange mechanism is the inversion of the hydrogen-bond donor–acceptor proprieties of the amide (Perrin, 1989). As pointed out by Perrin, this pathway is "more circuitous" than the N-protonation mechanism but is much more attractive as it avoids the protonation of the less basic site of the amide function (Perrin and Arrhenius, 1982; Perrin, 1989). Further support for the second mechanism is the fact that the large difference in basicity of the carbonyl oxygen and the nitrogen suggests that the oxygen is 10^6 times more likely to be protonated than the nitrogen (Martin and Hutton, 1973).

In summary, the base-catalyzed exchange of NHs in proteins and peptides occurs via the imidate anion (Fig. 2.1b, 1) whereas the O-protonation pathway appears to be predominant under acid-catalyzed conditions (Fig. 2.1b, 2ii). The fact that the imidic acid mechanism does not involve direct protonation of the nitrogen may provide access to potentially useful structural information since the measured acid-catalyzed exchange rates directly reflect the solvent accessibility of the amide carbonyl (Tüchsen and Woodward, 1985a, 1985b; Dempsey, 1992, 1995, 2001; Rohl and Baldwin, 1994). Similarly, the base-catalyzed exchange rates are related to the accessibility of the amide nitrogen. The environment of a particular peptide bond can thus be probed by measuring both base and acid-catalyzed exchange rates.

2.1.3   Factors Affecting Hydrogen Exchange

2.1.3.1   *pH Effects*   The relationship between the hydrogen exchange rate and the pH is of primary importance, and without this pH dependency, analysis of hydrogen exchange by mass spectrometry would not be possible (see Section 2.4). As described in Section 2.1.2, hydrogen exchange is both an acid- and a base-catalyzed reaction. Therefore, the exchange rate constant (k_{ex}) of freely exposed peptide hydrogens can be expressed as the sum of the acid- ($k_{H_3O^+}$), base- (k_{OH^-}), and water-catalyzed (k_{water}) contributions, respectively, as shown in Equation 2.2 (Woodward and Hilton, 1980; Englander and Kallenbach, 1984).

$$k_{ex} = k_{H_3O^+}[H_3O^+] + k_{OH^-}[OH^-] + k_{water} \qquad (2.2)$$

Exchange studies of alanine-based reference molecules, such as the random coil-like poly-DL-alanine peptide (PDLA), have yielded a quantitative description of pH effects on the NH exchange rate (Berger and Linderstrøm-Lang, 1957; Englander andPoulsen, 1969; Molday et al., 1972; Bai et al., 1993). The PDLA reference exchange rate constants $k_{H_3O^+}$, k_{OH^-}, and k_{water} were determined to be $41.7\,M^{-1}\,min^{-1}$, $1.12 \times 10^{10}\,M^{-1}\,min^{-1}$, and $3.16 \times 10^{-2}\,min^{-1}$, respectively, at 20°C and low salt conditions (Bai et al., 1993). The base-catalyzed exchange of NHs is therefore much more effective than acid catalysis by around eight orders of magnitude (Dempsey, 2001).

The pH dependence of several types of labile hydrogens found in proteins is presented in Fig. 2.2a. For NHs, a plot of $log(k_{ex})$ versus pH leads to the often-represented V-shaped curve with a minimum exchange rate value ($k_{ex,min}$) at pH 2.5–3.0 (pH$_{min}$), a point at which the rates for base- and acid-catalyzed exchange are equal. According to Leichtling and Klotz (1966), pH$_{min}$ reflects the ratio between the acid- and base-catalyzed exchange rate constants of the NH and can be expressed as follows:

$$pH_{min} = \frac{1}{2}\left[pK_{water} + log\left(\frac{k_{H_3O^+}}{k_{OH^-}}\right)\right] \qquad (2.3)$$

where K_{water} corresponds to the ionization constant for H_2O. On either side of pH$_{min}$, the slope $\Delta log(k_{ex})/\Delta pH$ is approximately equal to 1 for pH > pH$_{min}$ and to -1 for pH < pH$_{min}$, leading to a close to first-order pH dependence of k_{ex}. Thus, the NH exchange rate is mainly determined by OH$^-$ ion activity at pH values above pH$_{min}$, and by H_3O^+ ion activity at pH values below pH$_{min}$. Each shift of one pH unit from the pH$_{min}$ value modifies the exchange rate by 10-fold. The exchange rate of NHs is reduced by approximately four orders of magnitude when the pH is decreased from 7.0 to 2.5.

The contribution of water catalysis in the exchange process is generally considered to be insignificant around physiological pH ($\sim$7) and k_{water} is neglected in Equation 2.2. However, the contribution of water-catalyzed exchange becomes significant at pH values near pH$_{min}$ (Eq. 2.3). PDLA exchange studies in deuterium buffer at 25°C and pH $\sim$3 revealed that the contributions of acid-, base-, and water-catalyzed exchange account for 31%, 31%, and 38% of the global exchange rate, respectively (Gregory et al., 1983).

Most of the exchangeable hydrogens in side chains have pH_{min} values higher than NHs. Therefore, when the pH is adjusted to 2.5 to minimize exchange of NHs during MS analysis (see Section 2.4), most of the side chain hydrogens continue to exchange via acid-catalyzed reactions except for Arg(ε-NH) and Arg(η-NH), which are in the same range as NHs (Fig. 2.2a). The k_{ex} values for Arg(ε-NH) and Arg(η-NH) are

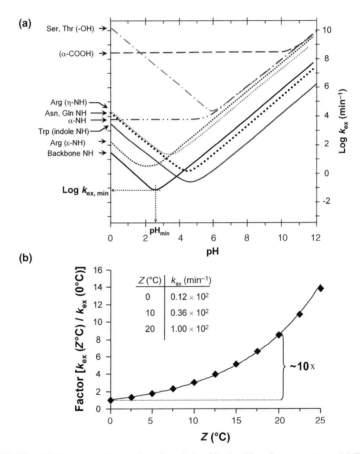

FIGURE 2.2 Hydrogen exchange is a function of both pH and temperature. (a) Exchange rate constants (k_{ex}) of several types of labile hydrogens found in proteins and peptides are shown as a function of solution pH (Creighton, 1993; Dempsey, 2001). A change of one pH unit modifies the exchange rate constant by approximately 10-fold. The exchange rate of backbone amide hydrogens (NHs) is at a minimum ($k_{ex,min}$) when the pH is approximately 2.5 (pH_{min}), depending on the sequence (see Section 2.1.3.4). (b) Effect of the temperature on the exchange rate constant of NHs at neutral pH. k_{ex} values were calculated for several temperatures (Z) using Equation 2.4 and normalized with the exchange rate value obtained at 0°C. Each change of 10°C affects the k_{ex} value by approximately threefold. The exchange rate is therefore decreased by approximately one order of magnitude when the temperature is reduced from 20 to 0°C.

close to their minima at pH 2.5–3.0, but they still exceed the $k_{ex,min}$ of NHs and will thus exchange more rapidly.

2.1.3.2 Temperature Effects

Hydrogen exchange rates are also temperature dependent. The rate constants $k(x)_T$ for acid-, base-, and water-catalyzed exchange at different temperatures T (in Kelvin) can be estimated using the following integrated form of the Arrhenius equation:

$$k(x)_T = k(x)_{293} \exp\left(-\frac{E_a(x)}{R}\left[\frac{1}{T}-\frac{1}{293}\right]\right) \tag{2.4}$$

where $k(x)_{293}$ corresponds to the reference rate constant $k_{H_3O^+}$, k_{OH^-}, or k_{water} at 20°C (see values in Section 2.1.3.1), R is the gas constant (8.134 J mol^{-1} K^{-1}), and $E_a(x)$ corresponds to the reference apparent activated energy for the acid- $(E_a(k_{H_3O^+}) = 14$ kcal mol$^{-1})$, base- $(E_a(k_{OH^-}) = 17$ kcal mol$^{-1})$, or water-catalyzed exchange $(E_a(k_{water}) = 19$ kcal mol$^{-1})$, respectively (Bai et al., 1993). Figure 2.2b shows the effects of temperature on the NH exchange rate at neutral pH. For this description, k_{ex} was calculated at several temperatures (ranging from 0 to 25°C) and normalized by using the estimated exchange rate value at 0°C. The ratio $[k_{ex}(Z°C)/(k_{ex}(0°C)]$ versus temperature plot gives rise to an exponential curve where the exchange rate changes approximately threefold for each 10°C increment. The NH exchange rate can therefore be reduced by approximately one order of magnitude when the temperature is adjusted from 20 to 0°C.

Changes in temperature mainly affect k_{ex} by altering catalyst concentrations (especially OH$^-$) through modifications of the ionization constant of water (Englander and Poulsen, 1969; Englander et al., 1972; Englander and Kallenbach, 1984). The diffusion–collision rate constant k_1 may also be affected (Fig. 2.1a). The number of hydrogen-bonded complexes leading to a successful transfer is reduced at lower temperatures as the viscosity of the solution highly depends on the temperature. Consequently, the probability that the proton donor A–H collides with the acceptor B decreases with the temperature.

2.1.3.3 Solvent and Pressure Effects

In addition to the temperature, solvent composition has significant effects on hydrogen exchange rates (Englander et al., 1972; Woodward et al., 1975; Englander and Kallenbach, 1984). Organic solvents mainly modify the OH$^-$ ion activity with little or no effect on acid-catalyzed exchange rates. Sensitivity to organic solvent occurs because the equilibrium constant K_{water}, which directly determines the OH$^-$ ion concentration at a given pH value ($[OH^-] = 10^{-(pK_{water}-pH)}$), is sensitive to solvent composition. Addition of organic cosolvents, such as dioxane, depresses K_{water} and lowers the concentration of OH$^-$ ions leading to a shift of the pH$_{min}$ to higher values (Eq. 2.3) (Leichtling and Klotz, 1966; Englander and Kallenbach, 1984; Maier and Deinzer, 2005).

Changes in pressure affect hydrogen exchange in folded proteins in solution. The base-catalyzed exchange rates in lysozyme, ribonuclease A, oxidized ribonuclease A, and poly-DL-lysine (PDLL) were all enhanced by an increase in pressure (Carter et al., 1978). The observed pressure dependence of hydrogen exchange can be due to either an enhancement of solvent penetration within folded proteins or pressure-induced denaturation. Although pressure significantly modifies exchange rates in folded proteins, changes are relatively insignificant in unstructured proteins and peptides (Carter et al., 1978). Engen and coworkers have recently shown that high pressure ($> 10,000\,\text{psi}$) did not cause an increase in deuterium loss in unstructured, highly deuterated model peptides (Wu et al., 2006).

2.1.3.4 *Side Chain and Ionic Strength Effects*   Amide hydrogen exchange rates in proteins and peptides are sensitive to inductive and steric blocking effects imposed by neighboring side chains (Molday et al., 1972; Kim and Baldwin, 1982; Bai et al., 1993). The effects of numerous substituents on NH exchange rates were first reported in several different amide models (Leichtling and Klotz, 1966; Sheinblatt, 1970; Kakuda et al., 1971; Englander et al., 1972). Inductive effects, imposed by the presence of electron-withdrawing substituents such as polar side chains, augment the acidity of the amide proton by lowering the pK for deprotonation. Therefore, the base-catalyzed exchange rate, which involves hydrogen abstraction and formation of the imidate anion (Fig. 2.1b), increases whereas acid catalysis decreases. Steric blocking effects, in contrast, are mainly induced by aliphatic and aromatic residues. The side chains of these residues sterically hinder the interaction between the amide hydrogen, the catalyst, and other aqueous species. As a result, the value of $k_{\text{ex,min}}$ is reduced with little change in pH_{min} (Bai et al., 1993).

The additivity of the inductive and blocking effects of side chains on NH exchange rates has been well characterized and calibrated by Molday et al. (1972) and Bai et al. (1993) by using random coil peptide models. Side chain specific acid (A) and base (B) factors were calculated for the 20 naturally occurring amino acid chains to correct $k_{H_3O^+}$ and k_{OH^-} values. These corrective factors are commonly used to predict the exchange behavior of unprotected NHs surrounded by particular neighboring residues using Equation 2.5:

$$k_{\text{ex}} = k_{H_3O^+}(A_{\text{Left}} \times A_{\text{Right}})[H_3O^+] + k_{OH^-}(B_{\text{Left}} \times B_{\text{Right}})[OH^-]$$
$$+ k_{\text{water}}(B_{\text{Left}} \times B_{\text{Right}}) \tag{2.5}$$

where X_{Left} and X_{Right} refer to the specific acid and base factors of the side chains located on the left or on the right side of the amide hydrogen, respectively. The theoretically corrected k_{ex} value is of particular interest as it allows the protection factor (PF) of specific amide hydrogens to be assessed. PF reflects the protection against exchange imposed by protein structure (through hydrogen bonding and/or solvent accessibility) and is expressed as the ratio between the calculated and the experimental k_{ex} value (Englander and Kallenbach, 1984; Dempsey, 2001).

The published residue-dependent correction factors to $k_{H_3O^+}$ and k_{OH^-} were estimated in the presence of salt to suppress coulombic effects. Charge effects are therefore neglected although they can have significant influences on NH exchange rates. The salt dependence of amide hydrogen exchange was first reported for the model compound N-methylacetamide (Schleich et al., 1968, 1971). Later, Kim and Baldwin (1982) reported on charge effects in the positively charged polypeptide PDLL. Both base- and acid-catalyzed exchange rates of PDLL were sensitive to salt concentration. Addition of NaCl (up to 2.0 M) enhanced the rate of acid-catalyzed exchange through electrostatic screening and diminished the rate of base-catalyzed exchange via counterion competition. As a result, the pH_{min} of PDLL shifted from 1.25 to 2.6 when the NaCl concentration increased from 0 to 2 M NaCl (Kim and Baldwin, 1982). Salt effects have also been seen in proteins such as the basic trypsin inhibitor BPTI and ribonuclease S (Kim and Baldwin, 1982; Matthew and Richards, 1983; Christoffersen et al., 1996). Salt effects on hydrogen exchange rates of 16 amide hydrogens near the solvent interface of BPTI showed considerable variation from site to site, suggesting that the exchange process depends on the local electrostatic field rather than the global net charge of the protein (Christoffersen et al., 1996). Salt concentration is therefore another parameter that should be considered in the hydrogen exchange process of proteins. Generally, in more dilute solutions (< 250 mM salt), the effects are minimal.

2.2 HX MECHANISMS IN PROTEINS

Backbone amide hydrogens in fully solvent-exposed peptides generally exchange rapidly at neutral pH with k_{ex} values ranging from 10^1 to $10^3 \, s^{-1}$ (Bai et al., 1993). However, the situation is quite different in folded proteins. Some NHs exchange quickly and others exchange much more slowly with half-lives of exchange ranging from minutes to months (Englander and Kallenbach, 1984; Engen and Smith, 2001; Hoofnagle et al., 2003; Wales and Engen, 2006a). The observed slowing of NH exchange in native proteins results from restricted access to the sites of exchange and/or intramolecular hydrogen bonding (NHs in folded proteins are all hydrogen bonded, either to water molecules or with other parts of the protein). It is this slowing of exchange rates that allows the use of hydrogen exchange methods for investigating protein conformation and dynamics. However, because both solvent accessibility and hydrogen bonding influence the exchange rate, it is generally not possible to assign secondary structure based on hydrogen exchange information; regions protected by strong hydrogen bonding in secondary structure elements may look similar (in terms of their exchange rates) to regions that have no secondary structure but are highly occluded from solvent.

Two exchange mechanisms have been developed to explain hydrogen exchange in proteins. In the first mechanism, exchange may occur directly from the folded state. In the second mechanism, partial or global unfolding occurs first, followed by exchange. The combined, observed rate constant for exchange at each NH position of a protein can therefore be described as

$$k_{obs} = k_f + k_u \qquad (2.6)$$

where k_{obs} is expressed as the sum of contributions of exchange from the folded (k_f) and unfolded states (k_u) (Woodward and Hilton, 1980; Kim et al., 1993; Kim and

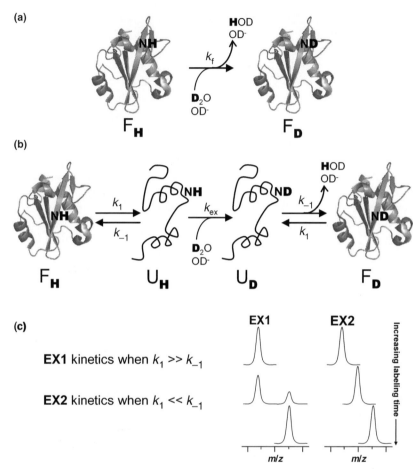

FIGURE 2.3 Principles of backbone amide hydrogen exchange in proteins. (a) In folded-state exchange, folded protein undergoes isotopic exchange with a rate constant k_f. (b) In exchange occurring by local unfolding, the protein unfolds, the backbone amide hydrogens are exposed to deuterated solvent and exchange with a rate constant k_{ex}. Rate constants k_1 and k_{-1} describe the unfolding and refolding reaction of the protein, respectively. Note that k_1 and k_{-1} here are different than k_1 and k_{-1} in Fig. 2.1. (c) EX1 and EX2 kinetic mechanisms have unique mass spectra. EX1 kinetics occurs when the unfolding rate is faster than the protein refolding $(k_1 \gg k_{-1})$ and gives rise to two distinct and separated mass envelopes. In EX2 kinetics, the protein refolding rate is much faster than the unfolding rate $(k_1 \ll k_{-1})$. The opening reactions must therefore occur multiple times before a successful exchange event takes place. EX2 kinetics leads to a gradual shift over the m/z range during the time course of deuteration.

Woodward, 1993). This two-process model of exchange is illustrated in Fig. 2.3a and b . There is experimental evidence for both processes, as described in detail in the following paragraphs. In both models, the transition from protiated to deuterated species (F_H–F_D and U_H–U_D transitions) is ordinarily considered to be irreversible as the D_2O concentration is purposely in vast excess of the H_2O concentration.

The rate of exchange from the folded state is given by Equation 2.7, where β is the probability (ranges from 0 to 1) that the amide hydrogen is simultaneously exposed to catalysts and water, and k_{ex} is the rate constant for NH exchange in random coil peptides (Kim and Woodward, 1993).

$$k_f = \beta\, k_{ex} \qquad (2.7)$$

Several hypotheses have been developed to explain how exchange can occur in folded proteins. One idea is referred to as the solvent penetration or diffusion model (Rosenberg and Enberg, 1969; Woodward and Rosenberg, 1971; Englander et al., 1972; Ellis et al., 1975; Woodward, 1977; Woodward and Hilton, 1980; Englander and Kallenbach, 1984; Miller and Dill, 1995). According to this model, exchange of amide hydrogens located near the surface of the protein or in close proximity to solvent channels proceeds with ease whereas exchange at buried positions occurs when catalysts enter the protein through transiently formed channels and cavities. Several mechanisms have been proposed to explain the formation of solvent channels within the protein (Englander and Kallenbach, 1984). One penetration mechanism proposes that these cavities arise from small and rapid molecular motions of protein atoms ranging from 0.1 to several angstroms. Another mechanism proposes that the formation of "active" channels arises from the fusion of preexisting cavities and might require hydrogen bond breakage. The exchange rate of buried NHs is therefore directly related to the average accessibility to solvent and reactivity with catalysts. A second model, the relayed imidic acid (or basic) mechanism, was developed by Tüchsen and Woodward to explain exchange in folded proteins (Tüchsen and Woodward, 1985a, 1985b). This model allows exchange of buried amide hydrogens via a charge delocalization process. This type of exchange may occur if amide hydrogens are connected to solvent via hydrogen bonds (Eriksson et al., 1995; Juranić et al., 1995; Dempsey, 2001). According to this view, the hydrogen exchange reaction takes place at the surface of the protein and does not require the diffusion of catalysts (OH^- or H_3O^+) inside the protein. Once the reaction is initiated at one end of the hydrogen-bonded peptide chain, for example, by O-protonation of the free carbonyl, the first acidified amide proton protonates the carbonyl oxygen of the amide proton to which it is hydrogen bonded and so on. The proton is thus relayed through the chain until one imidic acid intermediate exchanges its acidified hydrogen with a water molecule. Consequently, this mechanism only requires diffusion of water molecules into the exchange sites. The relayed mechanism has been used to explain the pH dependence of exchange for several buried amide hydrogens located at the end of hydrogen-bonding networks in proteins (Eriksson et al., 1995; Cotten et al., 1999).

Although exchange from the folded state generally requires small atomic movements to allow catalysts to access exchange sites, larger molecular motions are associated with exchange from the unfolded form. These motions may concern the entire protein (global unfolding), large segments of the protein (subglobal unfoldings, e.g., an α-helix), or very small and localized fragments of the protein (so–called protein "breathing") (Englander et al., 1996; Chamberlain and Marqusee, 1997). In general, exchange from the unfolded form, as described next, is the predominant view of how HX occurs in proteins.

Native proteins are considered to be "exchange incompetent" (F_H) and only become "exchange competent" (U_H) through unfolding events (Fig. 2.3b). This model has two components: (i) a structural factor corresponding to the rate of unfolding/refolding and (ii) an intrinsic chemical factor accounting for the exchange rate constant k_{ex} (Hvidt and Nielsen, 1966; Miller and Dill, 1995). The overall exchange rate constant from the unfolded state, k_u, is therefore given by Equation 2.8:

$$k_u = \frac{k_1 k_{ex}}{k_1 + k_{-1} + k_{ex}} = \left(\frac{k_1}{k_{-1} + k_{ex}}\right) k_{ex} \qquad (2.8)$$

where k_1 and k_{-1} are the rate constants for opening and closing of the native conformation, respectively, and k_{ex} is the intrinsic rate constant for exchange when there is no structure (as in unstructured peptides, see Section 2.1.3.4).

Proteins are ordinarily stable under native conditions so that the rate of unfolding is very small relative to the rate of refolding ($k_1 \ll k_{-1}$). The exchange rate of NHs thus depends on the unfolding dynamics of the protein, as well as on the rate constant k_{ex}. However, when the rate of unfolding is larger than the rate of refolding, $k_1 \gg k_{-1}$, the opening reaction is rate limiting. In this reaction, exchange rates of individual amide hydrogens located in the same region are considered to be equivalent. These two extremes of exchange ($k_1 \ll k_{-1}$ and $k_1 \gg k_{-1}$) were first described by Hvidt and Nielsen (1966) and are referred to as EX2 and EX1, respectively.

Exchange under EX2 kinetics, $k_1 \ll k_{-1}$, occurs when the refolding of the transient opening state (U_H) is fast compared to the intrinsic chemical exchange rate constant k_{ex} (Fig. 2.3b). In this case, the exchange rate of each individual amide hydrogen is proportional to the catalyst concentration and Equation 2.8 is reduced to

$$k_u = \left(\frac{k_1}{k_{-1}}\right) k_{ex} = K_{unfolding}\, k_{ex} \qquad (2.9)$$

where $K_{unfolding} = k_1/k_{-1}$ corresponds to the unfolding equilibrium constant. Since protein refolding is much faster than exchange, the probability of exchange during one unique unfolding event is small. The rate-limiting step of EX2 kinetics is thus the hydrogen exchange reaction (k_{ex}); the opening/closing process may occur many times before one hydrogen exchanges. Under EX2 kinetics, the unfolding rate constant (k_u) divided by the intrinsic chemical exchange rate (k_{ex}) yields the unfolding equilibrium constant. It is thus possible to obtain the equilibrium free energy of the unfolding reaction (ΔG^0_{open}) using Equation 2.10.

$$\Delta G^0_{open} = -RT \ln K_{unfolding} \qquad (2.10)$$

where R and T correspond to the gas constant and the temperature in Kelvin, respectively.

Exchange governed by EX1 kinetics, $k_1 \gg k_{-1}$, allows all the amide hydrogens in the unfolding unit to exchange before refolding, k_{-1}, occurs. EX1 is therefore often described as "cooperative exchange." Exchange of this type is much more rare than EX2 kinetics. Only a few proteins exhibit EX1 kinetics under native conditions (Engen et al., 1999; Wales and Engen, 2006b), and some proteins may contain distinct regions that undergo exchange via EX1 and EX2 kinetics simultaneously (Engen et al., 1997; Clarke and Itzhaki, 1998; Wales and Engen, 2006a). The EX1 regime is generally forced by using denaturing conditions such as high temperature, high pH or chaotropic reagents (Hilton et al., 1981; Bai et al., 1994; Qian and Chan, 1999; Sivaraman and Robertson, 2001; Houliston et al., 2002; Ferraro et al., 2004). EX1 and EX2 exchange mechanisms reflect different protein structural fluctuations and show differences in pH dependence: EX2 kinetics is characterized by a linear dependence on catalyst concentrations (Eq. 2.9) whereas EX1 kinetics is completely pH independent. The two kinetic regimes can be easily distinguished by mass spectrometry (Miranker et al., 1993). EX1 kinetics gives rise to two distinct and separated mass envelopes (the undeuterated and deuterated forms) whereas EX2 kinetics results in a gradual shift of the mass envelope with time (Fig. 2.3c).

To summarize, the overall observed exchange rate constant for exchange at each individual NH can be approximated by expanding Equation 2.6 into Equation 2.11:

$$k_{obs} = (\beta + K_{unfolding})k_{ex} \qquad (2.11)$$

The observed exchange rate constant under steady-state conditions corresponds to the sum of contributions of exchange from the folded state (Equation 2.7) and the unfolded state (Equation 2.9). Exchange from the unfolded state can occur by EX2, EX1 or a mixture of the two kinetics. Mass spectrometry can be used to measure the deuterium incorporation and can diagnose the kinetic regime, thereby assisting in the characterization of how the protein fluctuates in solution. It should be noted that this discussion refers to local unfolding events that occur in an ensemble of states; therefore, data from EX2 exchange should not be used to make interpretations about the overall folding of a protein.

2.3   DEUTERIUM INCORPORATION INTO PROTEINS

To measure how and where deuterium exchanges into a protein, the protein must be exposed to deuterium. While this sounds like a trivial concept, the way in which the exposure is performed can provide additional levels of insight. There are two basic kinds of labeling strategies: pulse labeling and continuous labeling (Deng et al., 1999a; Wales and Engen, 2006a) (Fig. 2.4). The choice of the labeling technique is primarily dictated by the type of information desired.

2.3.1   Continuous Labeling

In continuous labeling, the simplest labeling method, a fully protiated protein in a buffer containing 100% H_2O is diluted (15–20-fold is typical) with an identical buffer

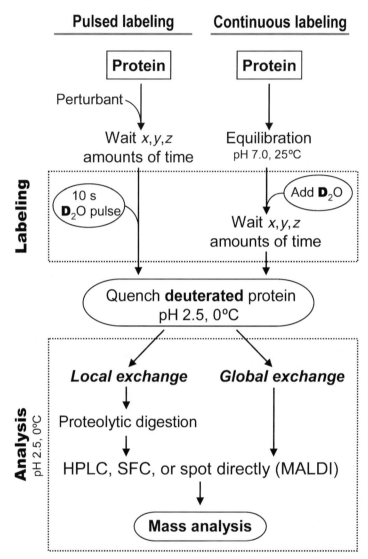

FIGURE 2.4 General scheme for HX MS experiments. In pulsed labeling (top, left) experiments, the protein is first exposed to a perturbant (ligands, denaturants, temperature, pH, etc.) and then labeled for a very brief period of time (generally 10 s) with a pulse of D_2O. After labeling, the exchange reaction is quenched by decreasing both the pH and the temperature to 2.5 and 0°C, respectively. For continuous labeling (top, right), the equilibrated protein is directly diluted in the deuterated buffer so that the final deuterium concentration is >95%. Aliquots are removed at various times and quenched. Samples can be analyzed directly either by online HPLC/SFC-ESI-MS or by MALDI-MS to obtain the global exchange behavior of the protein. To increase the spatial resolution (local exchange), the labeled and quenched protein is digested with an acid protease and the mass of each peptic fragment is determined.

containing 100% D_2O and the deuterium incorporation is monitored as a function of time (Fig. 2.4, top right). After a 15-fold or more dilution, the final deuterium concentration in the labeling solution exceeds 95%. The large excess of deuterium forces the hydrogen exchange reaction in only one direction (see Fig. 2.3b). In other words, once an amide hydrogen is replaced by a deuterium, it stays labeled. Continuous labeling proceeds until the maximum amount of exchange time desired has passed, for example, 24 h. At various times along the way, aliquots are removed from the labeling solution, quenched, and analyzed by mass spectrometry.

Continuous labeling experiments are usually performed under conditions where the native conformation of the protein is "stable." Thus, the monitored exchange rate at various incubation times provides information on the conformational dynamics of a protein under equilibrium conditions. However, protein stability may mean that, for example, 95% of the molecules exist in a given conformation at any one moment and the remaining 5% are in transition to another folded state. Transitioning between folded and unfolded species may be the result of natural motions of the protein but can also be induced in response to ligand binding, protein–protein complex formation, or by the addition of chaotropic agents (mild denaturing conditions). Because continuous labeling is performed while the populations of folded and unfolded proteins might be changing, if a protein (or a part of the protein) starts to unfold and make a transition, it becomes and stays labeled because the D_2O concentration is very high. The new species that underwent transition will therefore contain more deuterium than the rest of the population resulting in a unique mass spectrum. This kind of experiment is particularly useful to probe slow unfolding reactions (Engen et al., 1997; Wales and Engen, 2006b) and allow, for example, high energy state intermediates of a folding pathway to be studied (Krishna et al., 2004).

2.3.2 Pulse Labeling

In pulse labeling experiments, proteins are exposed to deuterium for a very brief period of time (Fig. 2.4, top left) after a protein has been forced to undergo structural changes via addition of a perturbant (Deng et al., 1999a; Konermann and Simmons, 2003). Perturbants are usually chaotropic agents (urea, guanidine hydrochloride) but can also be binding partners, changes in pH or temperature. After various times of interactions with the perturbing agent, samples are subjected to a "pulse" of deuteration buffer, typically 10 s or less (depending on pH). Since the labeling time is very short, only easily accessible and rapidly exchanging NHs (such as those in parts of proteins that are unfolded or on the surface and highly exposed to solvent) will become deuterated during the pulse.

As further described in Section 2.5, pulse labeling experiments have been extensively used to study protein folding mechanisms by mass spectrometry, as well as to identify transient intermediate states (Yang and Smith, 1997; Deng and Smith, 1998, 1999a; Deng et al., 1999a; Pan and Smith, 2003; Mazon et al., 2004; Pan et al., 2004). When pulsed labeling is done after protein denaturation, the monitored deuterium incorporation corresponds to a snapshot of the unfolded and folded proteins that existed at the time of the pulse (Deng and Smith, 1998, 1999b). The different mixing steps required for pulse labeling experiments may be performed manually or by using a quench-flow system (Yang and Smith, 1997; Konermann and Simmons, 2003; Pan et al., 2005). The use of an automated apparatus is required to follow rapid folding processes. Exposure times as short as 10 ms have been used to follow rapid folding of proteins (Heidary et al., 1997; Yang and Smith, 1997). The pulse length has to be short relative to the time of the conformational changes of interest, but is also a function of pH (Deng et al., 1999a).

2.3.3   Other Labeling Strategies

There are other techniques that permit selective labeling of protein sites undergoing structural changes: the kinetic and the functional labeling strategies (Rogero et al., 1986; Englander and Englander, 1994). In kinetic labeling experiments, the protein is exposed to labeling buffer for a limited exchange-in period so that all accessible NHs become deuterated. The protein is then exposed to 100% H_2O (exchange-out) for a time period generally shorter than the exchange-in time. Consequently, amide hydrogens with very fast exchange rates completely exchange out (return to H) while those with intermediate exchange rates stay deuterated. This method has been used in studies with myoglobin to measure the behavior of fast-exchanging amide hydrogens (Englander and Staley, 1969) and then extended to measure the effects of glycerol concentrations on the hydrogen exchange process (Calhoun and Englander, 1985).

As with kinetic labeling, functional labeling experiments start by exposing the protein to the labeling buffer for a defined period of time to deuterate all accessible amide hydrogens. At the end of the exchange-in period, the protein is "switched" to a slower exchanging form, for example, by adding a binding partner, and exchange-out is initiated by rapidly removing the deuterated buffer. Exchange-out proceeds for the same amount of time as exchange-in. Only sites with different HX rates in response to changing the conformation of the protein will remain labeled while all others will revert to hydrogen. This approach has been used, for example, to study the effects of the T to R allosteric transition on the hemoglobin structure (Ray and Englander, 1986; Louie et al., 1988; Englander et al., 2003).

2.4   MEASURING HX WITH MASS SPECTROMETRY

Hydrogen/deuterium (H/D) exchange can be monitored with any method that is sensitive to the unique properties of the isotopes of hydrogen. Initially, exchange was monitored after incorporation of tritium (T_2O) by measuring increasing radioactivity of the labeled proteins (Englander and Poulsen, 1969; Englander and Englander, 1972, 1978). NMR has also found great use in monitoring hydrogen exchange because deuterium is NMR silent whereas hydrogen is NMR active (Englander and Mayne, 1992). Replacement of amide hydrogens by deuterons causes the disappearance of the corresponding NMR peaks. The use of mass spectrometry to study H/D exchange in proteins began in the early 1990s when it became possible to analyze proteins by electrospray ionization. The first report of protein HX MS was by Katta and Chait (1991). The MS approach is based on the mass difference between hydrogen and deuterium. Replacement of hydrogen with a deuteron (or the inverse) results in a mass change of 1 Da, a variation easily observed by mass spectrometry.

2.4.1 Global Versus Local Exchange

The design of an HX MS experiment is dictated by the goals of the study. As shown in the lower half of Fig. 2.4, the deuterium uptake may be determined for all NHs in the protein (global exchange) or in short segments of the labeled protein (local exchange). The former is generally used to estimate the overall exchange behavior of the protein either alone (e.g., natural motions under equilibrium conditions) or in response to ligand binding, complex formation, activation, presence of denaturants, and so on. Global exchange has essentially no spatial resolution since the measured deuterium content corresponds to an average of all exchanged NHs present in the protein. Spatial resolution is increased by measuring the deuterium content of short peptide fragments generated by proteolysis of the labelled protein. This technique was first reported by Rosa and Richards (1979) and combined later with mass spectrometry analysis by Smith and colleagues (Zhang and Smith, 1993). Labeled proteins are digested under quench conditions using acidic proteases and the resulting peptide fragments can then be separated by online high-performance liquid chromatography (HPLC) (Fig. 2.4, bottom).

2.4.2 Back Exchange

A critical part of monitoring deuterium exchange by mass spectrometry is retaining the label. Before the mass can be measured, the labeled protein(s) may undergo digestion, desalting, concentration, and separation. Since protiated solvents are used during this process, deuterated positions may revert to hydrogen. This undesirable exchange, also referred to as back exchange, is unavoidable and must be minimized to retain as much of the label as possible. If totally deuterated solvents were used in the analysis steps instead of protiated solvents, there would still be a problem as there would then be artifactual forward exchange.

The primary way to minimize back exchange is to maintain a pH at which exchange is at its minimum and to keep everything cold. The pH and temperature of the labeled sample are adjusted to $\sim$2.5 and 0°C, respectively (see Fig. 2.2 and Sections 2.1.3.1 and 2.1.3.2). Under these quench conditions, the exchange rate constant of the unprotected NHs is decreased by approximately five orders of magnitude compared to what it was at pH 7 and 25°C.

The half-life for back exchange under quench conditions is between 30–120 min, depending on the sequence (Englander and Kallenbach, 1984; Bai et al., 1993; Smith et al., 1997). Consequently, the analysis steps must be completed as quickly as possible. One cannot perform a 1-h digestion and a 1-h HPLC gradient to completely digest and highly resolve all the peptic peptides. If that were the case, all the deuterium label would be gone by the time the peptides arrived at the mass spectrometer! If digestion and separation steps are kept under 12–15 min total, deuterium recovery can be around 85% or more. Although there are some losses in the ESI interface during spraying, these can usually be kept to less than 5%, depending on the conditions of the source.

As described in Section 2.1.3.1, side chain labile hydrogens exchange much faster than NHs at pH ∼2.5. Consequently, when labeled proteins or peptides are placed into the digestion and/or HPLC system in the presence of protiated solvents, any deuterium that exchanged into side chain positions very quickly reverts to hydrogen. The measured increases in mass, therefore, correspond to deuterium incorporation at backbone amide positions only. Because arginine has a side chain amide hydrogen with an exchange rate minimum close to that of the NHs (see Fig. 2.1a), proteins or peptides rich in arginine can complicate the analysis because all side chain deuterium may not be washed away during the analysis steps.

Adjustments can be made to correct the loss of deuterium during analysis. Zhang and Smith (1993) described a correction method that requires the analysis of totally deuterated version of the protein being analyzed. This correction essentially adjusts the measured deuterium levels by compensating for how much label is lost from a 100% deuterated version of the protein. It is often challenging to prepare such a totally deuterated version of a protein. Other correction methods have also been described (Hoofnagle et al., 2003, 2004). An alternative is to use relative deuterium measurements (Wales and Engen, 2006a) in which no back exchange correction is applied and losses are assumed to be constant because the experimental conditions are held constant.

Another way to minimize back exchange is to not perform the analysis steps in protiated solvents. The use of supercritical fluid chromatography (SFC) in place of HPLC has been described (Emmett et al., 2006). Preliminary SFC data obtained with a labeled myoglobin digest demonstrate that the back exchange can be greatly reduced with the use of nonexchanging CO_2 mobile phase. However, digestion must still be performed and because this cannot be accomplished in supercritical CO_2, there will still be losses of label during the digestion step.

It should also be noted that the label that is lost during analysis is lost randomly. This explains the Gaussian distribution of isotopes seen in deuterated peptides (explained in the appendix to Zhang and Smith (1993)). To determine the amount of deuterium incorporated, the average of the isotopic distribution must be obtained, as described in Section 2.4.4.

2.4.3 Proteolysis Before MS

The requirement for analysis under quench conditions of pH 2.5 and 0°C complicates the analysis. Quench conditions must be maintained at all times to minimize back exchange. Therefore, the HPLC system (i.e., columns, injection valves, and tubing) must be kept on ice or in refrigerated compartments. More importantly, digestion of the labeled protein must be performed at pH 2.5. Only a few proteases function at this low pH. To date, the best acid protease for digestion is pepsin (usually porcine) (Zhang and Smith, 1993; Smith et al., 1997). Relatively recently, two other acid proteases were used in combination with pepsin (Cravello et al., 2003). These acid proteases, from *Aspergillus saitoi* (type XIII) and *Rhizopus* sp. (type XVIII), are less efficient than pepsin under quench conditions. Like pepsin, neither of these enzymes is specific but they will generate the same peptic fragments when used under identical experimental conditions. A major downside to using acid proteases, therefore, is that they are usually nonspecific enzymes. While there are some general rules about where pepsin prefers to cleave, the cleavage sites cannot be predicted from sequence alone. It has become common practice to identify each peptic peptide produced under a given set of conditions, usually through a combination of exact mass analysis and MS/MS experiments. The combined use of more than one protease (each with slightly different specificity) creates more peptides, many of which are overlapping. The more peptides that can be produced during digestion, the higher the sequence coverage and spatial resolution of the deuterium exchange information (Cravello et al., 2003; Mazon et al., 2005).

2.4.4 Mass Measurements and Data Processing

Once the mass spectra of the labeled protein or the peptic fragments has been obtained, the deuterium uptake has to be determined. The isotope patterns observed during HX MS experiments are atypical, arising as a result of deuterium losses during analysis (see also Zhang and Smith, 1993; Weis et al., 2006a). It is therefore necessary to determine the average amount of deuterium incorporated at each exchange time point (Fig. 2.5a). The difference between unlabeled and labeled species ($\Delta 1$ and $\Delta 2$ in this example) corresponds to the average amount of deuterium incorporated after each incubation period.

For peptides, the deuterium levels that are obtained are the sum of deuterium incorporation at all the backbone amide positions in the peptide. Attempts have been made to use tandem mass spectrometry to determine deuterium levels at individual peptide amide linkages. The first studies were carried out using collision-induced dissociation (CID) with argon as the collision gas (Deng et al., 1999b; Kim et al., 2001). This fragmentation process apparently leads to intramolecular scrambling of amide deuteriums, especially in y_n ion species. Other studies using similar methods indicate that the b_n ions are also susceptible to scrambling Jorgensen et al., 2005. The use of electron capture dissociation (ECD) as a fragmentation technique may reduce deuterium scrambling in the gas phase (Kaltashov and Eyles, 2002a, 2002b). It is believed that the fragmentation of ions by ECD processes occurs much faster than the CID fragmentation caused when energy is absorbed and randomized until the weakest bond is cleaved (nonergodic process) (Zubarev et al., 1998). A recent study carried out with mellitin reveals that the calculated exchange rate of consecutive c_n ions correlates with NMR data for some amide hydrogens but evidence of deuterium scrambling in ECD was reported for z_n ions (Kweon and Hakansson, 2006). Therefore, the localization of deuterium at individual peptide amide position remains, for the moment, the only advantage enjoyed by NMR techniques.

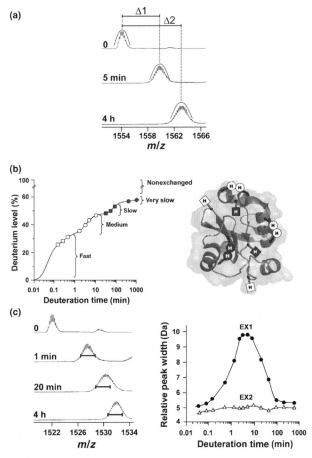

FIGURE 2.5   Hydrogen exchange mass spectrometry data processing. (a) The average amount of deuterium incorporation is measured by finding the centroid of each isotopic envelope at each exchange-in time. The monitored mass differences between the unlabeled and the labeled proteins in this example are Δ1 and Δ2 (not equivalent to 1 or 2 amu). (b) Deuterium level (Δ1, Δ2, etc. from part (a)) versus exchange time is plotted. Exchange rate constants can be extracted from such a graph by fitting a multiterm exponential equation (Eq. 2.12) to the data. The resulting fit (solid line) allows backbone amide hydrogens to be classified into different categories based on their exchange behavior. In this example data, each category is denoted by specific symbols (circles, squares, etc.) in the graph and in the accompanying structure. Backbone amide hydrogens with fast exchange rate constants are generally nonhydrogen bonded and at the surface of the protein whereas those with very slow exchange rates are buried in the protein and/or involved in hydrogen bonds. The "nonexchanged" category corresponds to highly protected amide hydrogens that are not deuterated during the timescale of the labeling experiment. (c) EX1 and EX2 kinetics can be easily characterized by measuring the peak width change of the isotopic cluster (left). Increases in peak width are directly related to EX1 kinetics (right). This approach allows EX1 and EX2 kinetics to be distinguished when mixed kinetics occurs simultaneously. All mass spectra presented in this figure were taken from Wales and Engen, (2006b).

It is possible to determine NH exchange rate constants for peptic fragments of a protein by using the equation

$$D = N - \sum_{i=1}^{N} \exp(-k_i t) \qquad (2.12)$$

where D is the deuterium level of the peptic fragment containing N peptide linkages, t is the deuterium incubation time and k_i the pseudo-first-order constant for deuterium exchange at each backbone amide linkage (Zhang and Smith, 1993; Smith et al., 1997). Although Equation 2.12 does not give the isotopic exchange rate at specific amide linkages, the resulting fit allows amide hydrogens of a specific peptide to be classified into categories (Fig. 2.5b, left). Amide hydrogens with fast exchange rate constants are generally nonhydrogen bonded and present at or near the surface of the protein while those with very slow exchange rates are hydrogen bonded and buried within the protein (Fig. 2.5b, right). Such correlations are particularly useful and can be used to obtain limited structural information of proteins that have not been or are difficult to crystallize (Yan et al., 2002, 2004).

As eluded to in Section 2.2 and Fig. 2.3c, MS can also be used to take a snapshot of the protein population at any one moment. In contrast to NMR, MS data are not averaged over all the molecules present in solution (Miranker et al., 1993). Consequently, different coexisting protein species can be detected by simple HX MS analysis, their corresponding exchange rates monitored, and unfolding/refolding rate constants determined. This particularity allows EX1 and EX2 regimes to be easily distinguished (Fig. 2.3c) even when mixed kinetics occur. Mixed kinetics, also referred to as EXX kinetics by some groups (Xiao et al., 2005), takes place when a protein undergoes hydrogen exchange by EX1 and EX2 regimes simultaneously (EX1 in one area and EX2 in another). EX1 kinetics can be characterized by measuring the peak width of the isotopic cluster (Weis et al., 2006a). Increases in peak width (Fig. 2.5c, left) are directly attributed to the EX1 regime. Therefore, the relative peak width versus deuteration time plot shows an increase in peak width for EX1 kinetics whereas no change is observed for EX2 (Fig. 2.5c, right). Much useful information can be obtained using this representation such as the half-life of the EX1 unfolding event (x-axis value when the apex occurs) or the relative number of hydrogens implicated in the EX1 kinetics (magnitude of the displacement in the y-axis). The analysis of peak width provides a simple diagnostic method for identifying EX1 kinetic events, and the process can be partially automated (Weis et al., 2006b).

2.5 CAPABILITIES OF HX MS IN STRUCTURAL BIOLOGY

Having explored the fundamental details of hydrogen exchange mass spectrometry in Sections 2.1–2.4, this section is provided to give an overview of the variety of applications of this technique. The reader will find more detailed applications in subsequent chapters of this book.

2.5.1 Protein Folding Studies

A powerful application of HX MS is the study of protein folding mechanisms. Mass spectrometry represents an excellent means to identify transient intermediates in a folding pathway since exchange MS data are not averaged over all the proteins present in solution. Different protein populations become labeled to different degrees and are therefore separated along the m/z scale. Such separation allows their respective exchange rates to be measured. Miranker et al. 1993 were the first to report the benefits of this experimental procedure in protein folding studies using the hen egg white lysozyme protein. Since then, the procedure has been extended to numerous other proteins such as interleukine 1β (Heidary et al., 1997), apo-myoglobin (Tsui et al., 1999), rabbit muscle aldolase and phosphotriose isomerase (Pan and Smith, 2003; Pan et al., 2004), cytochrome c (Yang and Smith, 1997), and malate dehydrogenase for which the folding process was directly followed inside the cavity of the chaperone GroEL–GroES (Chen et al., 2001).

In this kind of experiment, a protein is typically denatured in a protiated buffer containing a high concentration of denaturant. Refolding is initiated by mixing this solution with an excess of denaturant-free buffer. After a defined period of refolding, the sample is labeled with a pulse of deuterium and the exchange is rapidly quenched. Hydrogens located in folded regions will be protected from exchange during the deuterium pulse while those located in unfolded regions will become deuterated. As the refolding time increases, more parts of the protein become protected before the pulse and these remain fully protiated. An example of the resulting spectra is shown in Fig. 2.6a. A gradual mass shift to lower values with increasing folding time is observed. When no refolding time is allowed before the pulse, the entire population becomes deuterated (U_D) whereas after a sufficient refolding time the protein stays completely protiated (F_H). Between these two extremes, partially folded intermediates (I) may be observed in the presence of either fully protonated or fully deuterated species. (Note that it is possible to start the experiment with a deuterated buffer and to use H_2O as the labeling agent. However, it is advisable to follow the experiment as presented in Fig. 2.6a. Some aggregates or adducted species may appear during the refolding process, and it could be difficult to distinguish aggregates from intermediate forms using an H_2O pulse since both species are characterized by a mass shift to higher values). Protein folding studies by pulse labeling MS may be carried out under either equilibrium or kinetic conditions and in association with proteolytic fragmentation.

2.5.2 Quality Control

The exchange behavior of a folded protein is dictated by its structure. Consequently, each protein will behave slightly differently in a labeling experiment. Compact proteins with numerous secondary structure elements will generally exchange more slowly than proteins containing less secondary structural motifs. The monitored exchange rate as a function of time corresponds therefore to a signature of the protein in a defined folded state. This characteristic can be used to investigate the effects of

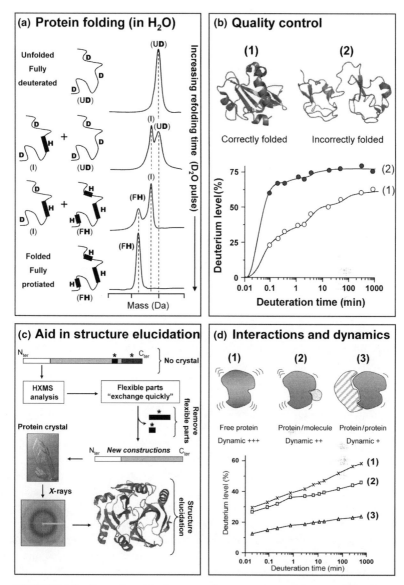

FIGURE 2.6 Examples of hydrogen exchange mass spectrometry applications.

mutations on protein structure and/or to determine if a recombinant protein is correctly folded (Brier et al., 2004, 2006a, 2006b; Pantazatos et al., 2004). HX MS may therefore be considered as a very potent quality control (QC) for correctly folded proteins. This could be applied to QC of preparations of therapeutic proteins, for example.

A simple comparison of the exchange behavior of a misfolded protein versus its correctly folded form (Fig. 2.6b) might indicate if the selected protein preparation method produces suitable folded protein. Such quality control could be performed either by pulse labeling or by continuous labeling. As described in Section 2.3, continuous labeling experiments provide information on the conformational dynamics of the protein under equilibrium conditions. As incorrectly folded regions might affect the overall protein dynamics, a modification of deuterium uptake would be observed. Continuous labeling experiments often require longer labeling periods (up to hours) and therefore would be more time consuming in an industrial setting. Pulse labeling might be more attractive in some situations as only very brief exposure to deuterium is needed (see Section 2.3). In addition, this labeling technique is easily automated allowing a precise control of pulse length and an increase in the number of samples that could be analyzed per hour.

2.5.3 Aid in Structure Elucidation

A major obstacle to determining protein structure by crystallography is obtaining crystals that diffract suitably. Unstructured regions of proteins can prevent crystallization by inhibiting the formation of stable crystal contacts. Removing these disordered parts may therefore greatly improve the probability of crystallization success. Algorithms predicting disorder can be used to localize unstructured regions, and this approach is reasonable when the protein to be crystallized shares structural homologies with known proteins. If the protein is completely unknown (weakly conserved structures motifs, novel folds, no existing structural homologues, etc.), analytical methods that are completely independent of structural comparison/homology are required. NMR spectroscopy can be used for this task but the protein quantity, the time of analysis, and the size of the proteins are often limiting factors.

Classically, crystallographers have used limited proteolysis to identify unfolded regions between domains as such unfolded regions are susceptible to cleavage. Hydrogen exchange mass spectrometry can also provide valuable information about disordered regions and aid in crystallography, as shown in Fig. 2.6c. Proteins that are difficult to crystallize are subjected to a brief D_2O pulse to selectively label the most rapidly exchanging NHs. After pepsin digestion, the deuterium content is determined by MS analysis. Differentiation between the structured/unstructured parts of the protein is based on the deuterium incorporation rates: amide hydrogens in unstructured regions are fully accessible to the solvent and will therefore exchange very quickly compared to those belonging to structured regions. Note that some nonhydrogen-bonded amide hydrogens found in folded regions may also display fast deuterium exchange rates.

Woods and colleagues have used this strategy to localize disordered regions (Pantazatos et al., 2004; Spraggon et al., 2004). The method was successfully applied to obtain crystals of two *Thermotoga maritima* proteins (Spraggon et al., 2004). Initial, MS-free attempts to obtain crystals of full-length versions of both proteins were unsuccessful. Based on the exchange results, disordered regions were removed and the truncated proteins reexpressed. The resulting constructs were all able to crystallize allowing the structure of both proteins to be determined at high resolution ($\sim$2 Å) (Spraggon et al., 2004).

2.5.4 Interactions and Dynamics

Another very important application of HX MS is to study protein interactions (Garcia et al., 2004; Borch et al., 2005; Komives, 2005; Kaveti and Engen, 2006). Upon complex formation, NHs located at protein interfaces may become protected from exchange. Consequently, a decrease of deuterium incorporation is observed by steric exclusion of the solvent. This characteristic, in ideal situations, may be used to identify protein–protein and protein–small molecule binding interfaces by comparing the protection pattern of the noncomplexed proteins with that of the complexed ones (Fig. 2.6d). Recent examples of protein–protein binding regions probed by HX MS include the extracellular signal-regulated protein kinase 2 (ERK2) with the mitogen-activated protein kinase phosphatase 3 (MKP3) (Zhou et al., 2006), the hexameric molecular motor P4 with the bacteriophage ϕ12 procapsid (Lisal et al., 2006), the SHEP1 protein with Cas (Derunes et al., 2006), and the intersubunit interactions of the protein kinase A (Hamuro et al., 2004) and the HIV-1 capsid proteins (Lanman et al., 2003). The precise localization of the binding interface is generally achieved by combining the exchange with proteolytic digestion.

In some circumstances, this method is useful for mapping the binding sites of small ligands (peptides or small molecules) on proteins. The binding region of the metalloproteinase inhibitor doxycycline on the active form of the matrilysin enzyme was recently investigated using this method (Garcia et al., 2005). Other examples include antimitotic drug binding to kinesins (Brier et al., 2006a, 2006), insulin-like growth factor I binding to IGFBP-I (Ehring, 1999), as well as substrate and inhibitor interactions with bacterial enzymes (Wang et al., 1997, 1998).

However, the observed modifications of deuteration upon complex formation must be treated extremely cautiously before drawing conclusions about the location of the interaction sites, especially with small molecules. The alterations to protein dynamics can be far from the site of ligand interaction. In many cases, the molecular motions of the protein are modified upon ligand binding, either locally or at long distances from the interaction site (Halgand et al., 1999; Nemirovskiy et al., 1999; Brier et al., 2004; Jorgensen et al., 2004; Frego and Davidson, 2006; Yao et al., 2006). These allosteric movements may slow down the rates of opening (k_1) and/or closing (k_{-1}) of several regions within the complexed protein leading to a decrease of deuterium uptake (Fig. 2.6d). In addition, all interactions do not cause changes in hydrogen exchange. For example, interactions driven principally by electrostatics may not alter exchange rates in highly ordered (e.g., alpha helices) secondary structural elements (Engen, 2003). To distinguish conformational changes from ligand protection, Komives and colleagues have examined the deuterium off-exchange rates of two protein–protein complexes after a short labeling time (10 min at 25°C) (Mandell et al., 2001). This approach allows rapidly exchanging surface amides to be preferentially labeled so that solvent accessibility changes are principally detected. Alternatively, mutagenesis experiments may also be used to discriminate both phenomena (Brier et al., 2006a).

Modifications of protein dynamics upon complex formation may also be used to probe protein–ligand interaction. In this binding assay, the capacity of a molecule to interact with a protein is estimated by its ability to modify the molecular motions. Binding of peptide ligands to several Src kinase proteins have been investigated with this method (Engen et al., 1997; Hochrein et al., 2006; Weis et al., 2006c).

ACKNOWLEDGMENT

This work was supported by funding from the National Institutes of Health, Grant R01-GM070590.

REFERENCES

Bai, Y., Milne, J. S., Mayne, L., and Englander, S. W., 1993. Primary structure effects on peptide group hydrogen exchange. *Proteins* 17 (1), 75–86.

Bai, Y., Milne, J. S., Mayne, L., and Englander, S. W., 1994. Protein stability parameters measured by hydrogen exchange. *Proteins* 20 (1), 4–14.

Berger, A. and Linderstrøm-Lang, K., 1957. Deuterium exchange of poly-DL-alanine in aqueous solution. *Arch Biochem Biophys* 69, 106–118.

Berger, A., Loewenstein, A., and Meiboom, S., 1959. Nuclear magnetic resonance and the proteolysis of N-methylacetamide. *J Am Chem Soc* 81, 62–67.

Borch, J., Jorgensen, T. J., and Roepstorff, P., 2005. Mass spectrometric analysis of protein interactions. *Curr Opin Chem Biol* 9 (5), 509–516.

Brier, S., Lemaire, D., Debonis, S., Forest, E., and Kozielski, F., 2004. Identification of the protein binding region of S-trityl-L-cysteine, a new potent inhibitor of the mitotic kinesin Eg5. *Biochemistry* 43 (41), 13072–13082.

Brier, S., Lemaire, D., DeBonis, S., Kozielski, F., and Forest, E., 2006a. Use of hydrogen/ deuterium exchange mass spectrometry and mutagenesis as a tool to identify the binding region of inhibitors targeting the human mitotic kinesin Eg5. *Rapid Commun Mass Spectrom* 20 (3), 456–462.

Brier, S., Lemaire, D., DeBonis, S., Forest, E., and Kozielski, F., 2006b. Molecular dissection of the inhibitor binding pocket of mitotic kinesin Eg5 reveals mutants that confer resistance to antimitotic agents. *J Mol Biol* 360 (2), 360–376.

Brier, S., Carletti, E., Debonis, S., Hewat, E., Lemaire, D., and Kozielski, F., 2006c. The marine natural product adociasulfate-2 as a tool to identify the MT-binding region of kinesins. *Biochemistry* 45 (51), 15644–15653.

Calhoun, D. B. and Englander, S. W., 1985. Internal protein motions, concentrated glycerol, and hydrogen exchange studied in myoglobin. *Biochemistry* 24 (8), 2095–2100.

Carter, J. V., Knox, D. G., and Rosenberg, A., 1978. Pressure effects on folded proteins in solution. Hydrogen exchange at elevated pressures. *J Biol Chem* 253 (6), 1947–1953.

Chamberlain, A. K. and Marqusee, S., 1997. Touring the landscapes: partially folded proteins examined by hydrogen exchange. *Structure* 5 (7), 859–863.

Chen, J., Walter, S., Horwich, A. L., and Smith, D. L., 2001. Folding of malate dehydrogenase inside the GroEL–GroES cavity. *Nat Struct Biol* 8 (8), 721–728.

Christoffersen, M., Bolvig, S., and Tuchsen, E., 1996. Salt effects on the amide hydrogen exchange of bovine pancreatic trypsin inhibitor. *Biochemistry* 35 (7), 2309–2315.

Clarke, J. and Itzhaki, L. S., 1998. Hydrogen exchange and protein folding. *Curr Opin Struct Biol* 8 (1), 112–118.

Cotten, M., Fu, R., and Cross, T. A., 1999. Solid-state NMR and hydrogen-deuterium exchange in a bilayer-solubilized peptide: structural and mechanistic implications. *Biophys J* 76 (3), 1179–1189.

Cravello, L., Lascoux, D., and Forest, E., 2003. Use of different proteases working in acidic conditions to improve sequence coverage and resolution in hydrogen/deuterium exchange of large proteins. *Rapid Commun Mass Spectrom* 17 (21), 2387–2393.

Creighton, T. E., 1993. Proteins: Structures and Molecular Properties. New York: W. H. Freeman.

Dempsey, C. E., 1992. Quantitation of the effects of an internal proline residue on individual hydrogen bond stabilities in an alpha-helix: pH-dependent amide exchange in melittin and [Ala-14]melittin. *Biochemistry* 31 (19), 4705–4712.

Dempsey, C. E., 1995. Hydrogen bond stabilities in the isolated alamethicin helix: pH-dependent amide exchange measurements in methanol. *J Am Chem Soc* 117 (28), 7526–7534.

Dempsey, C. E., 2001. Hydrogen exchange in peptide and proteins using NMR spectroscopy. *Prog Nucl Magn Reson Spectrosc* 39 (2), 135–170.

Deng, Y. and Smith, D. L., 1998. Identification of unfolding domains in large proteins by their unfolding rates. *Biochemistry* 37 (18), 6256–6262.

Deng, Y. and Smith, D. L., 1999. Rate and equilibrium constants for protein unfolding and refolding determined by hydrogen exchange-mass spectrometry. *Anal Biochem* 276 (2), 150–160.

Deng, Y. and Smith, D. L., 1999b. Hydrogen exchange demonstrates three domains in aldolase unfold sequentially. *J Mol Biol* 294 (1), 247–258.

Deng, Y., Zhang, Z., and Smith, D. L., 1999a. Comparison of continuous and pulsed labeling amide hydrogen exchange/mass spectrometry for studies of protein dynamics. *J Am Soc Mass Spectrom* 10 (8), 675–684.

Deng, Y., Pan, H., and Smith, D. L., 1999b. Selective isotope labeling demonstrates that hydrogen exchange at individual peptide amide linkages can be determined by collision-induced dissociation mass spectrometry. *J Am Chem Soc* 121 (9), 1966–1967.

Derunes, C., Burgess, R., Iraheta, E., Kellerer, R., Becherer, K., Gessner, C. R., Li, S., Hewitt, K., Vuori, K., Pasquale, E. B., Woods, V. L. Jr., and Ely, K. R., 2006. Molecular determinants for interaction of SHEP1 with Cas localize to a highly solvent-protected region in the complex. *FEBS Lett* 580 (1), 175–178.

Ehring, H., 1999. Hydrogen exchange/electrospray ionization mass spectrometry studies of structural features of proteins and protein/protein interactions. *Anal Biochem* 267 (2), 252–259.

Eigen, M., 1964. Proton transfer, acid-base catalysis, and enzymatic hydrolysis. Part I: elementary processes. *Angew Chem* 3 (1), 1–19.

Ellis, L. M., Bloomfield, V. A., and Woodward, C. K., 1975. Hydrogen–tritium exchange kinetics of soybean trypsin inhibitor (Kunitz): solvent accessibility in the folded conformation. *Biochemistry* 14 (15), 3413–3419.

Emmett, M. R., Kazazic, S., Marshall, A. G., Chen, W., Shi, S. D., Bolanos, B., and Greig, M. J., 2006. Supercritical fluid chromatography reduction of hydrogen/deuterium back exchange in solution-phase hydrogen/deuterium exchange with mass spectrometric analysis. *Anal Chem* 78 (19), 7058–7060.

Engen, J. R., 2003. Analysis of protein complexes with hydrogen exchange and mass spectrometry. *Analyst* 128 (6), 623–628.

Engen, J. R. and Smith, D. L., 2001. Investigating protein structure and dynamics by hydrogen exchange MS. *Anal Chem* 73 (9), 256A–265A.

Engen, J. R., Smithgall, T. E., Gmeiner, W. H., and Smith, D. L., 1997. Identification and localization of slow, natural, cooperative unfolding in the hematopoietic cell kinase SH3 domain by amide hydrogen exchange and mass spectrometry. *Biochemistry* 36 (47), 14384–14391.

Engen, J. R., Gmeiner, W. H., Smithgall, T. E., and Smith, D. L., 1999. Hydrogen exchange shows peptide binding stabilizes motions in Hck SH2. *Biochemistry* 38 (28), 8926–8935.

Englander, S. W., 2006. Hydrogen exchange and mass spectrometry: a historical perspective. *J Am Soc Mass Spectrom* 17 (11), 1481–1489.

Englander, S. W. and Englander, J. J., 1972. Hydrogen–tritium exchange. *Methods Enzymol* 26 (Pt C), 406–413.

Englander, S. W. and Englander, J. J., 1978. Hydrogen–tritium exchange. *Methods Enzymol* 49, 24–39.

Englander, S. W. and Englander, J. J., 1994. Structure and energy change in hemoglobin by hydrogen exchange labeling. *Methods Enzymol* 232, 26–42.

Englander, S. W. and Kallenbach, N. R., 1984. Hydrogen exchange and structural dynamics of proteins and nucleic acids. *Q Rev Biophys* 16 (4), 521–655.

Englander, S. W. and Mayne, L., 1992. Protein folding studied using hydrogen-exchange labeling and two-dimensional NMR. *Annu Rev Biophys Biomol Struct* 21, 243–265.

Englander, S. W. and Poulsen, A., 1969. Hydrogen–tritium exchange of the random chain polypeptide. *Biopolymers* 7 (3), 379–393.

Englander, S. W. and Staley, R., 1969. Measurement of the free and the H-bonded amides of myoglobin. *J Mol Biol* 45 (2), 277–295.

Englander, S. W., Downer, N. W., and Teitelbaum, H., 1972. Hydrogen exchange *Annu Rev Biochem* 41, 903–924.

Englander, S. W., Sosnick, T. R., Englander, J. J., and Mayne, L., 1996. Mechanisms and uses of hydrogen exchange. *Curr Opin Struct Biol* 6 (1), 18–23.

Englander, J. J., Del Mar, C., Li, W., Englander, S. W., Kim, J. S., Stranz, D. D., Hamuro, Y., and Woods, V. L. Jr., 2003. Protein structure change studied by hydrogen–deuterium exchange, functional labeling, and mass spectrometry. *Proc Natl Acad Sci USA* 100 (12), 7057–7062.

Eriksson, M. A., Hard, T., and Nilsson, L., 1995. On the pH dependence of amide proton exchange rates in proteins. *Biophys J* 69 (2), 329–339.

Ferraro, D. M., Lazo, N. D., and Robertson, A. D. 2004. EX1 hydrogen exchange and protein folding. *Biochemistry* 43 (3), 587–594.

Fersht, A. R., 1971. Acyl-transfer reactions of amides and esters with alcohols and thiols. A reference system for the serine and cysteine proteinases. Concerning the N protonation of amides and amide–imidate equilibria. *J Am Chem Soc* 93 (14), 3504–3515.

Frego, L. and Davidson, W., 2006. Conformational changes of the glucocorticoid receptor ligand binding domain induced by ligand and cofactor binding, and the location of cofactor binding sites determined by hydrogen/deuterium exchange mass spectrometry. *Protein Sci* 15 (4), 722–730.

Garcia, R. A., Pantazatos, D., and Villarreal, F. J., 2004. Hydrogen/deuterium exchange mass spectrometry for investigating protein–ligand interactions. *Assay Drug Dev Technol* 2 (1), 81–91.

Garcia, R. A., Pantazatos, D. P., Gessner, C. R., Go, K. V., Woods, V. L. Jr., and Villarreal, F. J., 2005. Molecular interactions between matrilysin and the matrix metalloproteinase inhibitor doxycycline investigated by deuterium exchange mass spectrometry. *Mol Pharmacol* 67 (4), 1128–1136.

Gregory, R. B., Crabo, L., Percy, A. J., and Rosenberg, A., 1983. Water catalysis of peptide hydrogen isotope exchange. *Biochemistry* 22 (4), 910–917.

Halgand, F., Dumas, R., Biou, V., Andrieu, J. P., Thomazeau, K., Gagnon, J., Douce, R., and Forest, E., 1999. Characterization of the conformational changes of acetohydroxy acid isomeroreductase induced by the binding of Mg^{2+} ions, NADPH, and a competitive inhibitor. *Biochemistry* 38 (19), 6025–6034.

Hamuro, Y., Anand, G. S., Kim, J. S., Juliano, C., Stranz, D. D., Taylor, S. S., and Woods, V. L. Jr., 2004. Mapping intersubunit interactions of the regulatory subunit (RIalpha) in the type I holoenzyme of protein kinase A by amide hydrogen/deuterium exchange mass spectrometry (DXMS). *J Mol Biol* 340 (5), 1185–1196.

Heidary, D. K., Gross, L. A., Roy, M., and Jennings, P. A., 1997. Evidence for an obligatory intermediate in the folding of interleukin-1 beta. *Nat Struct Biol* 4 (9), 725–731.

Hilton, B. D., Trudeau, K., and Woodward, C. K., 1981. Hydrogen exchange rates in pancreatic trypsin inhibitor are not correlated to thermal stability in urea. *Biochemistry* 20 (16), 4697–4703.

Hochrein, J. M., Lerner, E. C., Schiavone, A. P., Smithgall, T. E., and Engen, J. R., 2006. An examination of dynamics crosstalk between SH2 and SH3 domains by hydrogen/deuterium exchange and mass spectrometry. *Protein Sci* 15 (1), 65–73.

Hoofnagle, A. N., Resing, K. A., and Ahn, N. G., 2003. Protein analysis by hydrogen exchange mass spectrometry. *Annu Rev Biophys Biomol Struct* 32, 1–25.

Hoofnagle, A. N., Resing, K. A., and Ahn, N. G., 2004. Practical methods for deuterium exchange/mass spectrometry. *Methods Mol Biol* 250, 283–298.

Houliston, R. S., Liu, C., Singh, L. M., and Meiering, E. M., 2002. pH and urea dependence of amide hydrogen–deuterium exchange rates in the beta-trefoil protein hisactophilin. *Biochemistry* 41 (4), 1182–1194.

Hvidt, A. and Nielsen, S. O., 1966. Hydrogen exchange in proteins. *Adv Protein Chem* 21, 287–386.

Jorgensen, T. J., Gardsvoll, H., Dano, K., Roepstorff, P., and Ploug, M., 2004. Dynamics of urokinase receptor interaction with Peptide antagonists studied by amide hydrogen exchange and mass spectrometry. *Biochemistry* 43 (47), 15044–15057.

Jorgensen, T. J., Gardsvoll, H., Ploug, M., and Roepstorff, P., 2005. Intramolecular migration of amide hydrogens in protonated peptides upon collisional activation. *J Am Chem Soc* 127 (8), 2785–2793.

Juranić, N., Predrag, K. I., and Macura, S., 1995. Hydrogen bonding networks in proteins as revealed by the amide 1JNC' coupling constant. *J Am Chem Soc* 117 (1), 405–410.

Kakuda, Y., Perry, N., and Mueller, D. D., 1971. Hydrogen-deuterium exchange of a charged poly(methacrylamide) and its monomeric analog. *J Am Chem Soc* 93 (23), 5992–5998.

Kaltashov, I. A. and Eyles, S. J., 2002a. Crossing the phase boundary to study protein dynamics and function: combination of amide hydrogen exchange in solution and ion fragmentation in the gas phase. *J Mass Spectrom* 37 (6), 557–565.

Kaltashov, I. A. and Eyles, S. J., 2002b. Studies of biomolecular conformations and conformational dynamics by mass spectrometry. *Mass Spectrom Rev* 21 (1), 37–71.

Katta, V. and Chait, B. T., 1991. Conformational changes in proteins probed by hydrogen-exchange electrospray–ionization mass spectrometry. *Rapid Commun Mass Spectrom* 5 (4), 214–217.

Kaveti, S. and Engen, J. R., 2006. Protein interactions probed with mass spectrometry. *Methods Mol Biol* 316, 179–197.

Kim, P. S. and Baldwin, R. L., 1982. Influence of charge on the rate of amide proton exchange. *Biochemistry* 21 (1), 1–5.

Kim, K. S. and Woodward, C., 1993. Protein internal flexibility and global stability: effect of urea on hydrogen exchange rates of bovine pancreatic trypsin inhibitor. *Biochemistry* 32 (37), 9609–9613.

Kim, K. S., Fuchs, J. A., and Woodward, C. K., 1993. Hydrogen exchange identifies native-state motional domains important in protein folding. *Biochemistry* 32 (37), 9600–9608.

Kim, M. Y., Maier, C. S., Reed, D. J., and Deinzer, M. L., 2001. Site-specific amide hydrogen/deuterium exchange in *E. coli* thioredoxins measured by electrospray ionization mass spectrometry. *J Am Chem Soc* 123 (40), 9860–9866.

Komives, E. A., 2005. Protein–protein interaction dynamics by amide H/2H exchange mass spectrometry. *Int J Mass Spectrom* 240, 285–290.

Konermann, L. and Simmons, D. A., 2003. Protein-folding kinetics and mechanisms studied by pulse-labeling and mass spectrometry. *Mass Spectrom Rev* 22 (1), 1–26.

Krishna, M. M., Hoang, L., Lin, Y., and Englander, S. W., 2004. Hydrogen exchange methods to study protein folding. *Methods* 34 (1), 51–64.

Kweon, H. K. and Hakansson, K., 2006. Site-specific amide hydrogen exchange in melittin probed by electron capture dissociation Fourier transform ion cyclotron resonance mass spectrometry. *Analyst* 131 (2), 275–280.

Lanman, J., Lam, T. T., Barnes, S., Sakalian, M., Emmett, M. R., Marshall, A. G., and Prevelige, P. E. Jr., 2003. Identification of novel interactions in HIV-1 capsid protein assembly by high-resolution mass spectrometry. *J Mol Biol* 325 (4), 759–772.

Leichtling, B. H. and Klotz, I. M., 1966. Catalysis of hydrogen–deuterium exchange in polypeptides. *Biochemistry* 5 (12), 4026–4037.

Lisal, J., Kainov, D. E., Lam, T. T., Emmett, M. R., Wei, H., Gottlieb, P., Marshall, A. G., and Tuma, R., 2006. Interaction of packaging motor with the polymerase complex of dsRNA bacteriophage. *Virology* 351 (1), 73–79.

Louie, G., Englander, J. J., and Englander, S. W., 1988. Salt, phosphate and the Bohr effect at the hemoglobin beta chain C terminus studied by hydrogen exchange. *J Mol Biol* 201 (4), 765–772.

Maier, C. S. and Deinzer, M. L., 2005. Protein conformations, interactions, and H/D exchange. *Methods Enzymol* 402, 312–360.

Mandell, J. G., Baerga-Ortiz, A., Akashi, S., Takio, K., and Komives, E. A., 2001. Solvent accessibility of the thrombin-thrombomodulin interface. *J Mol Biol* 306, 575–589.

Martin, B. R. and Hutton, W. C., 1973. Predominant nitrogen-bound hydrogen exchange via oxygen-protonated amide. *J Am Chem Soc* 95 (14), 4752–4754.

Matthew, J. B. and Richards, F. M., 1983. The pH dependence of hydrogen exchange in proteins. *J Biol Chem* 258 (5), 3039–3044.

Mazon, H., Marcillat, O., Forest, E., Smith, D. L., and Vial, C., 2004. Conformational dynamics of the GdmHCl-induced molten globule state of creatine kinase monitored by hydrogen exchange and mass spectrometry. *Biochemistry* 43 (17), 5045–5054.

Mazon, H., Marcillat, O., Forest, E., and Vial, C., 2005. Local dynamics measured by hydrogen/ deuterium exchange and mass spectrometry of creatine kinase digested by two proteases. *Biochimie* 87 (12), 1101–1110.

Miller, D. W. and Dill, K. A., 1995. A statistical mechanical model for hydrogen exchange in globular proteins. *Protein Sci* 4 (9), 1860–1873.

Miranker, A., Robinson, C. V., Radford, S. E., Aplin, R. T., and Dobson, C. M., 1993. Detection of transient protein folding populations by mass spectrometry. *Science* 262 (5135), 896–900.

Molday, R. S. and Kallen, R. G., 1972. Substituent effects on amide hydrogen exchange rates in aqueous solution. *J Am Chem Soc* 94 (19), 6739–6745.

Molday, R. S., Englander, S. W., and Kallen, R. G., 1972. Primary structure effects on peptide group hydrogen exchange. *Biochemistry* 11 (2), 150–158.

Nemirovskiy, O., Giblin, D. E., and Gross, M. L., 1999. Electrospray ionization mass spectrometry and hydrogen/deuterium exchange for probing the interaction of calmodulin with calcium. *J Am Soc Mass Spectrom* 10 (8), 711–718.

Pan, H. and Smith, D. L., 2003. Quaternary structure of aldolase leads to differences in its folding and unfolding intermediates. *Biochemistry* 42 (19), 5713–5721.

Pan, H., Raza, A. S., and Smith, D. L., 2004. Equilibrium and kinetic folding of rabbit muscle triosephosphate isomerase by hydrogen exchange mass spectrometry. *J Mol Biol* 336 (5), 1251–1263.

Pan, J., Wilson, D. J., and Konermann, L., 2005. Pulsed hydrogen exchange and electrospray charge-state distribution as complementary probes of protein structure in kinetic experiments: implications for ubiquitin folding. *Biochemistry* 44 (24), 8627–8633.

Pantazatos, D., Kim, J. S., Klock, H. E., Stevens, R. C., Wilson, I. A., Lesley, S. A., and Woods, V. L. Jr., 2004. Rapid refinement of crystallographic protein construct definition employing enhanced hydrogen/deuterium exchange MS. *Proc Natl Acad Sci USA* 101 (3), 751–756.

Perrin, C. L., 1989. Proton exchange in amides: surprises from simple systems. *Acc Chem Res* 22 (8), 268–275.

Perrin, C. L. and Arrhenius, G. M. L., 1982. Mechanisms of acid-catalyzed proton exchange in *N*-methyl amides. *J Am Chem Soc* 104 (24), 6693–6696.

Pletcher, T. C., Koehler, S., and Cordes, E. H., 1968. Concerning the mechanism of hydrolysis of *N*-methylacetimidate esters. *J Am Chem Soc* 90 (25), 7072–7076.

Qian, H. and Chan, S. I., 1999. Hydrogen exchange kinetics of proteins in denaturants: a generalized two-process model. *J Mol Biol* 286 (2), 607–616.

Ray, J. and Englander, S. W., 1986. Allosteric sensitivity in hemoglobin at the alpha-subunit N-terminus studied by hydrogen exchange. *Biochemistry* 25 (10), 3000–3007.

Rogero, J. R., Englander, J. J., and Englander, S. W., 1986. Individual breathing reactions measured by functional labeling and hydrogen exchange methods. *Methods Enzymol* 131, 508–517.

Rohl, C. A. and Baldwin, R. L., 1994. Exchange kinetics of individual amide protons in 15N-labeled helical peptides measured by isotope-edited NMR. *Biochemistry* 33 (25), 7760–7767.

Rosa, J. J. and Richards, F. M., 1979. An experimental procedure for increasing the structural resolution of chemical hydrogen-exchange measurements on proteins: application to ribonuclease S peptide. *J Mol Biol* 133 (3), 399–416.

Rosenberg, A. and Enberg, J., 1969. Studies of hydrogen exchange in proteins. II. The reversible thermal unfolding of chymotrypsinogen A as studied by exchange kinetics. *J Biol Chem* 244 (22), 6153–6159.

Schleich, T., Gentzler, R., and Von Hippel, P. H., 1968. Proton exchange of *N*-methylacetamide in concentrated aqueous electrolyte solutions. I. Acid catalysis. *J Am Chem Soc* 90 (22), 5954–5960.

Schleich, T., Rollefson, B., and Von Hippel, P. H., 1971. Proton exchange of *N*-methylacetamide in concentrated aqueous electrolyte solutions. II. Acid catalysis in water-dioxane mixtures and base catalysis. *J Am Chem Soc* 93 (25), 7070–7074.

Sheinblatt, M., 1970. Determination of an acidity scale for peptide hydrogens from nuclear magnetic resonance kinetic studies. *J Am Chem Soc* 92 (8), 2505–2509.

Sivaraman, T. and Robertson, A. D., 2001. Kinetics of conformational fluctuations by EX1 hydrogen exchange in native proteins. *Methods Mol Biol* 168, 193–214.

Smith, D. L., Deng, Y., and Zhang, Z., 1997. Probing the non-covalent structure of proteins by amide hydrogen exchange and mass spectrometry. *J Mass Spectrom* 32 (2), 135–146.

Spraggon, G., Pantazatos, D., Klock, H. E., Wilson, I. A., Woods, V. L. Jr., and Lesley, S. A., 2004. On the use of DXMS to produce more crystallizable proteins: structures of the *T. maritima* proteins TM0160 and TM1171. *Protein Sci* 13 (12), 3187–3199.

Tsui, V., Garcia, C., Cavagnero, S., Siuzdak, G., Dyson, H. J., and Wright, P. E., 1999. Quench-flow experiments combined with mass spectrometry show apomyoglobin folds through and obligatory intermediate. *Protein Sci* 8 (1), 45–49.

Tüchsen, E. and Woodward, C., 1985a. Hydrogen kinetics of peptide amide protons at the bovine pancreatic trypsin inhibitor protein-solvent interface. *J Mol Biol* 185 (2), 405–419.

Tüchsen, E. and Woodward, C., 1985b. Mechanism of surface peptide proton exchange in bovine pancreatic trypsin inhibitor. Salt effects and O-protonation. *J Mol Biol* 185 (2), 421–430.

Wales, T. E. and Engen, J. R., 2006a. Hydrogen exchange mass spectrometry for the analysis of protein dynamics. *Mass Spectrom Rev* 25 (1), 158–170.

Wales, T. E. and Engen, J. R., 2006b. Partial unfolding of diverse SH3 domains on a wide timescale. *J Mol Biol* 357 (5), 1592–1604.

Wang, F., Blanchard, J. S., and Tang, X. J., 1997. Hydrogen exchange/electrospray ionization mass spectrometry studies of substrate and inhibitor binding and conformational changes of Escherichia coli dihydrodipicolinate reductase. *Biochemistry* 36 (13), 3755–3759.

Wang, F., Scapin, G., Blanchard, J. S., and Angeletti, R. H., 1998. Substrate binding and conformational changes of *Clostridium glutamicum* diaminopimelate dehydrogenase revealed by hydrogen/deuterium exchange and electrospray mass spectrometry. *Protein Sci* 7 (2), 293–299.

Weis, D. D., Wales, T. E., Engen, J. R., Hotchko, M., and Ten Eyck, L. F., 2006a. Identification and characterization of EX1 kinetics in H/D exchange mass spectrometry by peak width analysis. *J Am Soc Mass Spectrom* 17 (11), 1498–1509.

Weis, D. D., Engen, J. R., and Kass, I. J., 2006b. Semi-automated data processing of hydrogen exchange mass spectra using HX-express. *J Am Soc Mass Spectrom* 17 (12), 1700–1703.

Weis, D. D., Kjellen, P., Sefton, B. M., and Engen, J. R., 2006c. Altered dynamics in Lck SH3 upon binding to the LBD1 domain of *Herpesvirus saimiri* Tip. *Protein Sci* 15 (10), 2402–2410.

Woodward, C. K., 1977. Dynamic solvent accessibility in the soybean trypsin inhibitor–trypsin complex. *J Mol Biol* 111 (4), 509–515.

Woodward, C. K. and Hilton, B. D., 1980. Hydrogen isotope exchange kinetics of single protons in bovine pancreatic trypsin inhibitor. *Biophys J* 32 (1), 561–575.

Woodward, C. K. and Rosenberg, A., 1971. Studies of hydrogen exchange in proteins. VI. Urea effects on ribonuclease exchange kinetics leading to a general model for hydrogen exchange from folded proteins. *J Biol Chem* 246 (13), 4114–4121.

Woodward, C. K., Ellis, L. M., and Rosenberg, A., 1975. The solvent dependence of hydrogen exchange kinetics of folded proteins. *J Biol Chem* 250 (2), 440–444.

Wu, Y., Engen, J. R., and Hobbins, W. B., 2006. Ultra performance liquid chromatography (UPLC) further improves hydrogen/deuterium exchange mass spectrometry. *J Am Soc Mass Spectrom* 17 (2), 163–167.

Xiao, H., Hoerner, J. K., Eyles, S. J., Dobo, A., Voigtman, E., Mel'cuk, A. I., and Kaltashov, I. A., 2005. Mapping protein energy landscapes with amide hydrogen exchange and mass spectrometry: I. A generalized model for a two-state protein and comparison with experiment. *Protein Sci* 14 (2), 543–557.

Yan, X., Zhang, H., Watson, J., Schimerlik, M. I., and Deinzer, M. L., 2002. Hydrogen/deuterium exchange and mass spectrometric analysis of a protein containing multiple disulfide bonds: solution structure of recombinant macrophage colony stimulating factor-beta (rhM-CSFbeta). *Protein Sci* 11 (9), 2113–2124.

Yan, X., Watson, J., Ho, P. S., Deinzer, M. L., 2004. Mass spectrometric approaches using electrospray ionization charge states and hydrogen–deuterium exchange for determining protein structures and their conformational changes. *Mol Cell Proteomics* 3 (1), 10–23.

Yang, H. and Smith, D. L., 1997. Kinetics of cytochrome c folding examined by hydrogen exchange and mass spectrometry. *Biochemistry* 36 (48), 14992–14999.

Yao, Z. P., Zhou, M., Kelly, S. E., Seeliger, M. A., Robinson, C. V., and Itzhaki, L. S., 2006. Activation of ubiquitin ligase SCF(Skp2) by Cks1: insights from hydrogen exchange mass spectrometry. *J Mol Biol* 363 (3), 673–686.

Zhang, Z. and Smith, D. L., 1993. Determination of amide hydrogen exchange by mass spectrometry: a new tool for protein structure elucidation. *Protein Sci* 2 (4), 522–531.

Zhou, B., Zhang, J., Liu, S., Reddy, S., Wang, F., and Zhang, Z. Y., 2006. Mapping ERK2–MKP3 binding interfaces by hydrogen/deuterium exchange mass spectrometry. *J Biol Chem* 281 (50), 38834–38844.

Zubarev, R. A., Kelleher, N. L., and McLafferty, 1998. Electron capture dissociation of multiply charged proteins cations: a nonergotic process. *J Am Chem Soc* 120 (13), 3265–3266.

Covalent Labeling Methods for Examining Protein Structure and Protein Interactions

KEIJI TAKAMOTO and JANNA KISELAR

Center for Proteomics, Case Western Reserve University, Cleveland, OH, USA

3.1 INTRODUCTION

Methods that modify proteins covalently and probe their structure have been known for many years (Pauly used diazonium salts as a covalent reagent in 1904). Specific methods that use modification reagents to map "buried" versus "free" histidine side chains in proteins were introduced over 40 years ago by Hachimori and Horinishi (Hachimori et al., 1964; Horinishi et al., 1964). Footprinting of proteins by chemical modification was refined by Hanai and Wang in 1994 (Hanai and Wang, 1994) in which they mapped the solvent accessibility of lysine residues by determining their reactivity to modification by acetylation. The basic approach common to these methods is to generate modified sites within proteins using covalent adducts; thus, these sites can be used to probe conformational changes. However, the drawbacks of utilizing chemical reagents in protein footprinting experiments include preference of these reagents to a specific position within the protein, ultimately limiting the number of available reactive sites that can be probed or the large size of some reagents.

Another well-established footprinting method for probing protein structure in solution utilizes hydroxyl radicals as the reagent to modify macromolecules covalently. First, hydroxyl radical has been accepted as an ideal reagent for nucleic acid footprinting experiments after Tullius and Dombroski (1986) introduced Fe(II)-ethylendiaminetetraacetate (EDTA) Fenton–Haber–Weiss chemistry to generate radicals. The gamma rays were also used as a radiolysis source to generate hydroxyl radicals. The introduction of high-flux synchrotron X-rays for hydroxyl radical generation enabled millisecond timescale footprinting, which was used to examine Mg^{2+}-dependent RNA folding (Chance et al., 1997; Sclavi et al., 1998). The first

Mass Spectrometry Analysis for Protein–Protein Interactions and Dynamics, Edited by Mark Chance
Copyright © 2008 John Wiley & Sons, Inc.

report of protein footprinting method appeared in 1988 (Sheshberadaran and Payne, 1988). This early approach that utilized limited proteolysis of proteins followed by SDS polyacrylamide gel electrophoresis (SDS-PAGE) to separate the cleaved fragments resembled nuclease cleavage methods for DNA. Compared to nucleic acid gel methods, which have single-nucleotide resolution, protein gel approach provided less spatial resolution of structure. The use of tagged Fe(II)-EDTA (Rana and Meares, 1991) and free Fe(II)-EDTA Fenton chemistry (Heyduk and Heyduk, 1994) improved protein footprinting technique; however, nonspecific backbone cleavage with Fenton was relatively inefficient. The development of new analytical approaches for examining proteins that utilize mass spectrometry (MS) for the detection of cleaved fragments or irreversibly modified peptides after proteolysis improved the spatial resolution of the technique considerably (Suckau et al., 1992; Hanai and Wang, 1994). However, the intrinsic inefficiency of backbone cleavage is a serious drawback for the conduct of footprinting experiments. Analysis of the previous studies of the metal-catalyzed oxidation and radiolysis of peptides and proteins suggested that side chain modification by hydroxyl radical should be efficient and rapid (Garrison et al., 1970). The chemistry of amino acid and peptide oxidation using MS detection was investigated subsequent to radiolysis (Goshe and Anderson, 1999; Maleknia et al., 1999a). These experiments showed that oxidative modification of side chains is observed in considerable excess in dilute aqueous solution compared to backbone cleavage or cross-linking. Furthermore, tandem mass spectrometry method was found to be ideal for examining the specific sites of oxidation (Maleknia et al., 1999a; Chance 2001; Kiselar et al., 2002). Thus, the reactivity of a number of the side chains with hydroxyl radical and the changes of this reactivity as a function of ligand binding, unfolding, or macromolecular interactions result in changes in the surface accessibility at the defined sites. Knowledge of the radiolysis chemistry of amino acids available from the literature and the extensive studies of the radiolysis chemistry of amino acids, peptides, and proteins using MS-based detection in the past few years (Goshe et al., 2000; Nukuna et al., 2001; Xu et al., 2003, 2005a; Xu and Chance, 2004, 2005) have resulted in the development of hydroxyl-radical-mediated protein footprinting as an effective method for probing protein structure (Chance, 2001; Goldsmith et al., 2001; Guan et al., 2002; Kiselar et al., 2002, 2003a, 2003b; Liu et al., 2003; Rashidzadeh et al., 2003; Guan et al., 2004). In addition to radiolytic approaches, hydroxyl radical protein footprinting methods based on chemical generation of hydroxyl radicals by both Fenton chemistry (Sharp et al., 2003) and UV photolysis of hydrogen peroxide (H_2O_2) have been developed (Sharp et al., 2004, 2005).

3.2   CHEMISTRY OF HYDROXYL RADICAL FOOTPRINTING

Hydroxyl radicals used in the protein footprinting experiments have significant advantages over chemical reagents. First, these radicals have van der Walls area and solvent properties similar to those of water molecules. This makes hydroxyl radicals ideal as solvent accessibility probes for footprinting experiments. Second, they are highly reactive species that react with a number of surface accessible amino acid side chains and their chemical selectivity is well understood. These increase the number of target sites compared to chemical modifications, maximizing the structural information from the protein footprinting experiments. Third, they can be generated safely and conveniently under a wide range of solution conditions. In this chapter, we outline two main methods of hydroxyl radical generation, the reaction chemistry of hydroxyl radicals with different amino acid residue side chains, the relative reactivity of amino acid side chains, and principles of the protein footprinting approach.

3.2.1 Generation of Hydroxyl Radicals

Hydroxyl radicals can be generated by two main methods: transition metal-dependent chemical generation from peroxide and radiolysis of water (Fig. 3.1) (other methods can be used as well and are described in Chapter 4). Among the chemical methods, the most frequently used is Fe(II)-EDTA Fenton chemistry (Tullius and Dombroski, 1986), which requires commonly available chemicals that are not expensive and easy to handle. The EDTA chelate is used to neutralize the transition metal's positive charge; however, specific interactions of the chelate with macromolecules can bias the local reactivity observed. On the other hand, radiolysis of water using gamma or X-rays generates radicals isotropically in the solution and does not require the addition of any chemicals, although buffering of any solution for biochemical conditions of interest is desirable. In addition, radiolysis can also be used to carry out footprinting *in vivo* (Ottinger and Tullius, 2000). If high-flux X-rays from synchrotron sources are employed, millisecond exposures are sufficient for footprinting such that millisecond time resolution footprinting can be carried out (Kiselar et al., 2002; Guan et al., 2004).

The initial footprinting approach was based on the nonspecific cleavage of protein backbone using free hydroxyl radicals in solution in conjunction with SDS-PAGE to separate the cleavage products (Rana and Meares, 1991; Heyduk and Heyduk, 1994; Heyduk et al., 1996; Baichoo and Heyduk, 1997, 1999; Frank et al., 1998). The chemistry involves the hydrogen abstraction from the $C\alpha$ carbon and the subsequent

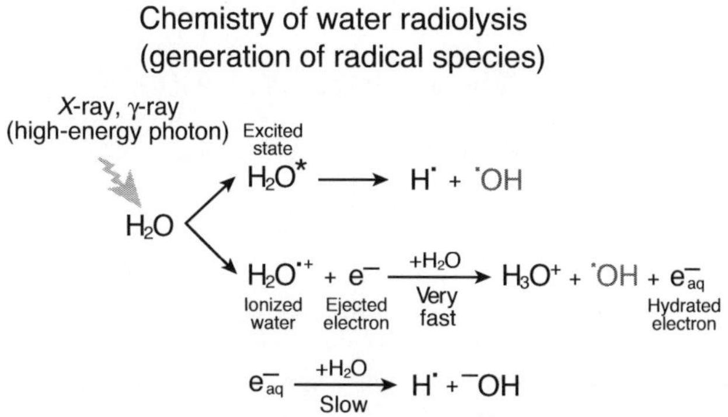

FIGURE 3.1 Schematic representation of hydroxyl radical generation during radiolysis of water upon ionizing radiations.

reaction of the radical species with oxygen leading to backbone cleavage, generating a new N-terminal end (Garrison et al., 1970, 1987). However, the reaction of side chains with hydroxyl radicals occurs at rates 10–1000 times faster than the abstraction of hydrogen from the Cα carbon (Garrison, 1987; Maleknia et al., 1999a; Xu and Chance, 2005). Thus, side chains are preferable as probes for the study of protein structure and dynamics. Use of hydroxyl radicals as probes of the solvent-accessible protein surface combined with quantitative mass spectrometric analyses of the covalently modified protein fragments provides a footprinting approach that can be used to probe protein structure, protein–ligand, and protein–protein interactions (Creighton, 1993; Guan et al., 2002, 2004, 2005; Kiselar et al., 2002, 2003a, 2003b; Liu et al., 2003; Rashidzadeh et al., 2003; Sharp et al., 2003, 2004, 2005). The side chain probe set for protein footprinting is revealed by examining the reactions of proteins and amino acids with radicals using mass spectrometry analysis under aerobic (Garrison, 1987; Maleknia et al., 1999a; Xu et al., 2003, 2004, 2005) or anaerobic conditions (Garrison, 1987; Maleknia et al., 1999a; Goshe et al., 2000; Nukuna et al., 2001).

3.2.2   Reactions of Hydroxyl Radical Products: Nature of Amino Acid Modifications

Major and some minor products from the reactions of amino acid side chains with hydroxyl radicals are shown in Fig. 3.2. In particular, hydroxyl radical reactions with aliphatic side chains of Leu, Ile, Val, and Ala are initiated by hydrogen abstraction from side chain carbon atoms. Under aerobic conditions, this carbon radical eventually reacts with oxygen and forms hydroxyl groups as the major products or aldehyde/ketone groups as the minor products (in Ala the side chain is converted to an aldehyde) resulting in a mass increase of 16 or 14 Da, respectively. In the case of amino acids with aromatic side chains, such as Phe, Tyr and Trp, hydroxyl radicals abstract hydrogen from the aromatic ring or insert directly. The former generates hydroxycyclohexadienyl radicals (in the case of phenylalanine and tyrosine) that further react with oxygen or hydroperoxyl radicals to yield hydroxylphenyl side chain groups. The addition of hydroxyl radical at the double bond of the indole heterocyclic ring of tryptophan generates a formylkinurenine product. Reactions of hydroxyl radicals with aromatic rings results in +16-Da or higher integer multiples of 16 (e.g., +32, +48) increases in mass. Acidic residues react with hydroxyl radical and lose CO_2, resulting in an aldehyde with an overall reduction in mass of 30 Da. Arginine reacts with hydroxyl radicals forming unique products, while the reaction of lysine is identical to that of aliphatic residues. Oxidation of proline produces hydroxyproline and pyroglutamine acid, while histidine undergoes a complicated set of reactions, including ring opening, which generates multiple products. Hydrogen abstraction of β- or γ-carbon of arginine upon hydroxyl radical attack results in products with +16- or +14-Da mass changes. However, when the δ-carbon is attacked, a guanidino elimination reaction follows, resulting in a product with a 43-Da reduction in mass. The loss of the strong positive charge in this reaction may influence the digestion with trypsin and make detection of reaction products difficult in positive-ion mode MS.

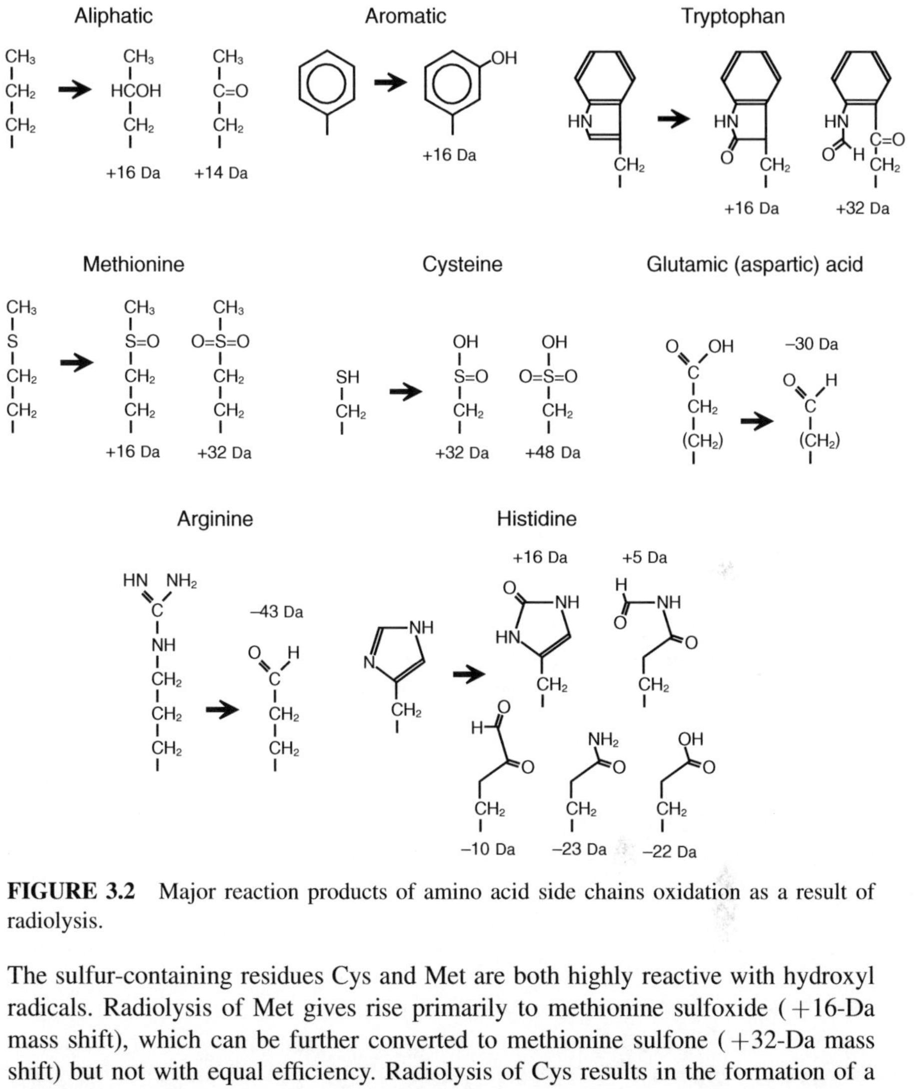

Aliphatic

Aromatic

Tryptophan

Methionine

Cysteine

Glutamic (aspartic) acid

Arginine

Histidine

FIGURE 3.2 Major reaction products of amino acid side chains oxidation as a result of radiolysis.

The sulfur-containing residues Cys and Met are both highly reactive with hydroxyl radicals. Radiolysis of Met gives rise primarily to methionine sulfoxide ($+16$-Da mass shift), which can be further converted to methionine sulfone ($+32$-Da mass shift) but not with equal efficiency. Radiolysis of Cys results in the formation of a primary oxidation product (sulfonic acid) with mass shift $+48$ Da. Oxidation of sulhydryl in Cys to sulfinic acid produces product with $+32$-Da mass shift with less efficiency. The formation of negatively charged sulfonic and sulfinic acid groups in these oxidation products decreases ionization efficiency in positive electrospray mode. The uniqueness of the chemical products can be utilized in some cases to identify the oxidation sites on the basis of characteristic mass signatures.

3.2.3 Relative Reactivity of Amino Acid Side Chains

The extent of radiolytic modification of amino acid side chains in proteins in solution depends on both its ability to react with hydroxyl radicals and its steric accessibility to solvent. The potential of an amino acid residue to serve as a valuable protein footprinting probe under typical aerobic conditions is determined by its relative reactivity and by the ability to detect oxidative products by mass spectrometry. The relative efficiency of reaction of the side chains under aerobic conditions based on MS detection was established as follows: Cys > Met > Trp > Tyr >Phe > Cystine > His > Leu ∼ Ile > Arg ∼ Lys ∼ Val > Ser ∼ Thr ∼ Pro > Gln ∼ Glu > Asp ∼ Asn > Ala > Gly. The low efficiency of detection for residues such as Gly, Ala, Asp, and Asn is due to the generation of minimal oxidized species (Gly is similar in reactivity to backbone carbon atoms) and/or the poor ionization efficiency of some nonstandard oxidation products (e.g., Arg, Asp, Glu). This translates into generation of weak ion signals that are not routinely detectible by mass spectrometry. Thus, these amino acids do not (so far) serve as informative probes in typical protein footprinting experiments. Ser and Thr, though as reactive as Pro, a residue seen to be very valuable in structural analysis, also have reaction products that are difficult to detect, presumably due to low ionization efficiency of the products. The sulfur-containing and aromatic amino acid residues (Met and Cys) are most susceptible to reaction with hydroxyl radicals, and the products of such a reaction can be easily detected in MS experiments. In addition, the sulfur-containing residues are susceptible to secondary oxidation after hydroxyl radical exposure; radiolysis can generate less reactive and long-lived radical species such as hydrogen peroxide and related peroxide radicals. If these species are not quenched, the observed extent of oxidation of peptides can increase during sample processing steps after radiolysis. In particular, methionine is known for its susceptibility to such oxidation. For example, in Fenton chemistry, hydrogen peroxide must be removed or quenched right after reaction completion. Thus, secondary oxidation can be avoided by the addition of reducing species such as methionine in its amino acid or amide forms (in excess) to compete with unwanted secondary oxidation of methionine and cysteine residues in digested peptides or protein species (Xu et al., 2005a). About half of the 20 residues are routinely used in protein footprinting experiments across many laboratories; the oxidation products from these amino acids can be relatively easily detected by mass spectrometry. In some cases, oxidation of other residues has been observed, such that the published set of oxidized residues in proteins that contribute to structural analysis (Xu et al., 2003) has been expanded. Overall, 14 of 20 residues have so far been used for protein footprinting. These 14 residues cover approximately 65% of the sequence of the typical protein, making structural resolution of the technique reasonable (Xu et al., 2003).

3.2.4   Principles of Hydroxyl Footprinting and Protein Integrity During Radiolysis

In the development of quantitative footprinting, which relies on modification-based methods, a number of principles have been laid down that need to be kept in mind when carrying out experiments of this type. First, the emphasis on first-order kinetics for the dose response has been stressed (see below). This is typically the case if the extent of protein modification, at the peptide level, is kept to the range

30–50%. The first-order behavior, extrapolated to 0% modified, is characteristic of a single class of side chain reactivity unaffected by dose within the observed range.

A second constraint to consider is what are the types and locations of the oxidative modifications and what effect are they likely to have on the protein structure. The modifications are typically oxygen additions, although some ring openings do occur. The oxidized residues are primarily located on the protein surface (with the exception of methionine). Making polar additions to surface residues is likely to have little effect on structure or stability. From numerous examples in the literature of introducing charged residues onto the surface of proteins, it is known that in some cases many residues can be introduced with minimal effect on protein structure (Creighton, 1993).

A third constraint to consider is the timescale on which the dose is introduced. If the dose is delivered faster than the protein has time to respond, then any damaging effect of the dose can be ignored. Protein structure changes can occur on the millisecond timescales. However, significant structural changes are often slower, on the order of seconds.

Finally, one needs to consider the method of data analysis, specifically, to what degree it is sensitive to any damage to the macromolecule. Our suggested method of examining the loss of unmodified material has several advantages in this respect. In particular, the rate constant derived from the analysis is sensitive entirely to the unmodified population, and "modified" molecules are not sampled. However, this is only true on a "local" scale. A protein molecule can be modified on one part of the protein, while another site within a different peptide could be unmodified and "sense" the modification elsewhere. For a peptide, being unmodified is a "local" determination, while for the protein, any modification moves the entire molecule out of the "unmodified" column. For the modification at one site to influence the reactivity of another site in another peptide, the "information" would need to be transmitted across the macromolecule, which is entirely conceivable. However, as we have seen, for an effect to be transmitted, it must change the protein structure and stability (not likely since the modifications are on the surface), must be transmitted quickly (since the dose is delivered within tens of milliseconds), must occur in a substantial fraction of the population (which is prevented by keeping the total amount of modification low), and must be significant enough to alter the accessibility of the affected residue. From the typical protein footprinting data the time points along the dose–response curve shows a constant rate of oxidation extrapolated to 0% modified. Any changes in reactivity induced by modification of the population would be reflected in increases or decreases in the rate of modification of a particular peptide. Thus, the method has internal controls to validate the results. The previous studies showed clearly that this method gives accurate structural data, as buried residues are not found to be reactive and accessible residues react in accordance with their solvent accessibility and intrinsic reactivity. If the above constraints are considered in the protein footprinting experiment, it is expected that the method can be a valuable tool for examining structure.

3.3 MASS SPECTROMETRY APPROACHES FOR QUANTITATIVE PROTEIN FOOTPRINTING

Mass spectrometric data of oxidized proteins correlate with the solvent accessibility of the residue side chains according to peptide reactivity and provide structural resolution at the level of single amino acid side chain using tandem MS. Accurate measure of the extent of oxidation provides the basis for sensitive determination of structural changes in ligand binding and in protein interactions. In this chapter, we outline the mass spectrometry approaches for quantitation of protein oxidation from hydroxyl radical footprinting experiments.

3.3.1 Quantification of Peptide Oxidation Using LC–MS

The irradiated protein is subjected to mass spectrometry analyses subsequent to protease digestion. As the radiolytic modifications of proteins are covalent and stable (unlike in deuterium exchange), subsequent analyses are relatively easy to carry out. For example, samples can be stored frozen after footprinting experiments for further analysis; a variety of specific and nonspecific proteases can be used at a range of temperatures and digestion times to generate peptide fragments of interest. In addition, reducing agents such as dithiothreitol(DTT), and tris (2-carboxyethyl) phosphine (TCEP) can be used to digest heavily disulfide-bonded proteins.

Proteolytic enzymes are selected for digestion to yield the highest protein coverage. Specifically, proteolysis is generally performed individually with a set of specific proteases such as trypsin, Asp-N, and/or Glu-C to maximize sequence coverage. Coverage of total protein sequence ranging from 80% to 95% is generally desirable. The protease digestion of the oxidized protein results in the formation of peptides and their stable oxidized products that can be separated by the gradient of reverse-phase chromatography and quantitated using mass spectrometry (Fig. 3.3). Because the proteolytic peptides and their oxidized products have in many cases very similar structure and a molecular mass that is shifted by $+16$ or $+32$ atomic mass units (corresponding to the addition of one or two oxygens, respectively), the ionization and detection efficiencies of these peptides are relatively comparable. However, since the oxygen adducts result in the formation of more polar peptides, the modified peptides are typically eluted from the reverse-phase column prior the unmodified peptides.

The peak area of each peptide is calculated from the total ion current chromatogram (TIC). In particular, specific m/z (mass-to-charge ratio) ions are extracted from TIC, and peak area is calculated by integration of the intensity values of this ion signal. Moreover, a set range of the retention time (determined empirically) within the chromatogram is used for each experiment to ensure consistency of quantitation for the modified and unmodified species. Importantly, all multiple modified species with single or multiple modifications that are present for the same peptide are quantified and added into the sum total modified. Then,

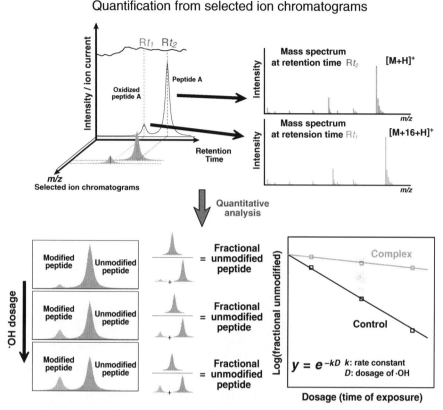

FIGURE 3.3   Representation of quantitative mass spectrometric analysis.

the fraction of unmodified peptide is calculated from the ratio of the peak area under the ion signal of the unmodified peptides to the peak area of the sum total modified. To provide oxidation rate data that emphasize the intact population, we monitor the loss of unmodified fraction. In contrast, the oxidation of already oxidized material would be directly measured by analyzing the rate of formation of the modified population. The dose-response curves are generated by plotting the fraction unmodified (on a log scale) for each peptide as a function of exposure time. Multiple independent experiments are globally fit to provide the oxidation rate constants using nonlinear regression. The data obey a pseudo-first-order reaction in the desired regime. Deviation from these kinetics is observed at increased exposure times due to protein degradation and secondary oxidations. Thus, the time points of oxidation have to be carefully chosen, and the time points at increased time that signal secondary oxidation should be removed from the fits in the data analysis.

3.3.2 Confirmation of Peptide Identity and Determination of Modification Sites by MS/MS

An important aspect of analysis is the confirmation that a peak with a particular m/z value corresponds to a peptide expected from the sequence based on the theoretically predicted digestion pattern. Moreover, structural resolution requires determination of the modification site(s) on the peptide. The fragmentation of peptides by collision-induced dissociation (CID) is typically used for this purpose (Hunt et al., 1986; Johnson et al., 1987; Tang et al., 1993; Tabb et al., 2004). The translational kinetic energy of selected ions is converted to internal energy of the ions (for low-energy collisions this energy is mostly vibrational), inducing cleavage of the peptide backbone. The fragments are termed a-, b-, and c-series (for N-terminal fragments) or x-, y-, and z-series (for C-terminal fragments) depending on the location of the bond cleavage (Fig. 3.4a) (Roepstorff and Fohlman, 1984). The most frequently observed ion types in low-energy collisions are b- and y-series ions, resulting from cleavage of the peptide bond with charge retention in the N- or C-terminal end of the peptide, respectively. By examining the mass differentials of ion peaks observed in the MS/MS spectrum, the peptide sequence can clearly be read (with exceptions for isobaric amino acids or posttranslational modifications). Thus, tandem mass spectrometry experiment for a particular m/z ion is routinely performed to confirm the identity of the peptide and determine the modification site generated by oxidation.

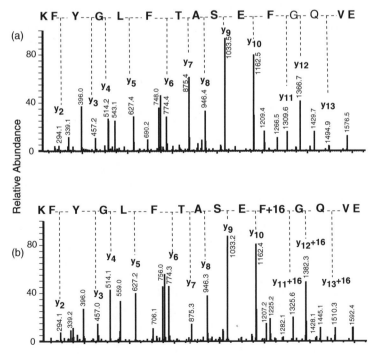

FIGURE 3.4 Representative tandem mass spectra utilized for peptide identification and determination of its modification sites.

Figure 3.4 shows representative ESI-MS/MS spectra upon selection of the doubly charged ion (m/z 862.0) of peptide with sequence EVQGFESATFLGYFK and its singly oxidized radiolytic product ($+16$). The y_n or b_n fragment ions observed represent cleavage products of the peptide bond with charge retention in the C- or N-terminal end of the peptide, respectively. The sequence for the unmodified peptide can clearly be confirmed from the spectrum in Fig. 3.4a. This is the control experiment for comparison so as to determine the location of the potentially modified (shifted) daughter ions in the case of the oxidized peptide. Specifically, m/z values for y_{2-13} give mass differences of 163, 57, 113, 147, and so on, that correspond to resides Y, G, L, F, and so on, respectively. The same analysis is performed for b-series of fragment ions (data not shown in Fig. 3.3). If the peptide has one oxidation adduct due to alcohol formation, the selected precursor ion will have m/z 870.0 for doubly charged ion (862.0 $+8$). The MS spectra of the singly oxidized peptide is shown in Fig. 3.4b, and the y-type series ions y_{2-10} are unchanged upon comparison to the spectrum of the unmodified peptide in panel B, while ions y_{11-13} are shifted by $+16$ Da. On the other hand, for the b-type series, ions b_{5-14} show mass shifts of $+16$ Da. No amount of the unmodified form for y_{11-13} and b_{5-14} ions is observed above the noise level. This indicates that phenylalanine at position 5 on the N-terminal end of the peptide is the only residue oxidized.

We can determine the modification site and confirm the identity of peptide; however, the CID process and spectrum are semiquantitative at best, so it will be risky to estimate the relative contribution of multiple residues in the case of multiple modification sites. Multiple enzymatic digestion can be used to determine the relative contribution of individual residues. If, as a result of using different enzymes, the digestion sites can isolate subfragments, it is sometimes possible to estimate the contribution from each site. However, it is highly sequence dependent and not always applicable.

3.4 EXAMPLES OF VARIOUS METHODS THAT GENERATE HYDROXYL RADICALS IN SOLUTION TO EXAMINE PROTEIN STRUCTURE

Hydroxyl-radical-mediated protein footprinting coupled to mass spectrometry has been developed rapidly over the last several years. It has matured into a viable approach for defining protein structure, subunit assembly, and conformational changes of macromolecules in solution based on measurements of reactivity of amino acid side chain groups with covalent modification reagents. Different groups have developed variations of the technique to adapt to specific biological problems of interest. In this section, we provide an overview of some of the protein footprinting approaches and their applications to probe protein structure and protein interactions.

3.4.1 Radiolytic Footprinting of Cytochrome *c*

Anderson's group has been developing an anaerobic radiolysis method that utilizes nonsolvent-exchangeable hydrogen/deuterium (H/D) exchange mediated by hydroxyl radical oxidation (Goshe et al., 2000; Nukuna et al., 2001; Xu et al., 2005a). Radiolysis leads to hydrogen abstraction from Cα (or side chain carbon atoms) and forms radicals at these positions. This radical is then repaired by dithiothreitol that has been equilibrated with deuterium oxide (D_2O) under anaerobic conditions. The –SH group has been replaced by –SD; thus, when the deuteron repairs the radical, a nonexchangeable deuterium addition event occurs providing a mass tag for the solvent-accessible site. These data were compared with the analysis of radiolytic modification of cytochrome c under aerobic conditions (Nukuna et al., 2004). The study also includes analyses of oxidation mechanisms of certain residues. Anderson and coworkers utilized 50% ^{18}O-labeled H_2O as a source of radicals generated by gamma-ray irradiation to differentiate the reaction directly by hydroxyl radical (contains ^{18}O) and through oxygen (mainly ^{16}O). As expected from previous literature (Goshe et al., 2000), no +18-Da oxidation of Tyr was observed, indicating that the oxidation source is molecular oxygen. They also found a +14-Da modification product of Tyr that was presumably the carbonyl form. They concluded that the oxidation of Tyr itself is mediated by the addition of radical but that carbonyl oxygen is exchanged with solvent during subsequent analyses.

3.4.2 Fenton Hydroxyl Radical Footprinting

Fenton's reaction was invented in the 1890s by H. J. H. Fenton when he discovered that ferrous iron(II) catalytically promoted the oxidation of tartaric acid by hydrogen peroxide. In Fenton reaction hydroxyl radicals are generated through the oxidation of Fe(II) to Fe(III) by H_2O_2. The Fenton system that includes three essential components Fe(II)-EDTA, H_2O_2, and ascorbate has become a standard method used for structural analyses of nucleic acids for years (Tullius and Dombroski, 1985, 1986; Wang and Padgett, 1989; Celander and Cech, 1990; Balasubramaniun et al., 1998; Hampel and Burke, 2001a, 2001b; Takamoto et al., 2002).

More recently, Hettich and coworkers applied Fenton chemistry-based hydroxyl radical footprinting to proteins. Hettich's group probed apomyoglobin solution structure using Fenton chemistry and analyzed the products by LC–MS/MS utilizing ES-FT-ICR-MS (electrospray Fourier transform ion cyclotron resonance mass spectrometry). They reported that apomyoglobin's structure by NMR is consistent with their footprinting experimental results. NMR studies showed that the overall structure of apomyoglobin is globular and rigid except in selected regions. From the crystal structure of native myoglobin, W7 and L11 amino acids are not solvent accessible, while Hettich and coworkers observed oxidative modifications for these residues. However, NMR analyses reported that the region that covers these two residues is disordered in solution. Thus, it is clear from footprinting data that these residues are much more accessible than those in the myoglobin crystal structure, consistent with the NMR data.

The drawback of the Fenton approach is that Fe^{2+} with EDTA and relatively high concentrations of ascorbate required for Fenton chemistry may affect the structure of proteins for some experiments. Also, H_2O_2 can react directly with S-containing residues; thus, these cannot be analyzed.

3.4.3   Laser Photolysis of H_2O_2

Knowing the disadvantages of the Fenton approach, Hettich's group has used a well-known photochemical method for generating hydroxyl radicals (Sharp et al., 2004, 2005). The basis for this approach is UV light-induced homolysis of hydrogen peroxide that can generate hydroxyl radicals. The UV photolysis of H_2O_2 has been widely used in industrial processes such as water treatment and medical equipment sterilization. More recently, it has been employed for nucleic acid footprinting studies and in protein surface oxidation studies. Sharp et al. (2004) applied this method to analyze the solvent-accessible surface of two model proteins, lysozyme and lactoglobulin. Protein samples irradiated by using UV lamp in the presence of 15% H_2O_2 were analyzed by FT-MS (Fourier transform mass spectrometry). The oxidation of amino acid side chains was found to be consistent with their solvent accessibility. In addition, they found a linear relationship between the accessible surface area and the extent of oxidation for both the entire protein (until cooperative oxidation is observed at high doses) and peptides digested by protease. Using FT instrument, they reported (Sharp et al., 2004) additional peptides that were missing in LC–MS/MS experiments and could be found in direct infusion ES-FT-MS due to the higher resolution and wider dynamic range.

Furthermore, Aye and coworkers applied a nanosecond laser-induced photochemical oxidation method for protein surface mapping to reduce the exposure time and concentration of H_2O_2 needed for protein surface modification (Aye et al., 2005). The oxidation of amino acid side chains was found to be consistent with their solvent accessibility. In this method, two model proteins, ubiquitin and apomyoglobin, were oxidized by hydroxyl radicals generated by exposure to a Nd:YAG pulse laser operating with a 3–5 ns pulse duration and a 2 mJ/pulse energy output at 266 nm for 1–100 laser shots. Immediately after irradiation, the samples were quenched by freezing in liquid nitrogen and subsequently lyophilizing in a vacuum to remove the residual H_2O_2. It is also important to perform the experiment with a single laser pulse to avoid oxidizing proteins after conformation change had occurred.

At the same time, Hambly and Gross developed laser flash photolysis of H_2O_2 to oxidize protein solvent-accessible residues on the microsecond timescale (Hambly and Gross, 2005). A 17-ns KrF excimer laser with a 50 mJ/pulse output (compared to a 2 mJ/pulse used by Aye) that operated at 248 nm was developed to minimize laser absorption of protein species. The exposed apomyoglobin solution was incubated with agarose-immobilized catalase to remove residual H_2O_2.

3.4.4   Radiolysis by High-Voltage Electric Discharge Within ESI Ion Source

An electrospray ionization (ESI) ion source was utilized as a hydroxyl radical generator by Maleknia et al. (1999). Specifically, a protein solution in micromolar concentration is infused at a rate of few microliters per minute into a conventional electrospray source of mass spectrometer. The high voltage given to the electrospray tip (8 kV) while using oxygen as a nebulizer gas in the ESI ion source (approximately $10 \, L \, min^{-1}$) produces hydroxyl radicals through generation of active oxygen species by electrochemical reactions. When a dilute protein solution is sprayed out of the needle, the protein molecules become oxidized on the surface with the bulk solution at the needle tip and within the sprayed droplets. The chemistry of the reactions with amino acids is the same as the chemistry of aerobic hydroxyl radical reactions. Oxidized proteins are then subjected to proteolysis and mass spectrometry analysis. This approach provides a snapshot of the structures of proteins and protein complexes in solution. The group successfully studied some examples that illustrate peptide–protein interactions (Wong et al., 2003, 2005).

The advantage of this technique is that no additional instrument except mass spectrometer is required for the radiolysis experiment. The major drawback of this approach is that the sprayed sample from the ESI ion source needle needs to be collected and condensed in order to retrieve the information of modification sites. It is also not clear whether the structure of proteins remains native under such a severe electrochemical condition within the aerosol. Nevertheless, the electrochemical footprinting approach can be useful for studying a protein structure.

3.4.5 Synchrotron X-Ray Footprinting for Protein Complexes and Assembly Studies: Probing Arp2/3 Complex Activation by ATP and WASp Binding Proteins

In synchrotron footprinting experiments, hydroxyl radicals are generated upon the radiolysis of protein solutions using synchrotron light. During X-ray irradiation, the energy of an incoming photon is transferred to an electron, which is then ejected from water molecules. These electrons deposit their energy in discrete ionization of other water molecule. Ionized water molecules then react with intact water molecules to produce $HO^{\bullet}$ radicals. Beamline X28C at the National Synchrotron Light Source (Brookhaven National Laboratory) is a national resource dedicated to radiolytic protein footprinting and is an excellent source of X-rays to conduct radiolysis experiments (Gupta et al., 2007). The beamline consists of a bending magnet producing white light over a range of energy from 3 to 30 keV with beam currents ranging from 300 to 150 mA over a 12-h injection cycle (Sclavi et al., 1998). At a ring energy of 2.8 GeV and a beam current of 250 mA, 5.5×10^{14} photons are absorbed per second by a 10-μL sample of $7 \, mm^2$ cross-sectional area. The photon flux that is dependent on the magnet strength, the energy, and the number of the electrons in the storage ring yields a steady-state concentration of $HO^{\bullet}$ of approximately 1 μM (Ralston et al., 2002). Milliseconds of X-ray dose exposure in the radiolysis experiments is controlled by an electronic X-ray shutter (Gupta et al., 2007). Thus, typical protein footprinting experiment is carried out on the millisecond timescale.

In protein complexes, the reactive residues buried between interfaces are protected from oxidation during radiolysis, making it possible to detect changes in surface accessibility. However, allosteric changes in protein conformation induced by ligand binding can also give rise to either protections or enhancements of oxidation.

Thus, confirmation of the proposed interface should include additional biochemical or structural data.

This approach has been successfully applied by Chance's group to large proteins and complexes such as gelsolin (85 kDa) (Kiselar et al., 2003), the transferring–transferrin receptor complex (a complex with total molecular mass of 330 kDa) (Xu et al., 2005b), and F-actin (a megadalton filament with a monomer size of 40 kD) (Guan et al., 2005); these prior experiments were carried out at 10–40 µM concentrations of protein with hundreds of picomoles of material. In this section, we present an overview of structure studies of the Arp2/3 complex that has the greatest complexity of any sample examined to date using this method (220 kDa of unique sequence). In addition, all experiments were carried out with only a few picomoles of material; this was necessitated in part by the need to fully occupy the WASp binding sites on Arp2/3 by using a 1 µM concentration of Arp2/3.

The Arp2/3 complex is thought to initiate actin polymerization as a branch on the side of an existing actin filament by forming a dimer of Arp2 and Arp3 arranged like subunits in an actin filament at the slow-growing pointed end of the new actin filament (Kelleher et al., 1995; Mullins et al., 1998; Robinson et al., 2001). The apo-complex requires ATP and nucleation-promoting cofactors such as SCAR/WASp family proteins and preexisting actin filaments to initiate growth of new filaments (Machesky et al., 1999; Blanchoin et al., 2001; Zalevsky et al., 2001). This tight regulation of Arp2/3 complex activity is crucial to the temporal and spatial regulation of filament assembly at the leading edge of motile cells. The high-resolution crystal structure of the apo-form of the complex shows a large separation between the Arp2 and Arp3 subunits, so formation of a dimer of Arps similar to two consecutive subunits in an actin filament requires substantial conformational rearrangements (Robinson et al., 2001). It has been postulated that nucleation-promoting factors may induce major conformational rearrangements in the Arp2/3 complex by bringing closer the two Arps (Robinson et al., 2001). However, the molecular mechanism leading to activation of the Arp2/3 complex remains primary hypothetical due to the lack of a high-resolution structure of an activated conformation as well as the lack of high-resolution, solution-based structural approaches to address these questions. Thus, methods such as oxidative protein footprinting become quite valuable in providing structural information for Arp2/3 activation by ATP and WASp in solution with resolution at the level of single side chains (Guan et al., 2004; Guan and Chance, 2005; Takamoto and Chance, 2006). Figure 3.5a shows the overall structure of the Arp2/3 complex apo-form with the modeled subdomains 1 and 2. Arp2/3 complex is an ensemble of seven conserved protein subunits, including two actin-related proteins, Arp2 and Arp3, and five additional subunits named ARPC1 p40, ARPC2 p34, ARPC3 p21, ARPC4 p20, and ARPC5 p16 (Baichoo and Heyduk, 1997, 1999; Balasubramanium et al., 1998; Mullins et al., 1998).

Kiselar et al. (2007) found that ATP binding on Arp2/3 complex induces conformational changes clustered within Arp3 (in peptide segments 5–18, 212–225, and 318–327) and Arp2 (within peptide segment 300–316) without any substantial changes in p40, p34, p21, p20, or p16. Tandem MS/MS analysis was used to identify

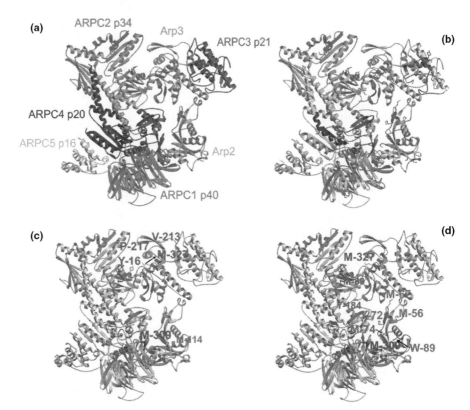

FIGURE 3.5   (a) Schematic representation of the structure of apo-form of the Arp2/3 complex with the modeled subdomains 1 and 2 of Arp2. Color codes for subunits are as follow: Arp3, orange; Arp2, red for subdomains 3 and 4, pink for the actin backbone model of subdomains 1 and 2; ARPC1 p40, green; ARPC2 p34, cyan; ARPC3 p21, magenta; ARPC4 p20, dark blue; and ARPC5 p16, yellow. (b) Peptides that were reactive in the absence of ATP are color coded as per (a); (c) Reactive peptides whose oxidation rate decreased in the presence of 1 mM ATP including 5–18, 212–225 and 318–329 (Arp3), and 300–316 (Arp2) are color coded as per (a); peptide 107–118 (Arp2) whose oxidation rate increased is also shown. (d) Peptides identified within Arp2/3 complex whose oxidation rate decreased on binding of WASp and 1 mM ATP including 80–91, 162–191 and 318–329 (Arp3), and 54–65, 66–80, 87–97, 300–316 (Arp2). Side chains of the oxidized peptide are shown. This figure was modified from Kiselar et al., (2007). (See the color version of this figure in Color Plates section.)

specific side chains sensitive to ATP binding (Fig. 3.5c). Specifically, residues within the nucleotide clefts of Arp2 and Arp3, including Tyr-16, Val-213, Pro-217, and Met-327 from Arp3, as well as Met-309 from Arp2, were less reactive to oxidation with bound nucleotide (Fig. 3.5c). For Arp3, the protection within the cleft between subdomains 1 and 4 indicates partial cleft closure (Beltzner and Pollard, 2003). These localized changes in the solution structure of Arp 2/3 complex are consistent with differences in the conformations of Arp3 in crystals with nucleotide soaked in it and without bound nucleotide (Nolen et al., 2004). Footprinting data also show that effects of nucleotide binding are confined to the clefts of the Arps, since the other 25 probe sites located throughout the complex did not change their reactivity. In addition, over 160 additional peptides that were routinely detected did not show any oxidation; thus, the surface areas of their potentially reactive residues were not appreciably exposed to solvent as a result of ATP binding.

The effect of WASp VCA binding on Arp2/3 complex was also studied. Seven of the 30 peptides exhibited slower oxidation in the presence of WASp VCA compared to ATP binding alone (Fig. 3.5b and d). These less reactive peptides are located in Arp3 including segments 80–91, 162–191, and 318–329 and in Arp2 including segments 54–65, 66–80, 87–97, and 300–316. The observed decrease in oxidation rate among the five regions ranged from two fold to five fold, whereas for peptides 54–65 and 318–329 the oxidation rate decreased by approximately 30%. Specific residues that are protected from oxidation in the Arp/WASp complex including Met-89, Tyr-184, and Met-327 in Arp3 and Met-56, Met-67, Tyr-72, Met-74, Trp-89, and Met-309 in Arp2 (Fig. 3.5d) were identified using tandem MS analysis. Binding of WASp VCA did not change the rate of oxidation of other subunits in the complex.

Several of the identified reactive residues were predicted to be useful for monitoring alteration of Arp2/3 complex by nucleotides and WASp. For example, sequence analysis and modeling studies (Beltzner and Pollard, 2003) predicted that Met-327 and Phe-328 on subdomain 3 of Arp3 should have altered solvent accessibility upon nucleotide binding. Furthermore, the bound nucleotide contains an additional amino group in position 6 of the adenine ring that can potentially interact with the A segment of WASp (Beltzner and Pollard, 2003). Probes on the back of the Arp2 subdomain 2 (e.g., Arg-42, Leu-55, Met-56, Leu-64, Ser-66, Asn-71, and Asp-90) may be candidates for binding WASp. Importantly, the residues Met-56, Met-67, Tyr-72, Met-74, Trp-89, and Met-327, which are accessible in the structure of apo-form of the Arp2/3 complex and have exhibited a protection from oxidation on WASp VCA binding, are overlapping with this proposed binding area. These findings are supported by numerous other biochemical and biophysical approaches, including cross-linking (Zalevsky et al., 2001; Weaver et al., 2002; Kreishman-Deitrick et al., 2005), NMR (Panchal et al., 2003; Kreishman-Deitrick et al., 2005), and fluorescence (Xu and Chance, 2005), that were used to examine the contact sites of WASp VCA on Arp2/3 complex. Specifically, chemical cross-linking studies (Weaver et al., 2002) favor binding of the A segment of WASp to Arp3. Furthermore, recently published NMR and cross-linking data suggest that a bivalent interaction of CA regions with the Arp2/3 complex stabilizes activated Arp2/3 complex through interactions with Arp2, Arp3, and ARPC3/p21 (Kreishman- Deitrick et al., 2005).

In addition, oxidative protein footprinting data showed that Arp2/3 complex binds to WASp within the C subdomain at residue Met 474 and within the A subdomain at Trp 500. Overall, studies of Arp2/3 complex activation upon ATP and WASp binding by hydroxyl radical protein footprinting suggest conformational changes of Arp2/3 complex resulting in a closed conformational state consistent with an "actin-dimer" model for the active state.

3.5 THE FUTURE: HYBRID APPROACHES THAT COMBINE EXPERIMENTAL AND COMPUTATIONAL DATA

Hydroxyl radical footprinting has been very successful in probing the protein structural changes in the context of different biological conditions, ligand binding, and macromolecular assembly such as protein/nucleic acid binding, protein/protein interactions, and large-scale protein complex formation (Heyduk et al., 1994, 1996; Baichoo and Heyduk, 1997, 1999; Frank et al., 1998; Maleknia et al., 1999a; Guan et al., 2002, 2004, 2005; Kiselar et al., 2002, 2003; Takamoto et al., 2002, 2006; Liu et al., 2003; Rashidzadeh et al., 2003; Xu et al., 2003, 2004, 2005; Gupta et al., 2004, 2007; Sharp et al., 2004, 2005). The success of these studies is based on the comparison between solved crystal structures and their solvent-accessible surface information. Although it is a very powerful technology, without structural information, footprinting studies are limited in their applicability. In other words, it is difficult to interpret the footprinting data without knowing the structure. Since we cannot calculate the solvent-accessible surface area information without solved structure, it becomes difficult task to interpret/understand the data. Thus, if there is no structural information about target protein available, we have to build the model structure consistent with hydroxyl radical footprinting data. The hydroxyl radical footprinting data basically provide surface accessibility information. In other words, they probe inside and outside of the protein molecule. This is tremendously useful information for protein structure prediction. As all proteins with well-defined structures have a hydrophobic core, the information about inside and outside amino acid residues is essential to define this core. The most useful probes so far observed and used for past studies are hydrophobic residues such as aliphatic and aromatic residues (Kiselar et al., 2002; Xu and Chance, 2004, 2005, 2007). If we observe aliphatic or aromatic modified residues (solvent accessible), we can rule out these residues from our candidate list for the hydrophobic core-forming residues.

There are many ways to predict structure, the two most popular being comparative modeling and *ab initio* modeling (Misura et al., 2006; Das et al., 2007; Melo and Sali, 2007). Although very powerful, comparative modeling will not be discussed here. This section will focus on cases where there is no crystal/NMR structures to be referenced for interpreting the footprinting data; thus, we even do not have useful template for comparative modeling. Thus, we have to take the path of *de novo* prediction. Fortunately, the accuracy of *de novo* prediction has been tremendously improved in the last decade. It even reached such a level of accuracy that the predicted structure can be successfully used for phasing by molecular replacement in favorable cases (Qian et al., 2007). However, there are cases where it predicts completely different fold from the actual protein, as these alternate predictions are of equal or lower energy from the actual structure. Thus, experimental data could discern among these predictions.

We are endeavoring to incorporate footprinting data into the highly successful Rosetta structure prediction package in collaboration with the Baker group. The experimental data can serve as a filter for the prediction algorithm to eliminate incompatible structures at an early stage of prediction. As Rosetta starts with low-resolution calculations in a decoy structure generation process, the decoy structures inconsistent with experimental data can be easily omitted from further steps (especially the computationally expensive full atom steps). This is expected to improve the speed of calculation. This means, for the same time of calculation, CPU time is spent on "productive" decoy structures that may be the right structure (at least they are consistent with observed experimental data) and allows Rosetta to explore larger conformational spaces, and thus makes it more likely to come across lower energy structures. On the other hand, as we have only the energy score to judge the decoy structures, it is extremely valuable to have experimental data to guide this process. Fortunately, as described above, low-energy but incorrectly folded structures are most likely ruled out by inconsistency with surface accessibility information, as it is very difficult to satisfy high-quality footprinting data with a structure having a different fold from actual one that is probed by footprinting.

Although it seems promising that Rosetta package will be a powerful tool for predicting proteins with unknown structures, ironically, the most difficult cases for footprinting studies are the visualization of structural change upon change in biological context surrounding the protein (such as ligand binding). The structure of Mg^{2+} form of G-actin has been proposed using footprinting data, despite the lack of tools. But it was a difficult process as whole processes had to be manually guided with consistency in all observed data. It is highly desirable to have tool to predict the structural shift upon these situations. In the case of structure change from solved structure (such as the case of G-actin), it would be possible to use the experimental data-guided molecular dynamics fitting (Jensen et al., 2007; Sener et al., 2007). These are likely to be the subject of many future studies.

REFERENCES

Aye, T. T., Low, T. Y., and Sze, S. K., 2005. Nanosecond laser-induced photochemical oxidation method for protein surface mapping with mass spectrometry. *Anal Chem* 77, 5814–5822.

Baichoo, N. and Heyduk, T., 1997. Mapping conformational changes in a protein: application of a protein footprinting technique to cAMP-induced conformational changes in cAMP receptor protein. *Biochemistry* 36, 10830–10836.

Baichoo, N. and Heyduk, T., 1999. Mapping cyclic nucleotide-induced conformational changes in cyclic AMP receptor protein by a protein footprinting technique using different chemical proteases. *Protein Sci* 8, 518–528.

Balasubramanian, B., Pogozelski, W., and Tullius, T., 1998. DNA strand breaking by the hydroxyl radical is governed by the accessible surface areas of the hydrogen atoms of the DNA backbone. *Proc Natl Acad Sci USA* 95, 9738–9743.

Beltzner, C. C. and Pollard, T. D., 2003. Identification of functionally important residues of Arp2/3 complex by analysis of homology models from diverse species. *J Mol Biol* 336, 551–565.

Blanchoin, L., Pollard, T. D., and Hitchcock-DeGregori, S. E., 2001. Inhibition of the Arp2/3 complex-nucleated actin polymerization and branch formation by tropomyosin. *Curr Biol* 11, 1300–1304.

Celander, D. W. and Cech, T. R., 1990. Iron(II)-ethylenediaminetetraacetic acid catalyzed cleavage of RNA and DNA oligonucleotides: similar reactivity toward single- and double-stranded forms. *Biochemistry* 29, 1355–1361.

Chance, M. R., 2001. Unfolding of apomyoglobin examined by synchrotron footprinting. *Biochem Biophys Res Commun* 287, 614–621.

Chance, M. R., Sclavi, B., Woodson, S. A., and Brenowitz, M., 1997. Examining the conformational dynamics of macromolecules with time-resolved synchrotron X-ray 'footprinting'. *Structure* 5, 865–869.

Creighton, T., 1993. Proteins: Structures and Molecular Properties. New York: W. H. Freeman.

Das, R., Qian, B., Raman, S., Vernon, R., Thompson, J., Bradley, P., Khare, S., Tyka, M. D., Bhat, D., Chivian, D., Kim, D. E., Sheffler, W. H., Malmström, L., Wollacott, A. M., Wang, C., Andre, I., and Baker, D., 2007. Structure prediction for CASP7 targets using extensive all-atom refinement with Rosetta@home. *Proteins* 69 (Suppl. 8), 118–128.

Frank, O., Schwanbeck, R., and Wisniewski, J. R., 1998. Protein footprinting reveals specific binding modes of a high mobility group protein I to DNAs of different conformation. *J Biol Chem* 273, 20015–20020.

Garrison, W. M., 1987. Reaction mechanisms in the radiolysis of peptides, polypeptides, and proteins. *Chem Rev* 87, 381–388.

Garrison, W. M., Kland-English, M. J., Sokol, H. A., and Jayko, M. E., 1970. Radiolytic degradation of the peptide main chain in dilute aqueous solution containing oxygen. *J Phys Chem* 74, 4506–4509.

Goldsmith, S. C., Guan, J. Q., Almo, S., and Chance, M. R., 2001. Synchrotron protein footprinting: a technique to investigate protein–protein interactions. *J Biomol Struct Dyn* 19, 405–418.

Goshe, M. B. and Anderson, V. E., 1999. Hydroxyl radical-induced hydrogen/deuterium exchange in amino acid carbon–hydrogen bonds. *Radiat Res* 151, 50–58.

Goshe, M. B., Chen, Y. H., and Anderson, V. E., 2000. Identification of the sites of hydroxyl radical reaction with peptides by hydrogen/deuterium exchange: prevalence of reactions with the side chains. *Biochemistry* 39, 1761–1770.

Guan, J. Q. Almo, S. C., and Chance, M. R., 2004. Synchrotron radiolysis and mass spectrometry: a new approach to research on the actin cytoskeleton. *Acc Chem Res* 37, 221–229.

Guan, J. Q. and Chance, M. R., 2005. Structural proteomics of macromolecular assemblies using oxidative footprinting and mass spectrometry. *Trends Biochem Sci* 30, 583–592.

Guan, J. Q., Takamoto, K., Almo, S. C., Reisler, E., and Chance, M. R., 2005. Structure and dynamics of the actin filament. *Biochemistry* 44, 3166–3175.

Guan, J. Q., Vorobiev, S., Almo, S. C., and Chance, M. R., 2002. Mapping the G-actin binding surface of cofilin using synchrotron protein footprinting. *Biochemistry* 41, 5765–5775.

Gupta, S., Mangel, W. F., McGrath, W. J., Perek, J. L., Lee, D. W., Takamoto, K., and Chance, M. R., 2004. DNA binding provides a molecular strap activating the adenovirus proteinase. *Mol Cell Proteomics* 3, 950–959.

Gupta, S., Sullivan, M., Toomey, J., Kiselar, J., and Chance, M. R., 2007. The Beamline X28C of the Center for Synchrotron Biosciences: a national resource for biomolecular structure and dynamics experiments using synchrotron footprinting. *J Synchrotron Radiat* 14, 233–243.

Hachimori, Y., Horinishi, H., Kurihara, K., and Shibata, K., 1964. States of amino acid residues in proteins. V. different reactivities with H_2O_2 of tryptophan residues in lysozyme, proteinases and zymogens. *Biochim Biophys Acta* 93, 346.

Hambly, D. M. and Gross, M. L., 2005. Laser flash photolysis of hydrogen peroxide to oxidize protein solvent-accessible residues on the microsecond timescale. *J Am Soc Mass Spectrom* 16, 2057–2063.

Hampel, K. J. and Burke, J. M., 2001a. A conformational change in the "loop E-like" motif of the hairpin ribozyme is coincidental with domain docking and is essential for catalysis. *Biochemistry* 40, 3723–3729.

Hampel, K. J. and Burke, J. M., 2001b. Time-resolved hydroxyl radical footprinting of RNA using Fe(II)-EDTA. *Methods* 23, 233–239.

Hanai, R. and Wang, J. C., 1994. Protein footprinting by the combined use of reversible and irreversible lysine modifications. *Proc Natl Acad Sci USA* 91, 11904–11908.

Heyduk, E. and Heyduk, T., 1994. Mapping protein domains involved in macromolecular interactions: a novel protein footprinting approach. *Biochemistry* 33, 9643–9650.

Heyduk, T., Heyduk, E., Severinov, K., Tang, H., and Ebright, R. H., 1996. Determinants of RNA polymerase alpha subunit for interaction with beta, beta', and sigma subunits: hydroxylradical protein footprinting. *Proc Natl Acad Sci USA* 93, 10162–10166.

Horinishi, H., Hachimori, Y., Kurihara, K., and Shibata, K., 1964. Sates of amino acid residues in proteins. 3. histidine residues in insulin, lysozyme, albumin and proteinases as determined with a new reagent of diazo-i-h-tetrazole. *Biochim Biophys Acta* 86, 477–489.

Hunt, D. F., Yates, J. R., III, Shabanowitz, J., Winston, S., and Hauer, C. R., 1986. Protein sequencing by tandem mass spectrometry. *Proc Natl Acad Sci USA* 83, 6233–6237.

Jensen, M. Q., Yin, Y., Tajkhorshid, E., and Schulten, K., 2007. Sugar transport across lactose permease probed by steered molecular dynamics. *Biophys J* 93 (1), 92–102.

Johnson, R. S., Martin, S. A., Biemann, K., Stults, J. T., and Watson, J. T., 1987. Novel fragmentation process of peptides by collision-induced decomposition in a tandem mass spectrometer: differentiation of leucine and isoleucine. *Anal Chem* 59, 2621–2625.

Kelleher, J. F., Atkinson, S. J., and Pollard, T. D., 1995. Sequences, structural models, and cellular localization of the actin-related proteins Arp2 and Arp3 from Acanthamoeba. *J Cell Biol* 131, 385–397.

Kiselar, J. G., Janmey, P. A., Almo, S. C., and Chance, M. R., 2003. Structural analysis of gelsolin using synchrotron protein footprinting. *Mol Cell Proteomics* 2, 1120–1132.

Kiselar, J. G., Janmey, P. A., Almo, S. C., and Chance, M. R., 2003. Visualizing the Ca^{2+}-dependent activation of gelsolin by using synchrotron footprinting. *Proc Natl Acad Sci USA* 100, 3942–3947.

Kiselar, J. G., Mahaffy, R., Pollard, T. D., Almo, S. C., and Chance, M. R., 2007. Arp2/3 activation mediated by binding of nucleotides and WASp: structural mass spectrometry approaches to large macromolecular complexes. *Proc Natl Acad Sci USA* 104, 1552–1557.

Kiselar, J. G., Maleknia, S. D., Sullivan, M., Downard, K. M., and Chance, M. R., 2002. Hydroxyl radical probe of protein surface using synchrotron X-ray radiolysis and mass spectrometry. *Int J Radiat Biol* 78, 101–114.

Kreishman-Deitrick, M., Goley, E. D., Burdine, L., Denison, C., Egile, C., Li, R., Murali, N., Kodadek, T. J., Welch, M. D., and Rosen, M. K., 2005. NMR analyses of the activation of the Arp2/3 complex by neuronal Wiskott-Aldrich syndrome protein. *Biochemistry* 44, 15247–15256.

Liu, R., Guan, J. Q., Zak, O., Aisen, P., and Chance, M. R., 2003. Structural reorganization of the transferrin C-lobe and transferrin receptor upon complex formation: the C-lobe binds to the receptor helical domain. *Biochemistry* 42, 12447–12454.

Machesky, L. M., Mullins, R. D., Higgs, H. N., Kaiser, D. A., Blanchoin, L., May, R. C., Hall, M. E., and Pollard, T. D., 1999. Scar, a WASp-related protein, activates nucleation of actin filaments by the Arp2/3 complex. *Proc Natl Acad Sci USA* 96, 3739–3744.

Maleknia, S. D., Brenowitz, M., and Chance, M. R., 1999a. Millisecond radiolytic modification of peptides by synchrotron X-rays identified by mass spectrometry. *Anal Chem* 71, 8965–8973.

Maleknia, S. D., Chance, M. R., and Downard, K. M., 1999b. Electrospray-assisted modification of proteins: a radical probe of protein structure. *Rapid Commun Mass Spectrom* 13, 2352–2358.

Melo, F. and Sali, A., 2007. Fold assessment for comparative protein structure modeling. *Protein Sci* 16 (11), 2412–2426.

Misura, K. M., Chivian, D., Rohl, C. A., Kim, D. E., and Baker, D., 2006. Physically realistic homology models built with ROSETTA can be more accurate than their templates. *Proc Natl Acad Sci USA* 103 (14), 5361–5366.

Mullins, R. D., Heuser, J. A., and Pollard, T. D., 1998. The interaction of Arp2/3 complex with actin: nucleation, high affinity pointed end capping, and formation of branching networks of filaments. *Proc Natl Acad Sci USA* 95, 6181–6186.

Nolen, B. J., Littlefield, R. S., and Pollard, T. D., 2004. Crystal structures of actin-related protein 2/3 complex with bound ATP or ADP. *Proc Natl Acad Sci USA* 101, 15627–15632.

Nukuna, B. N., Goshe, M. B., and Anderson, V. E., 2001. Sites of hydroxyl radical reaction with amino acids identified by (2) H NMR detection of induced (1) H/ (2) H exchange. *J Am Chem Soc* 123, 1208–1214.

Nukuna, B. N., Sun, G., and Anderson, V. E., 2004. Hydroxyl radical oxidation of cytochrome c by aerobic radiolysis. *Free Radic Biol Med* 37, 1203–1213.

Ottinger, L. M. and Tullius, T. D., 2000. High-resolution *in vivo* footprinting of a protein–DNA complex using gamma-radiation. *J Am Chem Soc* 122, 5901–5902.

Panchal, S. C., Kaiser, D. A., Torres, E., Pollard, T. D., and Rosen, M. K., 2003. A conserved amphipathic helix in WASP/Scar proteins is essential for activation of Arp2/3 complex. *Nat Struct Biol* 10, 591–598.

Qian, B., Raman, S., Das, R., Bradley, P., McCoy, A. J., Read, R. J., and Baker, D., 2007. High-resolution structure prediction and the crystallographic phase problem. *Nature* 450 (7167), 259–264.

Ralston, C. Y., Sclavi, B., Sullivan, M., Deras, M. L., Woodson, S. A., Chance, M. R., and Brenowitz, M., 2005. Time-resolved synchrotron X-ray footprinting and its application to RNA folding. *Methods Enzymol* 317, 353–368.

Rana, T. M., and Meares, C. F., 1991. Transfer of oxygen from an artificial protease to peptide carbon during proteolysis. *Proc Natl Acad Sci USA* 88, 10578–10582.

Rashidzadeh, H., Khrapunov, S., Chance, M. R., and Brenowitz, M., 2003. Solution structure and interdomain interactions of the *Saccharomyces cerevisiae* "TATA binding protein" (TBP) probed by radiolytic protein footprinting. *Biochemistry* 42, 3655–3665.

Robinson, R. C., Turbedsky, K., Kaiser, D. A., Marchand, J. B., Higgs, H. N., Choe, S., and Pollard, T. D., 2001. Crystal structure of Arp2/3 complex. *Science* 294, 1679–1684.

Roepstorff, P. and Fohlman, J., 1984. Proposal for a common nomenclature for sequence ions in mass spectra of peptides. *Biomed Mass Spectrom* 11, 601.

Sclavi, B., Woodson, S., Sullivan, M., Chance, M., and Brenowitz, M., 1998. Following the folding of RNA with time-resolved synchrotron X-ray footprinting. *Methods Enzymol* 295, 379–402.

Sener, M. K., Olsen, J. D., Hunter, C. N., and Schulten, K., 2007. Atomic-level structural and functional model of a bacterial photosynthetic membrane vesicle. *Proc Natl Acad Sci USA* 104 (40), 15723–15728.

Sharp, J. S., Becker, J. M., and Hettich, R. L., 2003. Protein surface mapping by chemical oxidation: structural analysis by mass spectrometry. *Anal Biochem* 313, 216–225.

Sharp, J. S., Becker, J. M., and Hettich, R. L., 2004. Analysis of protein solvent accessible surfaces by photochemical oxidation and mass spectrometry. *Anal Chem* 76, 672–683.

Sharp, J. S., Guo, J. T., Uchiki, T., Xu, Y., Dealwis, C., and Hettich, R. L., 2005. Photochemical surface mapping of C14S-Smllp for constrained computational modeling of protein structure. *Anal Biochem* 340, 201–212.

Sheshberadaran, H., and Payne, L. G., 1988. Protein antigen-monoclonal antibody contact sites investigated by limited proteolysis of monoclonal antibody-bound antigen: protein "footprinting". *Proc Natl Acad Sci USA* 85, 1–5.

Suckau, D., Mak, M., and Przybylski, M., 1992. Protein surface topology-probing by selective chemical modification and mass spectrometric peptide mapping. *Proc Natl Acad Sci USA* 89, 5630–5634.

Tabb, D. L., Huang, Y., Wysocki, V. H., and Yates, J. R., III, 2004. Influence of basic residue content on fragment ion peak intensities in low-energy collision-induced dissociation spectra of peptides. *Anal Chem* 76, 1243–1248.

Takamoto, K. and Chance, M. R., 2006. Radiolytic protein footprinting with mass spectrometry to probe the structure of macromolecular complexes. Annu Rev Biophys Biomol Struct. 35, 251–276.

Takamoto, K., He, Q., Morris, S., Chance, M. R., and Brenowitz, M., 2002. Monovalent cations mediate formation of native tertiary structure of the *Tetrahymena thermophila* ribozyme. *Nat Struct Biol* 9, 928–933.

Tang, X. J., Thibault, P., and Boyd, R. K., 1993. Fragmentation reactions of multiply-protonated peptides and implications for sequencing by tandem mass spectrometry with low-energy collision-induced dissociation. *Anal Chem* 65, 2824–2834.

Tullius, T. D. and Dombroski, B. A., 1985. Iron(II) EDTA used to measure the helical twist along any DNA molecule. *Science* 230, 679–681.

Tullius, T. D. and Dombroski, B. A., 1986. Hydroxyl radical "footprinting": high resolution information about DNA–protein contacts and application to lambda repressor and Cro protein. *Proc Natl Acad Sci USA* 83, 5469–5473.

Wang, X. D. and Padgett, R., 1989. Hydroxyl radical "footprinting" of RNA: application to pre-mRNA splicing complexes. *Proc Natl Acad Sci USA* 86, 7795–7799.

Weaver, A. M., Heuser, J. E., Karginov, A. V., Lee, W. L., Parsons, J. T., and Cooper, J. A., 2002. Interaction of cortactin and N-WASp with Arp2/3 complex. *Curr Biol* 12, 1270–1278.

Wong, J. W., Maleknia, S. D., and Downard, K. M., 2003. Study of the ribonuclease-S-protein–peptide complex using a radical probe and electrospray ionization mass spectrometry. *Anal Chem* 75, 1557–1563.

Wong, J. W., Maleknia, S. D., and Downard, K. M., 2005. Hydroxyl radical probe of the calmodulinmelittin complex interface by electrospray ionization mass spectrometry. *J Am Soc Mass Spectrom* 16, 225–233.

Xu, G. and Chance, M. R., 2004. Radiolytic modification of acidic amino acid residues in peptides: probes for examining protein–protein interactions. *Anal Chem* 76, 1213–1221.

Xu, G. and Chance, M. R., 2005. Radiolytic modification of sulfur-containing amino acid residues in model peptides: fundamental studies for protein footprinting. *Anal Chem* 77, 2437–2449.

Xu, G. and Chance, M. R., 2005. Radiolytic modification and reactivity of amino acid residues serving as structural probes for protein footprinting. *Anal Chem* 77, 4549–4555.

Xu, G. and Chance, M. R., 2007. Hydroxyl radical-mediated modification of proteins as probes for structural proteomics. *Chem Rev* 107, 3514–3543.

Xu, G., Kiselar, J., He, Q., and Chance, M. R., 2005a. Secondary reactions and strategies to improve quantitative protein footprinting. *Anal Chem* 77, 3029–3037.

Xu, G., Takamoto, K., and Chance, M. R., 2003. Radiolytic modification of basic amino acid residues in peptides: probes for examining protein–protein interactions. *Anal Chem* 75 (24), 6995–7007.

Xu, G., Liu, R., Zak, O., Aisen, P., and Chance, M. R., 2005b. Structural allostery and binding of the transferrin*receptor complex. *Mol Cell Proteomics* 4, 1959–1967.

Zalevsky, J., Lempert, L., Kranitz, H., and Mullins, R. D., 2001. Different WASP family proteins stimulate different Arp2/3 complex-dependent actin-nucleating activities. *Curr Biol* 11, 1903–1913.

Zalevsky, J., Grigorova, I., and Mullins, R. D., 2001. Activation of the Arp2/3 complex by the Listeria acta protein. Acta binds two actin monomers and three subunits of the Arp2/3 complex. *Biol Chem* 276, 3468–3475.

■■■■ **CHAPTER 4**

Complementary Methods for Structure Determination: Hydroxyl-Radical-Mediated Footprinting and Deuterium Exchange Mass Spectrometry as Applied to Serpin Structure

XIAOJING ZHENG

Case Center for Proteomics, Case Western Reserve University, Cleveland, OH, USA

PATRICK L. WINTRODE

Department of Physiology and Biophysics, Case Western Reserve University, Cleveland, OH, USA

4.1 INTRODUCTION

The most common applications of mass spectrometry (MS) to proteins involve the analysis of primary structure: high throughput sequencing and the analysis of posttranslational modifications. However, the power of mass spectrometry as a tool for probing protein secondary, tertiary, and quaternary structures is increasingly evident. In the past two decades, two structural mass spectrometry methodologies, hydroxyl-radical-mediated footprinting and hydrogen/deuterium (H/D) mass spectrometry, have been successfully and widely used for probing protein structure and dynamics by determining the solvent accessibility of side chains or stability and solvent accessibility of the protein backbone. These methodologies require picomoles of sample, are relatively rapid, and can be applied to extremely large proteins and macromolecular assemblies in aqueous solution that are difficult to study by other methods. Even when high-resolution X-ray structures are available, these methods can complement the static picture by providing valuable information on conformational dynamics in solution.

Mass Spectrometry Analysis for Protein–Protein Interactions and Dynamics, Edited by Mark Chance
Copyright © 2008 John Wiley & Sons, Inc.

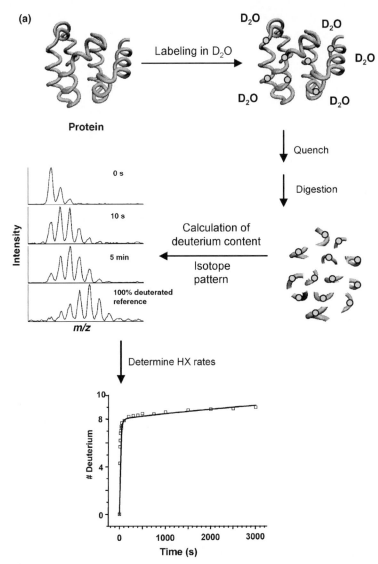

FIGURE 4.1 Schematic representations of synchrotron footprinting and H/D exchange mass spectrometry techniques. (a) By changing the solvent from H_2O to D_2O, amide protons in the backbone of protein exchange with the protons from the solvent. The exchange reaction is quenched by quickly changing solution pH to approximately 2.4 and usually combining with rapid freezing. Typically, the protein is digested with pepsin. The kinetics of amide H/D exchange is measured by mass spectrometry to provide essential dynamics information for protein. (b) When a protein is exposed to synchrotron X-ray, the hydroxyl radicals generated from water will modify side chains of the protein. After the X-ray exposure, the protein samples are digested by proteases and solvent accessibility information is provided by MS. The particular modification sites are determined by tandem MS and the side chain reactivity is accurately measured by quantitative liquid-chromatography-coupled mass spectrometry. Reproduced, with permission from Elsevier Ltd., from Zheng et al., 2008.

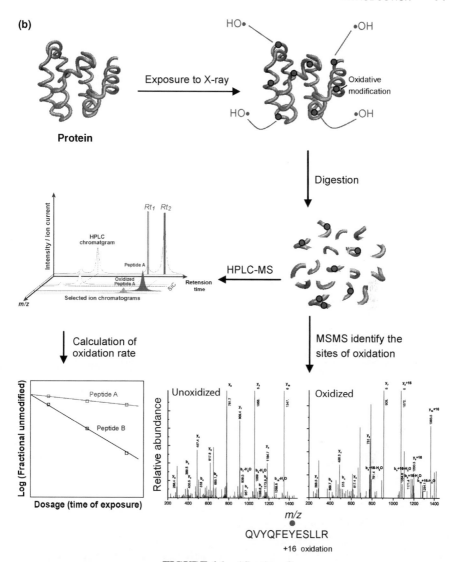

FIGURE 4.1   (*Continued*)

H/D exchange combined with mass spectrometry was developed in the early 1990s. The practicality of this method lies in the fact that amide hydrogens are sensitive probes of solvent accessibility, conformational lability, and secondary structure. A schematic representation of the H/D exchange technique is shown in Fig. 4.1(a). Each amino acid (except proline) has one amide hydrogen. Depending on their structural environment, backbone amide hydrogens will exchange with deuterium at different rates when a protein is incubated in D_2O. Amide hydrogens at the surface of proteins exchange very rapidly, while those that are buried or are involved in stable hydrogen bonds have much slower exchange rates (Busenlehner and Armstrong, 2005). Thus H/D exchange rates can be measured along the entire length of the protein backbone, providing a comprehensive measure of protein structure and solvent accessibility. Since backbone amide hydrogens are also involved in the formation of hydrogen bonds in protein secondary structures, their exchange rates are also a reflection of secondary structure and conformational stability (Englander et al., 1996).

"Hydroxyl-radical-mediated protein footprinting" refers to assays that examine protein structures and conformational changes, protein–ligand interaction, and protein–protein interaction by monitoring solvent accessibility using •OH modification (Xu and Chance, 2007). It has been well known for some time that •OH radicals can modify the side chains of proteins quite efficiently (Schuessler and Schilling, 1984). •OH radicals suitable for footprinting experiments can be generated by multiple methods such as using Fenton reagents from photo-oxidation of peroxide, using electrical discharge, and radiolysis of water (Guan and Chance, 2005). Figure 4.1b shows a schematic representation of synchrotron footprinting technology (SF), a hydroxyl-radical-mediated footprinting technique combining mass spectrometric detection that was developed by Chance and coworkers. Radiolysis of water using X-rays from synchrotron sources generates •OH radicals isotropically in solution without addition of chemicals. The •OH radicals generated through millisecond exposures of synchrotron X-rays can then react with proteins to yield stable oxidative modifications of solvent-accessible amino acid side chains. Subsequent to oxidation, proteins are digested by specific proteases to generate peptides for mass spectrometry analysis. Accurate measurements of side chain reactivity are achieved by quantitative liquid chromatography-coupled MS. Also the oxidized residues can be identified using tandem MS. The reactivity of side chains with •OH can give insights into protein structure and monitor conformational changes (e.g., due to ligand binding or macromolecular interactions) (Guan and Chance, 2005).

In this chapter, we illustrate the relative strengths and limitations of these two techniques, as well as their potential to complement each other, by reviewing recent work that applied both methods to the same protein (Zheng et al., 2008; Tsutsui et al., 2006), the serpin α_1-antitrypsin (α_1AT). Serpins are a large class of protease inhibitors that play important roles in regulating a variety of physiological processes, particularly those related to inflammation and tissue remodeling (Gettins, 2002). Serpins are unusual in that they do not fold to their true free energy minimum, but instead fold to a metastable state, and undergo a massive transition to a stable state when they inhibit their target proteases (Huntington et al., 2000; Dementiev et al., 2006; Ye et al., 2001). Conformational flexibility is therefore key to serpin function, making serpins especially attractive targets for techniques that can probe protein dynamics in solution. α_1AT has served as a canonical model for serpin structure and function. The major structural features of α_1AT as well as the inhibitory conformational change are shown in Fig. 4.2.

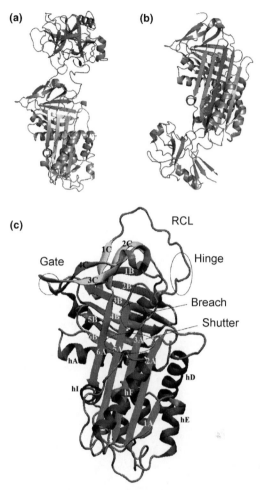

FIGURE 4.2   The structural and inhibitory properties of α_1-antitrypsin. (a) The encounter complex between a serpin and a target protease (PDBID 1K9O) (Ye et al., 2001). The protease is shown in orange. (b) The inhibitory complex of a serpin and a translocated protease (PDBID 1EZX) (Huntington et al., 2000). The inserted RCL is colored red. (c) The crystal structure of active α_1AT (PDB ID 1QLP) (Elliott et al., 2000). Functionally important regions as described by Whisstock et al. are indicated in circles. β-sheets A, B, and C are in green, purple, and yellow, respectively. All α-helices are in blue. The RCL is shown in red. Reproduced, with permission from Elsevier Ltd., from Zheng et al., 2008. (See the color version of this figure in Color Plates section.)

4.2   TECHNICAL COMPARISON OF HYDROXYL-RADICAL-MEDIATED FOOTPRINTING AND H/D EXCHANGE METHODOLOGIES

The examination of amino acid side chains by footprinting is complementary to the H/D exchange method that examines backbone structure. In this part, the advantages and disadvantages of the two techniques will be discussed.

Each method has different sample requirements such as buffer, pH, and additives. For both techniques, selection of buffer is critical as we want to optimize the stability and activity of the biological system. Additionally, the selected buffers also need to optimize the dose of radicals for footpinting experiments. Many commonly used biochemical buffers, such as Tris, HEPES, MOPS, CAPS, and citrate buffers are unsuitable for footprinting experiments due to the quenching of hydroxyl radicals. Sodium cacodylate or phosphate buffers are the typical choice for footprinting experiments because of their minimal quenching of ·OH. Different additives that are frequently presented in protein samples as cofactors (such as glycerol, ATP, ADP, EDTA, and $MgCl_2$) may also significantly impact footprinting because of their quenching effect (Xu and Chance, 2007). Recently, advancements in beamline instruments allow the examination of a wider range of biological buffers with the availability of the brighter beam (Gupta et al., 2007). Also the developments of assay methods using Alexa 488 dosimeter (fluorescence dye) provide a measure of the dose delivered to protein samples. Thus, the optimum conditions for radiolytic footprinting studies can be found and we can compare radiolysis under different conditions (e.g., with or without cofactors) to normalize experiments (Xu and Chance, 2007).

Compared with footprinting, an advantage of hydrogen exchange is that it allows for much greater flexibility regarding solution conditions. Any desired buffer can be employed, and a wide variety of additives and cofactors can be present during deuterium labeling. There are, however, some practical limitations regarding pH. The intrinsic (chemical) rate of amide H/D exchange is strongly pH dependent (Bai et al., 1993), and the extremely slow intrinsic exchange rates at low pH can create complications in designing and interpreting experiments. Most often, H/D exchange experiments are performed at pH >6.

Sequence coverage is critical for structural MS experiments because it determines how much information we are able to provide for a certain protein. For the 40 kDa α_1AT, both SF and H/D exchange techniques gave high coverage of its sequence (90% for SF and 89% for H/D exchange). Figure 4.3 indicates that 53 peptides (generated by trypsin, chymotrypsin, and Asp-N digestion) were identified in SF experiments, which cover ~90% of the α_1AT sequence. For the peptide mapping of H/D exchange experiments, the 29 peptic fragments analyzed by H/D MS are well distributed throughout the molecule, and the only significant gap consists in strand 2A (Fig. 4.3).

The oxidized products by hydroxyl radical exposure are very stable, which enables a wide range of sample sizes, proteases, and solution conditions to be examined. For example, three proteases (trypsin, chymotrypsin, and Asp-N) were used to maximize the α_1AT sequence coverage in footprinting experiments: the addition of chymotrypsin together with trypsin increases the coverage from 74% (trypsin only) to 90% because of the additional identification of three N-terminal fragments by chymotrypsin. Usage of multiple enzymes also efficiently segmented the α_1AT sequence into smaller units (Fig. 4.3), thus increased the resolution of footprinting data. In addition, usage of multiple proteases provided redundant information for SF on the reactivity of some peptides where modified residues overlap.

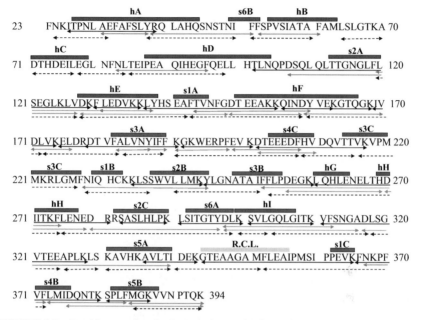

FIGURE 4.3   Peptide mapping for footprinting and H/D exchange mass analysis. Primary sequences corresponding to α-helices and β-strands are shown below blue and red bars, respectively. The RCL is shown below a yellow bar. Peptides digested by three proteases in footprinting experiments are indicated by colored double-headed arrows under the protein sequence. Black: trypsin; purple: chymotrypsin; and green: Asp-N. Peptic fragments for H/D exchange study are indicated by dashed double-headed arrows. (See the color version of this figure in Color Plates section.)

In contrast to footprinting, pepsin was the only protease used for proteolysis in H/D exchange experiments because the range of proteases and digestion conditions that can be employed with H/D exchange is limited due to the strict requirements of low pH and short (~5 min) digestion times. Back exchange is a major drawback of H/D exchange, which is an undesirable exchange of the deuterium for hydrogen, thus resulting in the loss of the label. The bulk of back exchange occurs during the process of proteolytic digestion and the subsequent analysis that is done in protonated solvents. To reduce the back exchange effect, the digestion of H/D experiments is performed at pH 2.4 on ice. As a result, the choice of a protease was narrowed and the digestion time was also limited (Kipping and Schierhorn, 2003). Porcine pepsin is used in H/D exchange experiments as it works the best at pH 2.4 and can tolerate a 0°C temperature. There are currently only a very few proteases other than pepsin that can digest proteins efficiently under conditions suitable for H/D MS.

Resolution is another important criterion for both techniques. For H/D exchange, typically the conformation change is only localized to the specific peptide fragment, while the stable modification of side chains in footprinting experiments allows a specific probe site to be identified using tandem mass spectrometry. The disadvantage of footprinting technique is that if a reactive side chain is absent in a particular peptide segment, there are no probes (Xu and Chance, 2007).

A handful of labs are working on overcoming the technical problem of both techniques. For example, back exchange attendant can be minimized by conducing online H/D exchange, online fragmentation and faster analyses, or by fragmenting intact proteins directly in the mass spectrometer to yield site-specific information (Yan et al., 2004). Recent advances in mass-resolving power and mass measurement accuracies of modern instruments will definitely enhance our ability to detect minimally abundant oxidative modifications, thus providing more probes. Also, the development of computational tools is undergoing in Chance's lab to provide robustness in data interpretation and structural determination.

4.3 STRUCTURAL MASS SPECTROMETRY DATA

4.3.1 H/D Exchange Data

H/D exchange studies of α_1AT were performed at pD 7.8 at 25°C followed by HPLC-MS to quantify the mass of each peptide fragment (Tsutsui et al., 2006). Based on the mass shift of the peptide isotopic envelops over time (Fig. 4.4a), deuterium incorporation was calculated and deuterium versus time curves were each fit to a sum of three exponentials. Figure 4.4b shows examples of normalized deuterium versus time curves for two typical peptides (160–171 and 325–338), along with best fits.

Exchange rate constants and the number of fast-, intermediate-, and slow-exchanging amide hydrogens in each peptide are summarized in Table 4.1. The fast-exchanging hydrogens are most likely amide hydrogen atoms that are fully solvent exposed. Half-lives for intermediate exchanging hydrogens range from 30 s to ~15 min, which are generally much slower than those of the completely unprotected hydrogens. These hydrogens are likely located in regions that are structured but are marginally stable and undergo frequent fluctuations. Half-lives for slowly exchanging hydrogens range from 488 s to 1.7×10^6 s indicating that these peptides are derived from regions that are both structured and stable.

4.3.2 Synchrotron Footprinting Data

The α_1AT protein sample at pH 7.0 was exposed to the synchrotron X-ray white beam for different time durations. Irradiated protein was subjected to protease digestion, and oxidized peptides were detected by LC-MS (Zheng et al., 2008).

Oxidative modifications in each peptide are detected by inspecting a selected peptide ion chromatogram (SIC) and its corresponding full mass spectrum. In Fig. 4.5a, the selected ion chromatograms of unoxidized and oxidized doubly charged peptide ions from residues 344–365 are given as an example. Peptides modified on both Met 351 and 358, or either Met, were eluted a few minutes

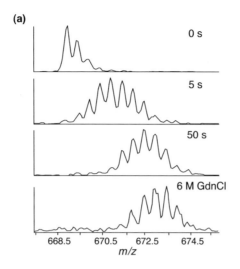

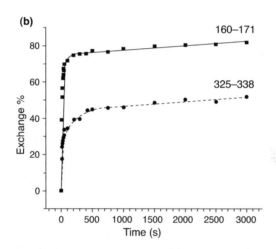

FIGURE 4.4 H/D exchange data analysis. (a) MS spectra of doubly charged peptide 160–171. Deuterium uptake of the peptide 160–171 at 0 (nondeuterated reference), 5, and 50 s and in 6 M guanidine deuterochloride (fully deuterated reference) is shown. (b) Normalized deuterium uptake versus time curves for two peptic fragments derived from α_1AT. Peptides are 160–171 (squares, solid line) and 325–338 (circles, dashed line). Reprinted with permission from Tsutsui et al., 2006. Copyright 2008 American Chemical Society.

earlier than the corresponding unoxidized peptide due to its greater hydrophobicity than the oxidized peptides. A full mass spectrum is also analyzed to further confirm oxidation of each peptide. An example is shown in Fig. 4.5b. Oxidative modification on peptide 202–217 increases its molecular weight by 16 Da due to an addition of an oxygen atom resulting m/z shifts of 5.2 and 8 for triply and doubly charged ions, respectively, compared to m/z of the corresponding unoxidized ions.

TABLE 4.1   Results of Fitting H/D Exchange Curves

Residues	N_{fast}	k_{fast} (s^{-1})	N_{int}	k_{int} (s^{-1})	N_{slow}	k_{slow} (10^{-5} s^{-1})
24–32	3	0.23	3	0.033	1	0.14
38–51	8	0.69	2	0.0020	3	3
38–60	6	0.40	3	0.0014	12	1
64–77	4	0.44	3	0.0026	6	3
64–84	10	0.42	1	8×10^{-4}	9	0.4
85–99	4	0.35	5	0.0038	4	2
120–142	8	0.36	7	0.0051	7	2
127–142	3	0.16	7	0.0033	5	7
131–142	3	0.085	4	0.0041	3	2
143–159	6	0.38	6	0.0014	4	3
160–171	7	0.17	2	0.0059	2	2
160–172	9	0.12	2	0.0038	1	0.9
173–182	6	0.69	2	0.024	1	0.23
183–208	9	0.29	6	0.0068	9	2
190–208	8	0.27	5	0.0035	4	0.04
209–227	7	0.23	2	0.0031	8	3
228–237	3	0.54	2	0.020	4	2
241–251	3	0.27	2	0.0015	4	0.9
251–275	9	0.23	6	0.0022	8	3
276–299	14	0.29	6	0.0019	2	7
297–303	3	0.37	1	0.0031	2	5
304–317	5	0.21	5	0.0019	4	1
318–338	7	0.20	4	0.0082	8	1
325–338	3	0.17	3	0.0031	6	0.6
339–351	10	1.0	1	0.0015	1	5
352–372	9	0.39	1	0.039	6	2
385–394	3	0.63	1	0.0044	4	0.6

Fourteen peptides that exhibited oxidation are shown in Table 4.2. Twelve of them are generated by trypsin digestion, while peptides composed of residues 159–170 and 270–279 are generated by Asp-N digestion. For each peptide in Table 4.2, the oxidative modified residues were identified by tandem MS spectra. Figure 4.5c shows a tandem MS spectrum of a tryptic peptide (residues 291–300) derived from strand 6A. Oxidation of a tyrosine residue in this peptide is evidenced by a peak difference between b_6 and b_7 ions or y_3 and y_4 ions. Other oxidized residues identified by tandem MS experiments are indicated within each peptide sequence in bold type (Table 4.2). The locations of these modified residues in α_1AT's crystal structure are shown in Fig. 4.6a and b. The solvent accessibility (SASA) of the oxidized side chains based on the crystal structure (PDB ID 1QLP) is shown below the one-letter codes (Table 4.2). Those peptides detected but not modified are colored yellow in Fig. 4.6a,b, indicating the absence or burial of reactive side chains in these regions. Oxidation rates for these 14 peptides were generated by analyzing LC-MS data and fitting dose–response curves. These curves

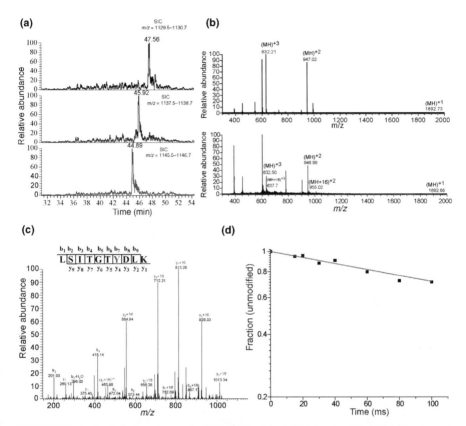

FIGURE 4.5 Footprinting data analysis. (a) SIC of unoxidized and oxidized doubly charged peptide ions from tryptic peptide 344–365 with an exposure time of 80 ms. Peptides with both Met (351 and 358) oxidized or single Met oxidized were eluted at a retention time (RT) of 44.89 min and 45.92 min, respectively. The corresponding unoxidized peptide was eluted at 47.56 min. (b) ESI-MS spectra of tryptic peptide 202–217 (top one) and its derivatives irradiated by synchrotron X-rays for 100 ms (bottom one). (c) MS/MS spectrum of oxidized doubly charged ion from tryptic peptide 291–300 at m/z 564.04 (+ 16 Da). The exposure time is 100 ms. (d) Dose–response curve for the radiolytic modification of tryptic peptide 26–39. Reproduced, with permission from Elsevier Ltd., from Zheng et al., 2008. The rate of radiolytic modification is listed in Table 4.1.

plot the fraction unmodified for each peptide as a function of exposure time (Fig. 4.5d). Data from duplicate experiments were averaged and shown in Table 4.2.

4.4 SOLVENT ACCESSIBILITY

Both SF and H/D exchange MS methods are capable of probing protein solvent accessibility. However, their capabilities are different. Synchrotron footprinting

TABLE 4.2 Rate Constants for the Modification of the Human α_1-Antitrypsin[#]

Protease	Residues	Peptide	Rate constant (s^{-1})
Trypsin	26–39	ITPNLAEFA **F**SLYR 6.83	2.9 ± 0.3
	102–129	TLNQPDSQLQLTTGNGLFLS**E**GLKLVDK 96.49	4.48 ± 0.06
	137–155	LYHSEA**F**TVNFGDTEEAKK 86.86	12.62 ± 0.28
	202–217	DTEEEDF**H**VDQVTTVK 74.33	3.28 ± 0.03
	244–259	YLGNATAIFFL **P**DEGK 4.28	1.46 ± 0.05
	291–300	LSITGT**Y**DLK 33.19	3.5 ± 0.1
	301–310	SVLG**Q**LGITK 37.22	6.11 ± 0.02
	311–328	VFSNGADLSGVTEEA**P**LK 133.51	13.04 ± 2.25
	336–343[*]	AVLTIDE**K** 55.61	≤10
	344–365	GTEAAGA **M**FLEAIP**M**SIPPEVK 189.45 204.55	102.2 ± 4.5
	366–380	FNKPFVFL**M**IDQNTK 0.51	10.95 ± 2.26
	381–387[*]	SPL**F**MGK 1.07	≤3
AspN	159–170	**D**YVEKGTQGKIV 22.99	3.51 ± 0.06
	270–279	DIITK**F**LENE 16.02	3.72 ± 0.03

[#]The modified amino acids are in bold, under which the accessible surface area (SASA) values are indicated.
[*]Oxidation products were observed at and above 100 ms exposure times. Reproduced, with permission from Elsevier Ltd., from Zheng et al., 2008.

examines the conformation by determining the solvent accessibility of side chain structures of proteins, while H/D exchange monitors the solvent exposure and the secondary structure of protein backbone.

In footprinting experiments, hydroxyl radicals react preferentially with the solvent-accessible reactive side chains of amino acid residues. Therefore, solvent

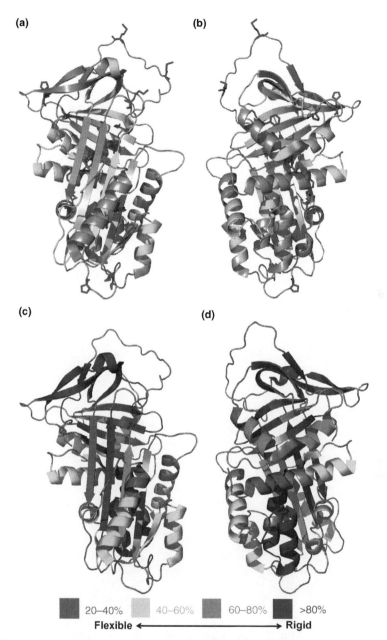

FIGURE 4.6 Comparison of the two structural MS methodologies. (a) The front view and
(b) the back view of the three-dimensional structure of active α_1AT (PDB ID 1QLP) with α-
helices and β-strands colored according to the ratio of the number of slow-exchanging
hydrogens/the number of protected hydrogens as described in *Results*. Loops are colored gray
in both figures. (c) The front view and (d) the back view of the structure of active α_1AT. Amino
acid side chains identified to be oxidized are shown in red. Peptides identified to be unmodified
are colored yellow. (See the color version of this figure in Color Plates section.)

accessibility and reactivity are the two criteria to determine whether a residue can be oxidized or not. Recent work has established the chemistry of radiolytic modification of different amino acid side chains and their relative reactivity (Xu and Chance, 2003; 2004; 2005a; 2005b). These studies proposed that at least 14 of the 20 amino acids are good footprinting probes that cover approximately 65% of the sequence of the typical protein. Their relative reactivity of the side chains using MS detection is as follows: Cys > Met > Trp > Tyr > Phe > Cystine > His > Leu, Ile > Arg, Lys, Val > Pro, Ser, Thr > Gln, Glu > Asp, Asn > Ala > Gly (Takamoto and Chance, 2006).

In terms of the two criteria mentioned above, the behavior of the 14 peptides that exhibited oxidation and the peptides that did not exhibit oxidation was generally consistent with the results of surface accessibility calculations on the native α_1AT molecule (Table 4.2). Figures 4.7a and b are two examples demonstrating that footprinting data accurately reflect the solvent accessibility of amino acid side chains consistent with the crystal structure. In peptide 202–217, His 209 but not Phe 208 is identified to be modified, although Phe is more reactive than His. In Fig. 4.7a, we can clearly see that His 209 (SASA 74 $\mathring{A}^2$) faces the solvent while Phe 208 (SASA 0 $\mathring{A}^2$) faces inside near sheet B. A second example demonstrating that footprinting data are consistent with crystallographic data is shown in Fig. 4.7b. The relative reactivity of Phe 35 (SASA 6.8 $\mathring{A}^2$) and Phe 33 (SASA 0 $\mathring{A}^2$) should be determined only by their structural environment. Although both Phe residues are located in helix A and adjacent to each other, they face in opposite directions. Phe 33 was not oxidized because it is completely buried in the space between helix A and sheet A. Thus, footprinting is a highly sensitive technique to probe solvent accessibility of side chains.

Generally, the level of exchanged hydrogens is a good reflection of solvent accessibility of backbone amide or hydrogen bonding strength of protein structure (Yan et al., 2004). In the H/D data shown in Table 4.1, the amide hydrogen atoms with fast exchange rates indicate that they are fully solvent exposed, whereas the slow exchanged hydrogens are likely located in the stable, buried regions of folded proteins. Also, regions containing significant α-helical or β-sheet content exchange more slowly than regions composed chiefly of loops or turns. For example, residues 51–78 encompassing helices B and C and residues 331–340 (strand 5A) show slow exchange, while the unstructured surface loop RCL shows rapid exchange.

However, one drawback of the H/D exchange is that it is not generally possible to achieve single-residue resolution. Some peptic fragments, especially the longer ones, may include multiple structure regions when mapped onto the crystal structure of the intact protein. Under such complicate situation, simply ranking peptides by their exchange rate may be misleading. For example, if a particular peptide encompasses 1/4 β strand and 3/4 surface loop, the peptide as a whole will display rapid exchange even if the β strand itself is stable.

Comparing the two structural MS methods, each has its respective strengths and limitations. Footprinting data have higher resolution than H/D exchange as it allows probing modifications on single amino acid residues, whereas H/D exchange is able to provide better sequence coverage than footpinting. Fourteen of 20 amino acids have been identified to be good footprinting probes, which cover approximately 65%

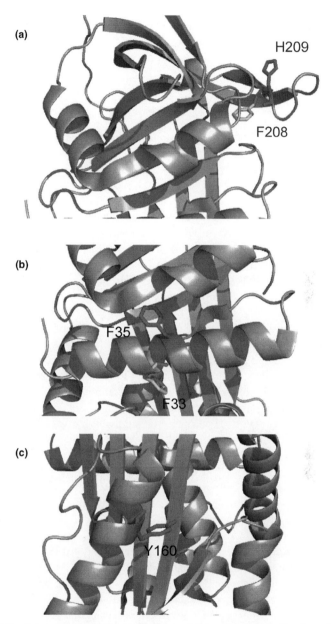

FIGURE 4.7 Solvent accessibility and dynamics information revealed by footprinting study. (a), (b) Oxidized probe residues are colored red and unmodified residues are colored yellow. (c) Oxidized probe residue Tyr 160 (red) is located in the C-terminus of helix F. Reproduced, with permission from Elsevier Ltd., from Zheng et al., 2008. (See the color version of this figure in Color Plates section.)

of the typical protein sequence. For a particular region, footprinting cannot provide structural information if a probe residue is absent. For H/D exchange, it probes almost the entire protein backbone as proline is the only amino acid without amide hydrogen.

4.5 DYNAMICS

H/D exchange is well established as a powerful and versatile tool for probing protein conformational dynamics. While amide hydrogens that are surface exposed and not involved in hydrogen bonding will exchange rapidly (within a few seconds at neutral pH), hydrogens that are hydrogen bonded or buried in the protein interior can exchange at variable rates. As discussed elsewhere (see previous chapters), the rate of exchange for a given hydrogen will depend on the frequency with which local interactions are broken by fluctuations, thus exposing the hydrogen to solvent and creating an opportunity for exchange. These structural fluctuations may be local, involving only a few nearby residues, or on a larger scale.

Conformational flexibility as determined by H/D exchange is mapped onto the structure of $\alpha_1 AT$ in Fig. 4.6c and d. For each element of secondary structure, the number of hydrogen-bonded amide hydrogens (as determined from the crystal structure) is compared with the number of slow-exchanging amide hydrogens (as determined from H/D exchange measurements) (Tsutsui et al., 2006). When these two numbers are equal, it indicates that the secondary structure element in question is stable. When the number of slow-exchanging hydrogens is significantly smaller than the number of hydrogen-bonded amide hydrogens, this indicates that the secondary structure is highly flexible and/or marginally stable. Where the ratio of slow-exchanging hydrogens to hydrogen-bonded hydrogens is >0.8, structures are colored blue (very rigid); where it is <0.2, structures are colored red (very flexible) with a scale of intermediate colors (Fig. 4.6c and d).

In general, the patterns of H/D exchange found in $\alpha_1 AT$ are consistent with what would be expected based on the crystal structure. Regions that consist largely of surface-exposed loops exchange quickly while regions that are buried or contain extensive structure exchange more slowly. There are, however, some exceptions. Particularly notable is the C-terminal half of helix F, which shows very little protection from exchange in spite of the fact that it is highly ordered in the crystal structure. From fitting of the H/D exchange data to a sum of three exponentials, it is found that the majority of the amide hydrogens from the peptide covering this region exchange at rates expected for completely exposed hydrogens. This suggests that the C-terminal half of the F helix is either unstructured in solution or undergoes frequent helix–coil transitions.

Another region that shows more rapid H/D exchange than expected from the crystal structure encompasses strands 2C, 6A, and half of helix H. From the pepsin digestion, the entire region is covered by a single peptide that is 23 residues in length. As a result of this lack of resolution, it is difficult to determine whether the structural flexibility indicated by rapid H/D exchange is localized to a particular region in this stretch of residues or evenly distributed throughout.

While hydroxyl-radical-mediated footprinting is well established as a technique for probing solvent accessibility, less is known about its ability to provide information on conformational dynamics. Results for several side chains in $\alpha_1 AT$ suggest that footprinting can, in fact, explain certain dynamic events in proteins. Tyr 160 is located at the C-terminus of helix F and is largely buried in the space between the helix F and sheet A. Of the total surface area of $140\,\mathring{A}^2$, only 22.9 is solvent accessible (Fig. 4.7c). The fact that Tyr 160 is readily oxidized suggests that its interactions with sheet A are in fact highly labile. This finding is consistent with the rapid H/D exchange seen in the C-terminal half of helix F. Both methods indicate that this region is weakly structured in solution and thus easily accessible to exchange or modification by chemical labels (deuterium or hydroxyl radicals).

Another example of how footprinting results reflect conformational dynamics can be seen from the oxidation of two methionine residues, Met 374 and Met 385. The total solvent-accessible surface area of Met 374 is $0.51\,\mathring{A}^2$, and its sulfur atom is totally buried. For Met 385, the total accessible surface area is $1.07\,\mathring{A}^2$, and its sulfur atom is likewise completely buried (Zheng et al., 2008). However, despite their sequestration from solvent, both methionines are rapidly oxidized during footprinting experiments. In the past, it has been found that oxidation of buried methionines is likely derived through secondary oxidation events mediated by peroxide generation during radiolysis (Maleknia et al., 2001; Kiselar et al., 2002) or peroxide addition as part of the oxidation mechanism (Sharp et al., 2003). However, more recent protocols have eliminated these side reactions (Xu and Chance, 2005a), and using these new methods methionine oxidation correlates well with solvent accessibility (Kamal et al., 2007; Kiselar et al., 2007). It is therefore likely that oxidation of Met 374 and Met 385 is made possible by transient structural fluctuations that expose their side chains to solvent. Only a very brief exposure would be required because the oxidation of sulfur containing side chains by hydroxyl radicals proceeds extremely rapidly (at the diffusion limit). This hypothesis is supported by molecular dynamics simulations (Zheng et al., 2008) that found that the sulfur atoms of both methionines are transiently exposed to solvent by local structural fluctuations (Fig. 4.8). During the simulations, the sulfurs of Met 374 and Met 385 achieved total accessible surface areas of 4.2 and $10.6\,\mathring{A}^2$, respectively (Fig. 4.8). The timescale of Met 374 and Met 385's momentary excursions is at the picosecond and nanosecond level (based on MD data), and the rate constant for reaction of methionine with hydroxyl radical is $10^9\,M^{-1}\,s^{-1}$ (Xu and Chance, 2007). Although the concentration of such radicals is low, radicals generated adjacent to a methionine residue that is experiencing a momentary excursion will immediately react, allowing these fast dynamics to be probed. Thus, under some circumstances hydroxyl radical footprinting can explain ultra fast (nano- and picosecond) dynamics in proteins. It is likely that this capability is limited to sulfur-containing residues, such as methionine, which are extremely reactive with hydroxyl radicals.

As mentioned above, H/D exchange results show that the region encompassing strands 2C and 6A and a portion of helix H shows very little protection from exchange despite containing extensive secondary structure. Because peptic digestion results in

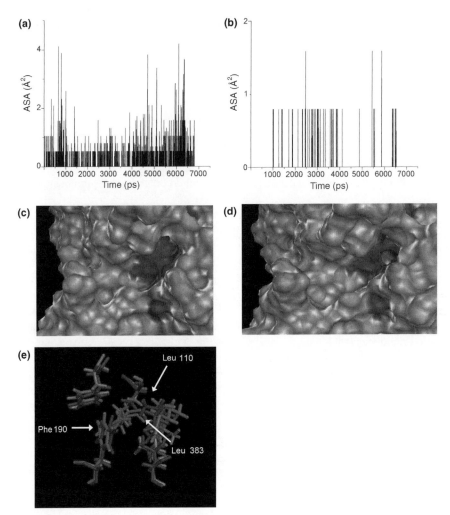

FIGURE 4.8 Molecular dynamics trajectories of Met 374. (a) Trajectory of Met 374 during 6.8 ns MD simulations. The average SASA for Met 374 residue is 0.13 ± 0.37 Å and the maximum SASA is 4.2 Å. (b) Trajectory of the sulfur atom of Met 374 during 6.8 ns MD simulation. Its average SASA is 0.01 ± 0.10 Å and the maximum SASA value is 1.6 Å. (c) and (d) show the surface of α_1AT at 5476 and 5478 ps of our MD simulation, respectively. Side chains involved in the burial/exposure of Met 374 are shown in blue, while Met 374 itself is shown in red. (e) Structures at the two time points in (c) and (d) are superimposed. It is clear that relatively small side chain displacements are sufficient to expose the sulfur atom of Met 374. Reproduced, with permission from Elsevier Ltd., from Zheng et al., 2008. (See the color version of this figure in Color Plates section.)

only a single long peptide covering this region (residues 276–299), it was not possible to localize the high flexibility to any subregion within this stretch of residues. Several reactive side chains, including His 287 and Pro 289, that are present in strand 2C are not modified by hydroxyl radicals, while Tyr 297 in strand 6A is modified despite its relatively low solvent accessibility (33 out of 140 Å^2). This suggests that the high flexibility indicated by H/D exchange is localized in strand 6A. This provides an example of how the ambiguities caused by low resolution in an H/D exchange experiment can be resolved by hydroxyl radical footprintng, and illustrates the advantages to be gained by combining both methods.

4.6 SIGNIFICANCE FOR SERPIN STRUCTURE AND FUNCTION

Because conformational mobility plays such a central role in serpin function, the distribution of flexibility in the active state has implications for the ability to efficiently inhibit target proteases. From the structures of the pre- and postinhibitory complexes (Fig. 4.1c), it is shown that helix F lies directly in the path of the reactive center loop (RCL) and target protease. Clearly, helix F must be displaced during translocation. Our findings that the C-terminal half of helix shows very rapid H/D exchange and that Tyr 160 is readily oxidized by hydroxyl radicals indicate that the top of the F helix is conformationally labile and that its contacts with the body of the molecule are weak. This would ensure that the F helix can be easily displaced from its position across the face of β-sheet A during inhibition and protease translocation.

It is generally believed that energy required for protease translocation is stored in the structure of serpins such as α_1AT in the form of conformational "strain" including suboptimal packing, steric clashes, and cavities (Seo et al., 2000). Interestingly, both Met 374 and Met 385 are located adjacent to cavities that are particularly important for α_1AT function (Zheng et al., 2008). It was found that stabilizing cavity filling mutations in these regions results in significant decreases in inhibitory activity (Lee et al., 2001). The fact that both buried methionines are readily modified by hydroxyl radicals indicates that the residues surrounding these cavities experience substantial structural fluctuations on a rapid timescale and points to an intriguing connection between local dynamics and serpin function.

Footprinting together with H/D exchange provides insight into the unique structural designs of human serpin protein and improves our understanding of its unusual inhibition mechanism.

4.7 SUMMARY

In this chapter, we have used the example of the serpin α_1AT to illustrate the types of information that can be obtained when hydrogen exchange mass spectrometry and radiolytic footprinting are both applied to the same protein. H/D exchange MS data revealed the distribution of flexibility and rigidity in functionally important regions of α_1AT (Tsutsui et al., 2006). While footprinting has revealed that several buried side chains are readily oxidized despite their apparent inaccessibility to solvent in the X-ray crystal structure. This indicates that the environments of these side chains are dynamic in solution and, significantly, these side chains are located in regions previously identified by mutagenesis as being important for the metastability of the α_1AT structure (Zheng et al., 2008).

Also, examining footprinting results together with the H/D exchange results demonstrates that the combination of both techniques provides a more complete picture of protein conformation and conformational dynamics than either technique provides alone. SF and H/D exchange are two complementary structural MS techniques to study protein structural dynamics, one probing amino acid side chains and the other probing backbone amide hydrogens, respectively. H/D exchange can probe dynamics on millisecond timescale using pulse labeling and rapid quench techniques (Busenlehner and Armstrong, 2005). In previous stop–flow footprinting experiments, millisecond dynamics was also probed (Shcherbakova et al., 2004). The results on α_1AT show that footprinting is also capable of probing ultra fast dynamics, even to the nanosecond level. Therefore, protein structural dynamics can be monitored not only at the backbone and the side chain levels but also at wide ranging timescales when both techniques are used.

Structural mass spectrometry method is an increasingly important tool to study protein structures and dynamics because of its relatively fast analysis, minimal sample requirements, and capability of providing essential dynamic information for protein segmental motions that are important for biological function. The comparison of these two techniques will enable investigators to better understand the complementarity of the two MS methods and assist them to efficiently employ these methods into their studies.

ACKNOWLEDGMENT

This work is supported in part by grants from NIBIB (P41-EB-01979) and NIH HL085469.

REFERENCES

Bai, Y., Milne, J. S., Mayne, L., and Englander, S. W., 1993. Primary structure effects on peptide group hydrogen exchange. *Proteins* 17, 75–86.

Busenlehner, L. S. and Armstrong, R. N., 2005. Insights into enzyme structure and dynamics elucidated by amide H/D exchange mass spectrometry. *Arch Biochem Biophys* 433, 34–46.

Dementiev, A., Dobo, J., and Gettins, P. G., 2006. Active site distortion is sufficient for proteinase inhibition by serpins: structure of the covalent complex of α1-proteinase inhibitor with porcine pancreatic elastase. *J Biol Chem* 281, 3452–3457.

Elliott, P. R., Pei, X. Y., Dafforn, T. R., and Lomas, D. A., 2000. Topography of a 2.0 Å structure of alpha1-antitrypsin reveals targets for rational drug design to prevent conformational disease. *Protein Sci.* 9, 1274–1281.

Englander, S. W., Sosnick, T. R., Englander, J. J., and Mayne, L., 1996. Mechanism and uses of hydrogen exchange. *Curr Opin Struct Biol* 6, 18–23.

Gettins, P. G., 2002. Serpin structure, mechanism and function. *Chem Rev* 102, 4751–4803.

Guan, J. Q. and Chance, M. R., 2005. Structural proteomics of macromolecular assemblies using oxidative footprinting and mass spectrometry. *Trends Biochem Sci* 30, 583–592.

Gupta, S., Sullivan, M., Toomey, J., Kiselar, J., and Chance, M. R., 2007. The beamline X28C of the Center for Synchrotron Biosciences: a national resource for biomolecular structure and dynamics experiments using synchrotron footprinting. *J Synchrotron Radiat* 14, 233–243.

Huntington, J. A., Read, R. J., and Carrell, R. W., 2000. Structure of a serpin–protease complex shows inhibition by deformation. *Nature* 407, 923–926.

Kamal, J. K., Benchaar, S. A., Takamoto, K., Reisler, E., and Chance, M. R., 2007. Three-dimensional structure of cofilin bound to monomeric actin derived by structural mass spectrometry data. *Proc Natl Acad Sci USA* 104, 7910–7915.

Kipping, M. and Schierhorn, A., 2003. Improving hydrogen/deuterium exchange mass spectrometry by reduction of the back-exchange effect. *J Mass Spectrom* 38, 271–276.

Kiselar, J. G., Maleknia, S. D., Sullivan, M., Downard, K. M., and Chance, M. R., 2002. Hydroxyl radical probe of protein surfaces using synchrotron X-ray radiolysis and mass spectrometry. *Int J Radiat Biol* 78, 101–114.

Kiselar, J. G., Mahaffy, R., Pollard, T. D., Almo, S. C., and Chance, M. R., 2007. Visualizing Arp2/3 complex activation mediated by binding of ATP and WASp using structural mass spectrometry. *Proc Natl Acad Sci USA* 104, 1552–1557.

Lee, C., Maeng, J. S., Kocher, J. P., Lee, B., and Yu, M. H., 2001. Cavities of α_1-antitrypsin that play structural and functional roles. *Protein Sci* 10, 1446–1453.

Maleknia, S. D., Ralston, C. Y., Brenowitz, M. D., Downard, K. M., and Chance, M. R., 2001. Determination of macromolecular folding and structure by synchrotron X-ray radiolysis techniques. *Anal Biochem* 289, 103–115.

Schuessler, H. and Schilling, K., 1984. Oxygen effect in the radiolysis of proteins. Part 2. Bovine serum albumin. *Int J Radiat Biol Relat Stud Phys Chem Med* 45, 267–281.

Seo, E. J., Im, H., Maeng, J. S., Kim, K. E., and Yu, M. Y., 2000. Distribution of the native strain in human α_1-antitrypsin and its association with protease inhibitor function. *J Biol Chem* 275, 16904–16909.

Sharp, J. S., Becker, J. M., and Hettich, R. L., 2003. Protein surface mapping by chemical oxidation: structural analysis by mass spectrometry. *Anal Biochem* 313, 216–225.

Shcherbakova, I., Gupta, S., Chance, M. R., and Brenowitz, M., 2004. Monovalent ion-mediated folding of the Tetrahymena thermophila ribozyme. *J Mol Biol* 342, 1431–1442.

Takamoto, K. and Chance, M. R., 2006. Radiolytic protein footprinting with mass spectrometry to probe the structure of macromolecular complexes. *Annu Rev Biophys Biomol Struct* 35, 251–275.

Tsutsui, Y., Liu, L., Gershenson, A., and Wintrode, P. L., 2006. The conformational dynamics of a metastable serpin studied by hydrogen exchange and mass spectrometry. *Biochemistry* 45, 6561–6569.

Whissock, J. C., Skinner, R., Carrell, R. W., and Lesk, A. M., 2000. Conformational changes in serpins: I. The native and cleaved conformations of alpha(1)-antitrypsin. *J Mol Biol* 295, 651-665.

Xu, G. and Chance, M. R., 2003. Radiolytic modification of basic amino acid residues in peptides: probes for examining protein–protein interactions. *Anal Chem* 75, 6995–7007.

Xu, G. and Chance, M. R., 2004. Radiolytic modification of acidic amino acid residues in peptides: probes for examining protein–protein interactions. *Anal Chem* 76, 1213–1221.

Xu, G. and Chance, M. R., 2005a. Radiolytic modification and reactivity of amino acid residues serving as structural probes for protein footprinting. *Anal Chem* 77, 4549–4555.

Xu, G. and Chance, M. R., 2005b. Radiolytic modification of sulfur-containing amino acid residues in model peptides: fundamental studies for protein footprinting. *Anal Chem* 77, 2437–2449.

Xu, G. and Chance, M. R., 2007. Hydroxyl radical-mediated modification of proteins as probes for structural proteomics. *Chem Rev* 107, 3514–3543.

Yan, X., Watson, J., Ho, P. S., and Deinzer, M. L., 2004. Mass spectrometric approaches using electrospray ionization charge states and hydrogen–deuterium exchange for determining protein structures and their conformational changes. *Mol Cell Proteomics* 3, 10–23.

Ye, S., Cech, A. L., Belmares, R., Bergstrom, R. C., Tong, Y., Corey, D. R., Kanost, M. R., and Goldsmith, E. J., 2001. The structure of a Michaelis serpin–protease complex. *Nat Struct Biol* 8, 979–983.

Zheng, X. J., Wintrode, P. L., and Chance, M. R., 2008. Complementary structural mass spectrometry techniques reveal local dynamics in functionally important regions of a metastable serpin. 2008. *Structure* 16, 38–51.

■■■■■ CHAPTER 5

Deuterium Exchange Approaches for Examining Protein Interactions: Case Studies of Complex Formation

ELIZABETH A. KOMIVES

Department of Chemistry and Biochemistry, University of California, San Diego, CA, USA

In this chapter, we will demonstrate state-of-the-art applications of amide hydrogen/deuterium (H/D) exchange coupled to mass spectrometry (MS) to understand the biophysics of protein–protein interactions in protein kinase A (PKA) and nuclear factor kappa B (NF-κB). It will demonstrate how such MS data are used in combination with docking, biochemical, and genetic data to provide a comprehensive view of protein complex formation.

5.1 INTERACTIONS OF REGULATORY AND CATALYTIC SUBUNITS OF PROTEIN KINASE A

Many extracellular signals result in the activation of adenylate cyclase, which produces the second messenger cAMP. Signaling by cAMP largely depends on its ability to activate protein kinase A, which phosphorylates hundreds of target proteins upon activation (Meinkoth et al., 1993). The protein kinase A holoenzyme consists of a catalytic (kinase) bound to one of the several different regulatory subunits. The regulatory subunits contain a dimerization domain that also docks to A-kinase anchoring proteins. Each regulatory subunit contains two cAMP-binding domains that together bind a single catalytic domain. Thus, the holoenzyme complex contains two regulatory and two catalytic domains, which are inactive. When four molecules of cAMP bind, the complex dissociates releasing two active kinase domains (Fig. 5.1).

Mass Spectrometry Analysis for Protein–Protein Interactions and Dynamics, Edited by Mark Chance
Copyright © 2008 John Wiley & Sons, Inc.

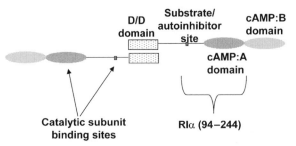

FIGURE 5.1 A schematic showing the domain organization of intact RIα. The dimerization/docking domain (10–61) is hatched, the pseudosubstrate region (94–97) is indicated in red, and the cAMP-binding domain A is in cyan and B is in blue. The region encompassing the two binding sites, the pseudosubstrate and the unknown site in the cAMP-binding A domain, is marked with arrows, and the truncated RIα(94–244) is indicated with a bracket. (See the color version of this figure in Color Plates section.)

Site-directed mutagenesis experiments had led to the model that cAMP binds first to the cAMP-binding B domain and then to the cAMP-binding A domain followed by dissociation of the regulatory subunit in either one or two steps (Taylor et al., 2005).

5.1.1 Interaction of the Catalytic Subunit with a Pseudosubstrate

Our goal was to use amide H/D exchange experiments to understand how the regulatory domains interact with the catalytic domains, and subsequently how cAMP dissociates the holoenzyme complex. We took a piece-by-piece approach based on previous genetic data. In one of the very first studies on using H/D exchange to study protein–protein complexes, we reported that addition of an inhibitor peptide with similar sequence to the pseudosubstrate/inhibitor region of the regulatory subunit (residues 94–114) "protected" amides on the surface of the catalytic subunit (Mandell et al., 1998). These experiments were done by first deuterating the ATP complex of the catalytic subunit for 10 min in D_2O buffer, then adding the peptide, and after allowing binding to occur, diluting 10-fold into H_2O buffer. Samples were taken over time during this off-exchange reaction and quenched by rapidly chilling to 0°C and decreasing the pH by addition of 2% trifluoroacetic acid (TFA). The quenched sample was digested with immobilized pepsin, and the resulting peptides were analyzed by MALDI-TOF (matrix-assisted laser desorption ionization time of flight) mass spectrometry. Several peptides from the catalytic subunit retained deuterium only when the inhibitor peptide was bound. These marked the surface of the catalytic subunit that was no longer solvent accessible when the inhibitor peptide was bound. The inaccessible surface was consistent with the crystal structure of the inhibitor peptide-kinase structure (Knighton et al., 1991).

5.1.2 Interaction of the Catalytic Domain with the RIα Regulatory Domain

The regulatory domain was known to have at least two binding sites for the catalytic subunit; one corresponded to the pseudosubstrate/inhibitor region, but the location of the other binding site was not known. For the RIα isoform of the regulatory subunit, a truncated version was shown to bind the catalytic subunit with full affinity; so, next we studied this protein, which only contained the pseudosubstrate/inhibitor sequence and the cAMP-binding A domain. It was missing the dimerization domain at the N-terminus and the cAMP-binding B domain at the C-terminus (Gibson et al., 1997). A crystal structure of this protein had recently been solved, but no structure of the complex was available (Wu et al., 2004a, 2004b). Because the complex was essentially irreversible, we used an on-exchange experimental format. In this experiment, deuterium exchange into the surface amides of the complex is compared to deuterium exchange into the surface amides of each individual protein. In this case, the interface is identified as those regions for which less deuteration is observed in the complex as compared to the individual proteins. Using MALDI-TOF MS, we were able to analyze all the peptides from the pepsin digest of the complex between the RIα (94–244) and the catalytic subunit at once with very little overlap of peptides from each of the proteins (Fig. 5.2a). When the H/D exchange results from the free proteins were compared to those obtained from the protein complex, the same regions of the catalytic subunit that we had previously identified as the pseudosubstrate/inhibitor binding site incorporated less deuterium in the complex as compared to the free protein. Simple binding sites such as the pseudosubstrate/inhibitor binding at the active site give equivalent surface mapping results whether the experiment is set up in the on-exchange or off-exchange format. If one suspects conformational changes, it is sometimes possible to obtain additional insights into such events by doing both on-exchange and off-exchange experiments. In addition to the pseudosubstrate/inhibitor binding site, a new contiguous surface in the large lobe of the kinase showed decreased surface solvent accessibility where on-exchange was less in the complex than in the free protein (Fig. 5.2b and c).

Concomitantly, a contiguous surface on the regulatory subunit that also incorporated less deuterium in the complex as compared to the free regulatory subunit was observed (Fig. 5.3a). Other groups found a similar surface when the full-length regulatory subunit was studied (Hamuro et al., 2004).

5.1.3   Combination of H/D Exchange Data and Computational Docking

Faced with trying to put these two surfaces together in a meaningful structure of the complex, we turned to molecular docking. Attempts at molecular docking of the regulatory and catalytic subunits had proven unsuccessful (Fig. 5.3b). We thought that if we could filter the docked solutions based on the H/D exchange results, we might "find" the correct solutions, and indeed we did (Fig. 5.3c) (Anand et al., 2003).

The docked structure gave a good idea of where the interface between the regulatory and catalytic subunits was located, but the H/D exchange experiments did not reveal the large conformational change in the regulatory subunit that was later found to occur upon complex formation (Kim et al., 2005). Once the conformational change was known from the structure of the complex, we went back to see how it was possible that the H/D exchange data did not reveal such a large change. The conformational change involved the straightening of the bent helix, and remarkably the helix lost some interactions and gained others so that the change was not seen in

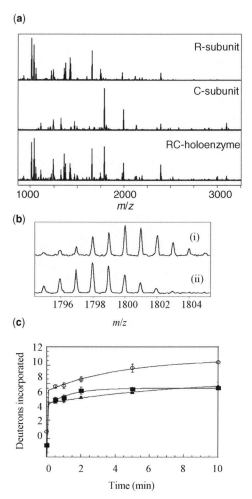

FIGURE 5.2   (a) MALDI-TOF mass spectrum of the peptic digest from RIα (top), from the C-subunit (middle), and from the RIα$_2$-C$_2$ holoenzyme (bottom). (b) Region of the MALDI-TOF mass spectrum showing one of the peptides (m/z = 1793.97, residues 247–261) from the C-subunit that experienced slowed exchange in the RIα$_2$-C$_2$ holoenzyme complex. (i) The isotopic envelope for the peptide from the free C-subunit bound to MgATP after 10 min of deuteration. (ii) The isotopic envelope for the same peptide from the RIα$_2$-C$_2$ holoenzyme after 10 min of deuteration. (c) Plot of deuterium incorporation into the amide positions of the region of C-subunit for the peptide fragment, m/z = 1793.97, from the C-subunit (+MgATP) (○), RIα$_2$-C$_2$ holoenzyme (■), and RIα (94–244)-C holoenzyme (▲).

the surface amide exchange. These results showed that it was possible to combine H/D exchange interface mapping with rigid body molecular docking to build up structural models of protein–protein complexes from the structures of the individual protein components with the caveat that if there are large conformational changes, these may not be detected by the H/D exchange.

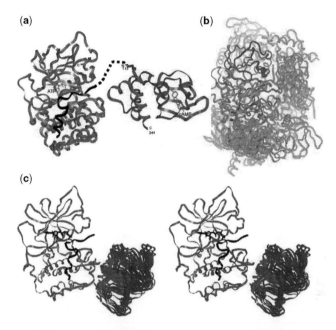

FIGURE 5.3 (a) Structure of the protein kinase A catalytic subunit (gray) showing the inhibitor/pseudosubstrate peptide (black). Those regions of the protein identified by amide H/D exchange that are in the inhibitor/pseudosubstrate interface are shown in gold and the newly observed interface with the regulatory subunit are shown in red. The structure of the regulatory subunit (residues 94–244) is shown in blue with the interface that was observed with the catalytic subunit shown in red. The connection between the inhibitor peptide and the cAMP-binding A domain is dashed because this region is disordered in the structure of the regulatory subunit. The cAMP bound in the A domain is shown as gold sticks. (b) The top 100 structures from the rigid body docking of the regulatory domain (residues 94–244) (green) with the catalytic domain (blue). The catalytic domain is in the same orientation as in (a). (c) Stereoview of the ensemble of structures obtained after filtering the top 100,000 molecular docking solutions using the amide H/D exchange results. (See the color version of this figure in Color Plates section.)

5.2 ALLOSTERY IN PROTEIN–PROTEIN INTERACTIONS REVEALED BY H/D EXCHANGE

5.2.1 Allostery within the Regulatory Subunit Revealed by H/D Exchange

Another very interesting finding came from the studies on the regulatory subunit. This protein can exist in three states: with nothing bound, with cAMP bound, and with the catalytic subunit bound. It is not possible for the regulatory subunit to bind both cAMP and the catalytic subunit simultaneously. Indeed, this is how the signaling process starts. When a cell receives a signal, adenylate cyclase is activated and cAMP is produced. One of the first responders to cAMP binding is protein kinase A, which binds the cAMP causing the catalytic subunits to dissociate, releasing two active kinase molecules. Interestingly, crystal structures did not reveal any conformational differences between the cAMP-bound and cAMP-free regulatory subunits (Wu et al., 2004a, 2004b). However, amide exchange revealed communication between the cAMP-binding site and the catalytic subunit-binding site. In the cAMP-bound form, the region of the regulatory subunit that binds the catalytic subunit was observed to exchange more deuterium. Conversely, in the catalytic subunit-bound form, the cAMP-binding pocket exchanged more deuterium (Fig. 5.4) (Anand et al., 2002). Because amide exchange is sensitive to protein conformation and dynamics, it can reveal both types of allostery in proteins.

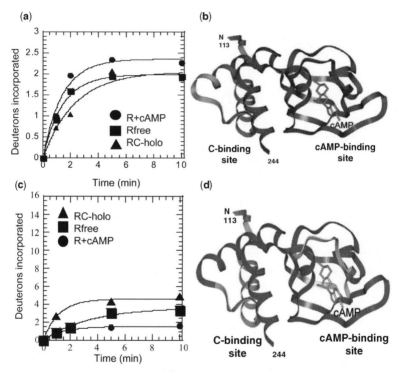

FIGURE 5.4   (a) The time course of deuterium incorporation into the peptide m/z = 1594.73, corresponding to residues 136–148 of the A-helix, shows that the most deuterium is incorporated in the R + cAMP state (●), while somewhat less is incorporated into the free R-subunit (■), and the least is incorporated when the C-subunit is bound (▲). (b) Structure of the regulatory (R) subunit (blue) showing the location of residues 136–148 (gold). (c) The time course of deuterium incorporation into the peptide m/z = 2115.27, corresponding to residues 202–221 of the phosphate binding cassette, shows that the most deuterium is incorporated into the R + C state (▲), while less is incorporated into the free regulatory subunit (■), and the least is incorporated when the cAMP is bound (●). (d) Structure of the regulatory (R) subunit (blue) showing the location of residues 202–221 (gold). (See the color version of this figure in Color Plates section.)

5.2.2 Allostery in the Thrombin–Thrombomodulin Interaction

In studies of the thrombin–thrombomodulin interaction, amide H/D exchange experiments have revealed a similar line of communication between anion binding exosite 1 and the active site (Mandell et al., 2001). As it turns out, only cofactor-active thrombomodulin fragments cause the full complement of loop tightening in the active site (Koeppe et al., 2005). Others have observed this type of dynamic allostery as well (Shi et al., 2006). It may turn out that the real power of amide exchange experiments is in revealing these subtle dynamic changes upon protein–protein binding as much as in identification of interfaces.

5.3 INTERACTIONS OF THE INHIBITOR IκBα WITH THE TRANSCRIPTION FACTOR NF-κB

NF-κB proteins are a family of eukaryotic transcription factors that regulate more than 150 target genes involved in a wide variety of cellular functions (Pahl, 1999). Inflammatory cytokines, growth factors, and some bacterial and viral products are among the signals that activate NF-κB (Ghosh et al., 1998). NF-κB regulates gene transcription in stress and immune response, cellular differentiation, and proliferation (Ghosh et al., 1998; Pahl, 1999; Li and Stark, 2002; Bonizzi and Karin, 2004). Misregulation of NF-κB occurs in many diseases, including heart disease, Alzheimer's disease, diabetes, AIDS, and cancers (Greten and Karin, 2004; Kumar et al., 2004).

The NF-κB family includes five different subunits (p50, p52, p65/RelA, RelB, and cRel) that form homo- and heterodimers, but the canonical form is the p50/p65 heterodimer (Ghosh et al., 1998). The NF-κB transcriptional activity is regulated by a family of inhibitor proteins, IκBα, IκBβ, and IκBε (Verma et al., 1995; Hoffmann et al., 2002). In resting cells, IκB binds NF-κB, causing the complex to be sequestered in the cytosol by masking the nuclear localization sequence (Baeuerle and Baltimore, 1988; Baldwin, 1996). When the cell receives an activating signal, the IκB kinases (IKKs) are activated. These phosphorylate the NF-κB-bound IκB, causing subsequent ubiquitination and proteasomal degradation of the IκB (Traenckner et al., 1994; Brown et al., 1995; Traenckner and Baeuerle, 1995; Chen et al., 1996; Karin and Ben-Neriah, 2000). The freed NF-κB translocates to the nucleus, binds to DNA, and activates transcription of target genes (Pahl, 1999). One of the genes that is activated by NF-κB is the IκBα gene (Brown et al., 1993; de Martin et al., 1993; Scott et al., 1993; Sun et al., 1993). Newly synthesized IκBα enters the nucleus and binds to NF-κB. The NF-κB·IκBα complex is exported to the cytosol and shuts down the transcriptional activation. Studies in IκBβ$^{-/-}$ and IκBε$^{-/-}$ cells show an oscillatory NF-κB response, due to rapid activation of NF-κB transcriptional activity by signal-induced IκBα degradation and strong negative feedback by NF-κB-induced IκBα production (Hoffmann et al., 2002).

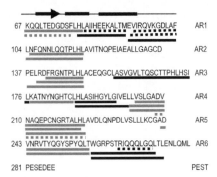

FIGURE 5.5 The coverage map showing all of the peptides resulting from pepsin cleavage of IκBα (25 peptides, bars below the sequence), which cover 74% of the sequence. Six peptides (dashed lines) can only be analyzed qualitatively. The peptides cover each of the β-hairpin regions in all six ARs (gray bars). A schematic of an AR above the sequence shows the α-helices (bars) and β-sheets (arrows).

5.3.1 H/D Exchange of IκBα Reveals Partially Unfolded Regions

Amide H/D exchange experiments were performed on the free and NF-κB-bound IκBα. IκBα is a repeat protein, and fortuitously, pepsin cleaved it in the same place in each repeat. In this way, the amide exchange of the β-hairpin from each ankyrin repeat could be compared to the amide exchange of each of the other repeats (Fig. 5.5).

5.3.2 H/D Exchange Reveals IκBα Folds upon Binding to NF-κB

For the study of the IκBα in complex with NF-κB, pepsin cleavage of the entire complex resulted in many peptides from NF-κB that overlapped with IκBα peptides obscuring the results. To circumvent this problem, an N-terminal cysteine residue was introduced into the p65 subunit that could be specifically biotinylated. The p65 subunit was immobilized on streptavidin beads via the biotin so that the NF-κB (all of the p65 and most of the p50) could be removed from the H/D exchange mixture prior to pepsin digestion. This "trick" was used previously to map a protein–antibody interface and it worked extremely well in both cases (Baerga-Ortiz et al., 2002).

The kinetic plots for deuterium incorporation into the free and NF-κB-bound IκBα are shown in Fig. 5.6.

The most striking result is the very large difference in amides exchanging into the β-hairpins of the fifth and sixth ankyrin repeats. More than 10 amides that were rapidly exchanging in the free protein exchanged slowly in the bound protein. In fact, the amide exchange into the β-hairpins of the fifth and sixth ankyrin repeats became comparable to the low exchange seen for the second, third, and fourth repeats in the free protein (Fig. 5.7).

Typically, on-exchange experiments of protein complexes reveal regions of each protein that exchange less deuterium in the complex as compared to in the free proteins. If the on-exchange is done for short periods of time, one can

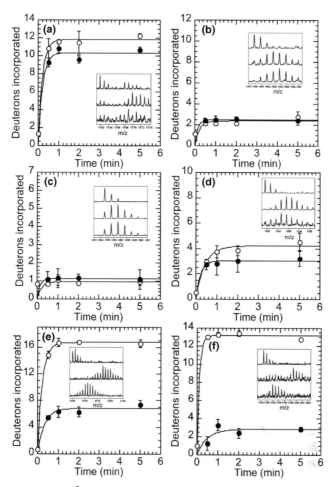

FIGURE 5.6 Amide H/²H exchange in IκBα β-hairpins with and without NF-κB. (a) Deuterium incorporation in the β-hairpin in AR 1 peptide MH+ 1761.85 shows only small differences in the extent of exchange in free (○) and NF-κB-bound IκBα (●) that may be due to protection at the IκBα·NF-κB interface. (b) Deuterium incorporation in the β-hairpin of AR 2 peptide MH+ 1679.87 was low in both the free and the bound state. (c) Deuterium incorporation in the β-hairpin of AR 3 peptide MH+ 1054.58 was low in both the free and the bound state. (d) Deuterium incorporation in the β-hairpin of AR 4 peptide MH+ 1221.56 showed only a small change in the extent of exchange between free and bound IκBα again likely due to interface protection. (e) Deuterium incorporation into the β-hairpin of AR 5 peptide MH+2165.08 shows a large decrease in amide exchange in the bound form that is much larger than expected for protection at the IκBα·NF-κB interface. (f) Deuterium incorporation into the β-hairpin of AR 6 peptide MH+1788.89 shows a large decrease in amide exchange similar to that seen for the fifth repeat. Error bars represent the standard deviation of triplicate reactions, and the y-axis maximum corresponds to the total number of exchangeable amide protons in the peptide, except for panel (f), which only has 13 amide protons. Insets show MALDI mass envelopes in nondeuterated controls (top), free IκBα after 2 min of exchange (middle), and NF-κB-bound IκBα after 2 min of exchange (bottom).

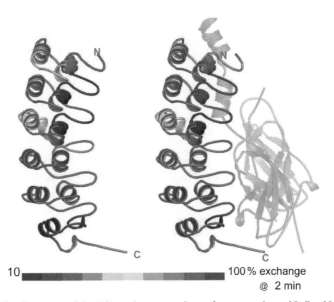

FIGURE 5.7   Summary of the H/D exchange results on free versus bound IκBα. Nearly all of the amides in the β-hairpins of the fifth and sixth ankyrin repeats exchange rapidly in the free state, but in the bound state, they show similar low exchange as was observed for the second, third, and fourth repeats. (See the color version of this figure in Color Plates section.)

generally conclude that the amides exchanging in the free proteins are near the surface of the protein, and the decreases in the complex are due to decreased surface solvent accessibility. However, there are other reasons why amide exchange may slow upon protein–protein binding and these include changes in backbone dynamics, changes in conformation, and changes in the foldedness of one of the binding partners. In the case of the fifth and sixth ankyrin repeats, a very large number of amides were no longer exchanging in the protein–protein complex. To understand how such a large decrease in amide exchange could come about, we correlated the amide exchange results with the computed solvent-accessible surface area of the IκBα molecule from the crystal structure of the NF-κB·IκBα complex (Huxford et al., 1998; Jacobs and Harrison, 1998). Since there is no structure of free IκBα, the solvent-accessible surface area for the free protein was computed from the structure of IκBα taken from the structure of the complex (Truhlar et al., 2006a, 2006b). It is important to emphasize that the H/D exchange experiments were done for short periods of time, and the exchange at 2 min was used in the correlation. Under these conditions, the correlation between computed solvent-accessible surface area and surface amide exchange is extremely high (Fig. 5.8).

Clearly, if exchange is allowed to occur for hours, this correlation will break down. This analysis allowed us to conclude that the large decrease in amide exchange into the β-hairpins of the fifth and sixth repeats could not be accounted

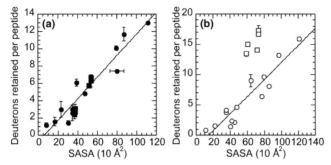

FIGURE 5.8   (a) A strong correlation was observed between the solvent-accessible surface area computed from the crystal structure of the NF-κB/IκBα complex and the results from amide H/D exchange of the surface of the complex for 2 min (●), including for the fifth and sixth ankyrin repeats (■). (b) A strong correlation was also observed for the solvent-accessible surface area computed for the structure of IκBα taken from the crystal structure of the NF-κB/IκBα complex (○) except for the fifth and sixth repeats, which show much more exchange than expected from the structure (□).

for by interface protection. In contrast, the small decreases in amide exchange into the β-hairpins of the first and fourth ankyrin repeats could be accounted for by interface protection perfectly. This analysis allowed us to make the definitive conclusion that the fifth and sixth repeats of IκBα fold upon binding to NF-κB (Truhlar et al., 2006b). This case highlights how the use of known structural information can help with interpretation of amide H/D exchange data.

REFERENCES

Anand, G. S., Hughes, C. A., Jones, J. M., Taylor, S. S., and Komives E. A., 2002. Amide H/2H exchange reveals communication between the cAMP and catalytic subunit-binding sites in the R(I)alpha subunit of protein kinase A. *J Mol Biol* 323 (2), 377–386.

Anand, G. S., Law D., Mandell, J. G., Snead, A. N., Tsigelny, I., Taylor, S. S., Ten Eyck, L., and Komives, E. A., 2003. Identification of the protein kinase A regulatory RI–catalytic subunit interface by amide H/²H exchange and protein docking. *Proc Natl Acad Sci USA* 100, 13264–13269.

Baerga-Ortiz, A., Hughes, C. A., Mandell, J. G., and Komives, E. A., 2002. Epitope mapping of a monoclonal antibody against human thrombin by H/D-exchange mass spectrometry reveals selection of a diverse sequence in a highly conserved protein. *Protein Sci* 11 (6), 1300–1308.

Baeuerle, P. A. and Baltimore, D., 1988. I-Kappa-B: a specific inhibitor of the Nf-kappa-B transcription factor. *Science* 242 (4878), 540–546.

Baldwin, A. S., 1996. The NF-kappa-B and I-kappa-B proteins: new discoveries and insights. *Annu Rev Immunol* 87, 13–20.

Bonizzi, G. and Karin,M., 2004. The two NF-kappaB activation pathways and their role in innate and adaptive immunity. *Trends Immunol* 25 (6), 280–288.

Brown, K., Park, S., Kanno T., Franzoso, G., and Siebenlist, U., 1993. Mutual regulation of the transcriptional activator NF-kappa B and its inhibitor, I kappa B-alpha. *Proc Natl Acad Sci USA* 90 (6), 2532–2536.

Brown, K., Gerstberger, S., Carlson, L., Franzoso, G., and Siebenlist, U., 1995. Control of I kappa B-alpha proteolysis by site-specific, signal-induced phosphorylation. *Science* 267 (5203), 1485–1488.

Chen, Z. J., Parent, L., and Maniatis, T., 1996. Site-specific phosphorylation of IkappaBalpha by a novel ubiquitination-dependent protein kinase activity. *Cell* 84 (6), 853–862.

de Martin R., Vanhove B., Cheng Q., Hofer E., Csizmadia V., Winkler H., and Bach F. H., 1993. Cytokine-inducible expression in endothelial cells of an I kappa B alpha-like gene is regulated by NF kappa B. *EMBO J* 12 (7), 2773–2779.

Ghosh, S., May, M. J., and Kopp, E. B., 1998. NF-kappa, B and Rel proteins: evolutionarily conserved mediators of immune responses. *Annu Rev Immunol* 16, 225–260.

Gibson, R. M., Ji-Buechler, Y., and Taylor, S. S., 1997. Interaction of the regulatory and catalytic subunits of cAMP-dependent protein kinase: electrostatic sites on the type Ia regulatory subunit. *J Biol Chem* 272, 16343–16350.

Greten, F. R. and Karin,M., 2004. The IKK/NF-kappaB activation pathway: a target for prevention and treatment of cancer. *Cancer Lett* 206 (2), 193–199.

Hamuro, Y., Anand, G. S., Kim, J. S., Juliano C., Stranz, D. D., Taylor, S. S., and Woods, V. L. J., 2004. Mapping intersubunit interactions of the regulatory subunit (RIalpha) in the type I holoenzyme of protein kinase A by amide hydrogen/deuterium exchange mass spectrometry (DXMS). *J Mol Biol* 340 (5), 1185–1196.

Hoffmann, A., Levchenko, A., Scott, M. L., and Baltimore, D., 2002. The IkappaB-NF-kappaB signaling module: temporal control and selective gene activation. *Science* 298 (5596), 1241–1245.

Huxford, T., Huang, D. B., Malek, S., and Ghosh, G., 1998. The crystal structure of the IkappaBalpha/NF-kappaB complex reveals mechanisms of NF-kappaB inactivation. *Cell* 95, 759–770.

Jacobs, M. D. and Harrison, S. C., 1998. Structure of an IkappaBalpha/NF-kappaB complex. *Cell* 95, 749–758.

Karin, M. and Ben-Neriah,Y., 2000. Phosphorylation meets ubiquitination: the control of NF-[kappa]B activity. *Annu Rev Immunol* 18, 621–663.

Kim, C., Xuong, N. H., and Taylor, S. S., 2005. Crystal structure of a complex between the catalytic and regulatory (RIalpha) subunits of PKA. *Science* 307 (5710), 690–696.

Knighton, D. R., Zheng, J. H., Ten Eyck, L. F., Xuong, N. H., Taylor, S. S., and Sowadski, J. M., 1991. Structure of a peptide inhibitor bound to the catalytic subunit of cyclic adenosine monophosphate-dependent protein kinase. *Science* 253 (5018), 414–420.

Koeppe, J. R., Seitova, A., Mather, T., and Komives, E. A., 2005. Thrombomodulin tightens the thrombin active site loops to promote protein C activation. *Biochemistry* 44, 14784–14791.

Kumar, A., Takada, Y., Boriek, A. M., and Aggarwal, B. B., 2004. Nuclear factor-kappaB: its role in health and disease. *J Mol Med* 82 (7), 434–448.

Li X. and Stark G. R., 2002. NFkappaB-dependent signaling pathways. *Exp Hematol* 30 (4), 285–296.

Mandell, J. G., Falick, A. M., and Komives, E. A., 1998. Identification of protein:protein interfaces by decreased amide proton solvent accessibility. *Proc Natl Acad Sci USA* 95 (25), 14705–14710.

Mandell, J. G., Baerga-Ortiz A., Akashi, S., Takio, K., and Komives, E. A., 2001. Solvent accessibility of the thrombin–thrombomodulin interface. *J Mol Biol* 306 (3), 575–589.

Meinkoth, J. L., Alberts, A. S., Went W., Fantozzi, D., Taylor, S. S., Hagiwara, M., Montminy, M., and Feramisco, J. R., 1993. Signal transduction through the cAMP-dependent protein kinase. *Mol Cell Biochem* 127–128, 179–186.

Pahl, H. L., 1999. Activators and target genes of Rel/NF-kappaB transcription factors. *Oncogene* 18, 6853–6866.

Scott, M. L., Fujita, T., Liou, H. C., Nolan, G. P., and Baltimore, D., 1993. The p65 subunit of NF-kappa B regulates I kappa B by two distinct mechanisms. *Genes Dev* 7 (7A), 1266–1276.

Shi, Z., Resing, K. A., and Ahn, N. G., 2006. Networks for the allosteric control of protein kinases. *Curr Opin Struct Biol* 16, 686–692.

Sun, S. C., Ganchi, P. A., Ballard, D. W., and Greene, W. C., 1993. NF-kappa B controls expression of inhibitor I kappa B alpha: evidence for an inducible autoregulatory pathway. *Science* 259 (5103), 1912–1915.

Taylor, S. S., Kim, C., Vigil, D., Haste, N. M., Yang, J., Wu, J., and Anand, G. S., 2005. Dynamics of signaling by PKA. *Biochim Biophys Acta* 1754, 25–37.

Traenckner, E. B. and Baeuerle, P. A., 1995. Appearance of apparently ubiquitin-conjugated I kappa B-alpha during its phosphorylation-induced degradation in intact cells. *J Cell Sci Suppl* 19, 79–84.

Traenckner, E. B., Wilk S., and Baeuerle, P. A., 1994. A proteasome inhibitor prevents activation of NF-kappa B and stabilizes a newly phosphorylated form of I kappa B-alpha that is still bound to NF-kappa B. *EMBO J* 13 (22), 5433–5441.

Truhlar, S. M. E., Torpey, J., Koeppe, J. R., Croy, C. H., and Komives, E. A., 2006a. Solvent accessibility of protein surfaces by amide H/2H exchange MALDI-TOF mass spectrometry. *J Am Soc Mass Spectrom* 17, 1490–1497.

Truhlar, S. M. E., Torpey, J. W., and Komives, E. A., 2006b. Regions of IkBa that are critical for its inhibition of NFkB·DNA interaction fold upon binding to NF-kB. *Proc Natl Acad Sci USA* 103, 18951–18956.

Verma, I. M., Stevenson, J. K., Schwarz, E. M., Van, Antwerp, D., and Miyamoto, S., 1995. Rel/NF-kappa B/I kappa B family: intimate tales of association and dissociation. *Genes Dev* 9 (22), 2723–2735.

Wu, J., Brown, S., Xuong, N. H., and Taylor, S. S., 2004a. RIalpha subunit of PKA: a cAMP-free structure reveals a hydrophobic capping mechanism for docking cAMP into site B. *Structure* 12 (6), 1057–1065.

Wu, J., Jones, J. M., Nguyen-Huu X., Ten Eyck, L. F., and Taylor, S. S., 2004b. Crystal structures of RIalpha subunit of cyclic adenosine 5′-monophosphate (cAMP)-dependent protein kinase complexed with (Rp)-adenosine 3′,5′-cyclic monophosphothioate and (Sp)-adenosine 3′,5′-cyclic monophosphothioate, the phosphothioate analogues of camp. *Biochemistry* 43 (21), 6620–6629.

CHAPTER 6

Hydrogen/Deuterium Exchange Studies of Viruses

SEBYUNG KANG and PETER E. PREVELIGE JR.

Department of Microbiology, University of Alabama at Birmingham, Birmingham, AL, USA

6.1 OVERVIEW OF VIRUS LIFECYCLES

Viral capsids are protein shells that contain and protect the viral genome from external insult. Viral capsids may appear static but they are actually dynamic supramolecular structures that undergo a programmed series of conformational transformations during their replication cycle. The replication cycle is generally composed of four recognizable steps: entry, viral protein and nucleic acid synthesis, assembly, and exit. Entry can be accomplished either via the formation of a pore through which the nucleic acid translocates, via fusion between the viral envelope and the host plasma membrane leading to internalization of the viral capsid, or via endocytosis and release. If the nucleic acid is not injected, some disassembly is generally required to liberate the nucleic acid. A variety of postentry strategies are employed by viruses to coopt the biosynthetic machinery of the infected cell into replicating their proteins and nucleic acid. The replicated proteins and nucleic acid assemble into viral capsids, which are subsequently released from the host cell by either budding or cell lysis. The released particles can be enveloped by either a host cell-derived lipid bilayer or naked protein shells. Frequently, the immediate product of assembly is noninfectious, either because the nucleic acid has not yet been packaged or because the viral capsid needs to undergo a maturation event. Maturation of these intermediates, termed procapsids, is often induced by either or both genome packaging and proteolysis.

6.2 STRUCTURAL INVESTIGATIONS OF VIRAL CAPSIDS

X-ray crystallography and cryo-electron microscopy (cryo-EM) coupled with image analysis are powerful tools for determining the structure of viral capsids. Viral

Mass Spectrometry Analysis for Protein–Protein Interactions and Dynamics, Edited by Mark Chance
Copyright © 2008 John Wiley & Sons, Inc.

capsids are generally assembled from hundreds of copies of a small number of protein subunits using repeated bonding interactions (Caspar and Klug, 1962). This strategy results in the formation of a symmetrical capsid, and both helically and icosahedrally symmetric viruses are commonly observed.

The atomic structures of a number of viruses have been determined crystallographically, and in a few cases crystal structures of the virus in various stages of the life cycle are available. However, obtaining the atomic structures of viruses in multiple states is challenging. An alternative "divide and conquer" strategy takes advantage of the fact that the resolution routinely achievable in a cryo-EM reconstruction (currently 8–10 Å) is sufficient to identify elements of secondary structure and often track conformational changes during the life cycle either directly or by fitting the known crystal structure of the subunit into the reconstructions. Although many viral capsids are amenable to either crystallographic or cryo-EM-based studies, not all capsids can be studied in this manner. Enveloped capsids have proven refractory to crystallography but can be approached by electron microscopy provided the underlying capsid geometry is regular, that is, all the particles in an ensemble are structurally identical. This requirement stems from the need to obtain projection images from multiple angles to reconstruct the three-dimensional structure. Should the particles not be identical, as is the case for HIV, for example, tomographic reconstructions in which tilted images of a single particle are analyzed can provide three-dimensional reconstructions. Tomographic reconstructions are limited to approximately 50–90 Å in resolution by radiation damage, arising from the need to repeatedly image the same particle at different tilt angles. One limitation to the approach of merging crystal structures with cryo-EM reconstructions is that the resolution of cryo-EM reconstructions has generally been insufficient to reveal the detailed side chain interactions involved in assembly and maturation.

6.3 DYNAMICS OF VIRAL CAPSIDS

The first indication of the dynamic nature of viral capsids came from studies on poliovirus. The capsid of poliovirus is made of four proteins: VP1, VP2, VP3, and VP4. The crystal structure of poliovirus demonstrated that VP1, VP2, and VP3 make up the outer capsid and VP4 is located on the inside of the capsid. However, antibodies directed against VP4 were capable of neutralizing the virus, suggesting that VP4 was dynamic and fluctuating between the inside and outside of the capsid at physiological temperature (Li et al., 1994). Subsequent evidence for capsid dynamics was obtained by mass spectrometry-based analysis of the kinetics of trypsin cleavage. Flock house virus is an RNA-containing insect virus. Viral capsids can be produced that contain either the viral RNA or the nonspecific cellular RNA. The crystal structures of the two capsids are identical. However, an analysis of the kinetics of trypsin digestion by mass spectrometry revealed that the particles that contained the viral RNA were significantly stabilized relative to the particles that contained host RNA (Bothner et al., 1998, 1999). This striking finding suggested that in addition to carrying genetic information, the viral nucleic acid may play a structural role and influence the dynamics of the viral capsid.

6.4 HYDROGEN/DEUTERIUM EXCHANGE STUDIES OF VIRUS CAPSID STRUCTURE

Hydrogen/deuterium exchange studies of viruses have provided key insights into the structures and structural changes that viral capsids undergo in circumstances where the techniques of X-ray diffraction or electron microscopy have been unable to provide full answers. While performing hydrogen/deuterium exchange studies on viral capsids is in principle similar to performing exchange studies on other systems, they introduce some unique technical considerations. These will be discussed at the end of this chapter. The following section will offer a review of the systems and key results where exchange studies have been applied to the study of viruses.

6.4.1 Bacteriophage P22

The assembly of the *Salmonella typhimurium* bacteriophage P22 is a multistep process. The initial assembly product is a procapsid composed of 415 copies of the external coat protein arranged as a $T = 7$ lattice, and approximately 300 copies of the internal scaffolding protein are required to ensure high-fidelity assembly. The genomic DNA is packaged into the procapsid through the action of a complex of proteins that constitute a molecular motor. DNA packaging results in the release of scaffolding protein and a 10% expansion of the coat protein lattice. The expansion is accompanied by stabilization, and cryo-electron micrographic reconstructions of the procapsid and mature forms at approximately 9 Å resolution indicates that the stabilization arises from the formation of new intersubunit contacts upon maturation (Fig. 6.1a) (Prevelige and King, 1993).

Raman spectroscopy, a vibrational spectroscopy, can detect hydrogen/deuterium exchange (HDX) in proteins through the change in vibrational frequencies arising from the mass change upon deuteration. Raman spectroscopy-based hydrogen/deuterium exchange studies indicated that as the 431-residue-long coat protein subunit progressed through the assembly pathway from an isolated subunit, to a procapsid incorporated subunit, to the final mature form, the size of the exchange-protected core increased from 14% to 25% (Tuma et al., 1998a). These data suggested that the subunit underwent a progressive coupled folding and association pathway in which the initial state more resembled a protein folding intermediate and the final state more resembled a typical globular folded protein. While Raman spectroscopy provides a sensitive measure of the extent of the change in protection, and can identify the secondary structure elements that undergo change through their unique vibrational frequencies, it cannot identify the particular regions of the protein undergoing the change in protection as mass spectrometry can.

A series of biochemical experiments were performed to identify the regions of the coat protein that became more stable upon assembly and maturation. Protease digestion experiments indicated that the coat protein was composed of N- and C-terminal proteolytic domains (Lanman et al., 1999). To examine their behavior, the individual domains were cloned and purified (Kang and Prevelige, 2005). Hydrodynamic and spectroscopic studies demonstrated that the N-terminal domain was largely unfolded and extended in solution while the C-terminal domain formed

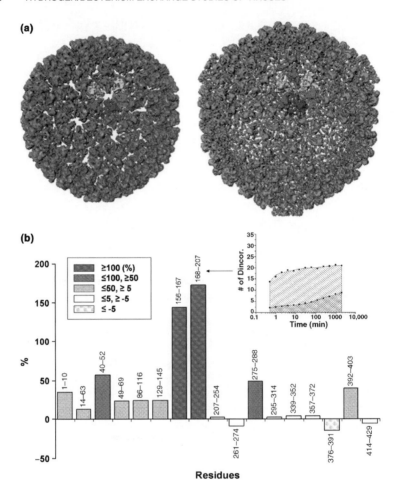

FIGURE 6.1 Hydrogen/deuterium exchange of bacteriophage P22 capsid. (a) Three-dimensional reconstruction of bacteriophage P22 procapsids (left) and phage (right) from cryo-electron micrographs. Note the increase in diameter, angular appearance, and generally more closed structure of the mature phage lattice. The seven "quasi-equivalent" subunits that comprise an asymmetric unit of the capsid are colored. (b) The relative change in exchange protection upon capsid expansion. The number of deuterons incorporated at each time point was plotted and the area under the curve calculated (inset). The percent change was then plotted for each identified peptide. The N-terminal domain spans residues 1–145 and the C-terminal domain spans residues 207–429. The loop region spans residues 156–207. Part (b) of this figure reproduced, with permission from Elsevier Ltd., from Kang and Prevelige, 2005. (See the color version of this figure in the Color Plates section.)

compact dimers. To characterize the regions undergoing change upon maturation, hydrogen/deuterium exchange experiments were performed on the coat protein in the procapsid and capsid lattices (Kang and Prevelige, 2005). As is frequently the case with viral capsids, it was necessary to include a denaturing agent, in this case 8 M urea, in the acid quench to disrupt the lattice.

Exchanged, quenched, and dissociated procapsids and capsids were digested in solution with pepsin. Approximately, 85% sequence coverage was achieved. Comparing the exchange protection in the procapsid and mature capsid forms revealed that the C-terminal domain underwent relatively little change and was relatively protected in both the immature and mature forms (Fig. 6.1b). In contrast, the N-terminal domain underwent an overall increase in exchange protection upon maturation. This suggested that the flexibility in the N-terminal region detected in the isolated domain was part of an adaptive region that formed increasingly stable contacts during the assembly and maturation pathway. The N-terminal domain might make the new stabilizing contacts that accompany maturation.

Particularly noteworthy was the approximately 50-residue peptide that linked the N- and C-terminal domains. This region was 80% exchanged within 3 min in the procapsid form but remained approximately 70% unexchanged in the mature form even after several days. These data are consistent with the hypothesis that this region is an exposed loop prior to maturation but forms new stabilizing contacts upon maturation. On the basis of these data, a cysteine residue was introduced into the middle of the loop (Kang et al., 2006). The cysteine residue slowly but spontaneously formed a disulfide bond with the cysteine residue in an adjacent subunit. The effect of disulfide bond formation was the increase in the rate of the maturation expansion. Interestingly, there was only one peptide (residues 376–391 of the C-terminal domain) that became less protected upon maturation. The results suggest that the C-terminal domain represents the stable folded core of the subunit that is capable of undergoing biomolecular recognition and self-association. This core drives procapsid formation. The N-terminal domain represents a more flexible, perhaps partially unfolded, domain that is remodeled upon assembly to form new, strong intra- and intermolecular interactions, providing the capsid with the stability necessary to survive the harsh enteric environment in which bacteriophage P22 replicates.

6.4.2 HIV

Similar to the case with bacteriophage P22, the immediate product of HIV assembly is an immature viral capsid. The immature capsid is composed of approximately 3000–5000 molecules of the 55-kDa Gag polyprotein. In the immature virus, the polyprotein is radially arranged with the myristoylated N-terminus embedded in the viral membrane. Budding from the infected cell activates a viral-encoded protease packaged within the capsid. The protease cleaves the Gag polyprotein into three structural domains: a 17-kDa matrix (MA) protein, a 24-kDa capsid (CA) protein, and a 7-kDa nucleocapsid (NC) protein. The matrix protein, which is N-terminally myristoylated, remains associated with the viral membrane, and the CA protein collapses to form a conical core that encompasses a complex of NC and two molecules of the viral RNA (Freed, 1998). The formation of the conical core is crucial for the virus to become infectious. Drugs (the protease inhibitors) that inhibit the protease and thereby prevent cleavage-induced maturation have proven to be very effective clinically. Furthermore, mutations within the CA protein itself, which allow cleavage but prevent proper core formation, invariably show compromised infectivity (von Schwedler et al., 2003). The block in replication in these mutants appears to be at the step of reverse transcription (Tang et al., 2001).

Because the structural rearrangement accompanying proteolytic cleavage is central to infectivity, and therefore represents a therapeutic target, significant effort has been put into characterizing the rearrangement at the molecular level. This effort has been complicated by the fact that both immature and mature virions are membrane enveloped and pleomorphic, factors which make them largely refractory to the traditional approaches of X-ray crystallography and cryo-electron microscopy. High-resolution X-ray crystallographic and NMR studies suggest that the CA subunit is composed of distinct N- and C-terminal domains connected by a four-residue flexible linker (Gitti et al., 1996; Momany et al., 1996; Gamble et al., 1997; Berthet-Colominas et al., 1999; Worthylake et al., 1999). The challenge in understanding maturation therefore lies in building a detailed model of how these CA molecules pack together to form both the immature and mature capsid lattices.

Recombinant CA can be assembled *in vitro* and forms both cones and a family of long, regular helical tubes (Li et al., 2000). These tubes are thought to be good mimics of the mature lattice at the molecular level. Cryo-EM analysis of these helical tubes indicated that they were built on a hexameric lattice (Fig. 6.2a) (Li et al., 2000). The N-terminal domain of CA formed the hexameric florets and the florets were tied together by dimerization of the C-terminal domain. While the C-terminal domain had been demonstrated to form dimers in solution, no N-terminal domain self-association has been observed. This raised the question of how the NTD:NTD interactions were stabilized in the hexamers that make up the cones and helical tubes.

The cryo-electron micrographic reconstructions of the helical tubes displayed limited resolution with regard to the C-terminal domain and it was possible that crucial interactions therefore escaped detection. To detect the changes in intersubunit interaction upon assembly and maturation, comparative hydrogen/deuterium exchange studies were performed on unassembled and assembled CA subunits in the helical form (Lanman et al., 2003). Three major regions of increased protection were identified: helices I and II in the N-terminal domain, helix IX in the C-terminal domain, and a region composed of the base of helix IV and the "flexible" linker region (Fig. 6.2b). Based on fitting the crystallographic structure of the N-terminal domain into the reconstructed helical tubes, it was proposed that helices I and II would mediate NTD:NTD homotypic interactions. The H/D exchange protection observed in these helices was consistent with this proposal. Similarly, based on the high-resolution crystal structure of the CTD dimer, it was evident that helix IX mediates dimerization, and the protection observed in this helix upon assembly is consistent with that model. In contrast, protection of the base of the fourth helix and the linker region was unexpected. To identify the basis for this protection, the tubes were chemically cross-linked, dissociated, digested with trypsin, and the cross-linked peptides identified. A cross-link between lysine 70 in the N-terminal domain of one subunit and lysine 182 in the C-terminal domain of another subunit was identified. Knowledge of the protected region of the NTD and its proximity to the CTD of an adjacent subunit suggested a model of how the hexamer might be stabilized in the mature form (Fig. 6.2c).

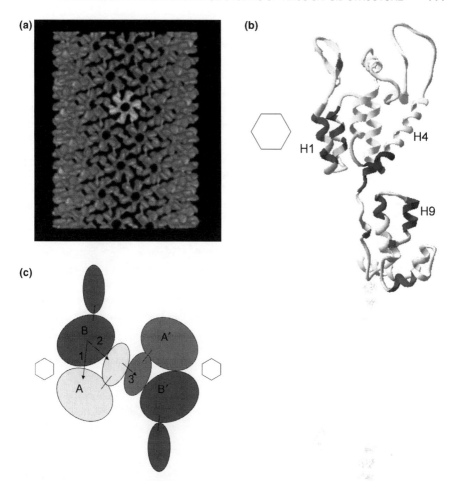

FIGURE 6.2   Hydrogen/deuterium exchange of HIV CA. (a) Three-dimensional reconstruction of CA helical tube from cryo-electron micrographs. The central hexameric florets are composed of the N-terminal domains of six subunits. Each floret is tied to its six nearest neighbors via dimerization of the C-terminal domain. (b) The difference in exchange protection upon CA polymerization into tubes mapped onto the three-dimensional structure of CA protein. The three shades of increasingly dark blue represent 10×, 100×, and 1000× increased protection upon assembly, respectively. Note that the relative disposition of the N- and C-terminal domain is arbitrary. (c) Schematic representation illustrating how an interaction between the N-terminal domain of subunit B and the C-terminal domain of subunit A can promote formation of the hexamer. Note that the N-terminal domain:C-terminal domain interaction (interaction 2) can occur only if the C-terminal domain has formed a dimer (interaction 3). This figure reproduced, with permission from Nature Publishing Group (part a) and Elsevier Ltd. (parts b and c), from Li et al., 2000 and Lanman et al., 2003, respectively. (See the color version of this figure in the Color Plates section.)

In this model, dimerization of the C-terminal domain induces a conformational change that creates a binding site for the N-terminal domain. By harvesting the bonding energy of both an NTD:NTD interaction and an NTD:CTD interaction, a subunit can be stably incorporated into the hexamer.

To investigate the changes in intersubunit interactions in actual virus particles, clones were constructed that when transfected into 293T cells produced enveloped virus-like particles lacking only the envelope glycoprotein (Env) and the reverse transcriptase (RT) enzyme (Lanman et al., 2004). One construct also carried a mutation in the active site of the protease, resulting in the production of immature particles blocked in cleavage and maturation. Budded viral particles were harvested from the media and purified by differential centrifugation. To initiate exchange, the purified particles were diluted into deuterated PBS buffer and then sampled and acid quenched at appropriate time points.

When the global exchange into intact CA was examined in the mature particles, it was apparent that they displayed a bimodal profile (Fig. 6.3a). The area under the two peaks was roughly evenly distributed, and both peaks exchanged more slowly than did free CA in solution. This result indicated that the CA protein within viral particles existed in two distinct exchange-protected states, one more protected than the other by approximately 35 residues. Both peaks eventually ended up fully exchanged but

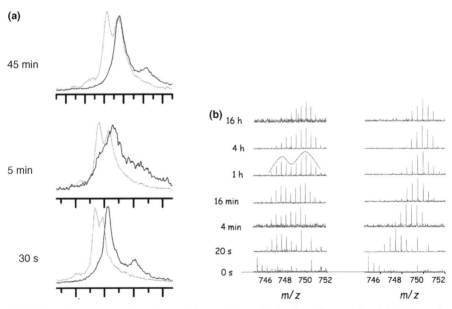

FIGURE 6.3 Hydrogen/deuterium exchange of HIV CA in virus-like particles. (a) Hydrogen/deuterium exchange of intact CA protein in mature virus-like particles and free in solution. Note that CA in particles is protected relative to free CA but that the profile is bimodal. (b) Hydrogen/deuterium exchange profile of peptide spanning N-terminal domain residues 55–68 that are part of the N-terminal domain:C-terminal domain interface in the mature (left) and immature (right) virus-like particles. Note that the peptide displays bimodal exchange in the mature form but not in the immature form. This figure reproduced from Lanman et al., 2004.

did so without altering their relative areas, indicating that the two forms of CA were not interconverting. Approximately 17 residues were protected in the faster exchanging peak relative to CA free in solution. This number is consistent with the number of residues buried in the CTD dimer interface and suggests that all the CA present in the virion is in the form of CTD dimers. An additional 35 residues were protected in the slower exchanging peak. This suggests that both NTD:NTD and NTD:CTD interactions, as well as the CTD:CTD dimer interaction, have been formed in this subset of the CA molecules. To test this hypothesis the viral particles were pepsin digested and the peptide corresponding to the region involved in the NTD:CTD interaction analyzed. This peptide proved to be bimodal, indicating that only 50% of the subunits in the mature virion had been incorporated into the conical core (Fig. 6.3b). In contrast, in the immature virus this peptide was unimodal and less protected. This result suggested that the NTD:CTD interaction is formed only upon maturation and, furthermore, that only 50% of the molecules are incorporated into the core. Interestingly, a small molecule inhibitor of maturation has been identified. NMR footprinting experiments suggest that, this molecule binds to the NTD in the region of the protein that forms the new NTD:CTD interface (Tang et al., 2003). Binding of this small molecule apparently blocks the formation of this interaction resulting in aberrant maturation and a consequent reduction in infectivity. Thus, hydrogen/deuterium exchange experiments have provided key information about the maturation of HIV and a rationale for understanding the mechanism of a potential new class of antiviral compounds.

6.4.3 Brome Mosaic Virus

The RNA-containing brome mosaic virus (BMV) and cowpea chlorotic mottle virus (CCMV) are simple 180-subunit icosahedra whose subunits are structurally similar. Both virus capsids undergo pH-induced swelling over the pH 5–7. In the case of CCMV, the structural basis of the swelling transformation has been defined using a combination of X-ray crystallography and cryo-electron microscopy (Speir et al., 1995). In the capsids of both CCMV and BMV, chemically identical subunits form both hexameric and pentameric clusters called capsomeres. The pentameric capsomeres are positioned at the 12 vertices of the icosahedron, while the hexamers are threefold coordinated. The swelling induced at high pH results in little change internal to the capsomers, rather the pentameric and hexameric capsomeres move outward along the fivefold and quasi-sixfold axes, respectively, with the result that gaps are opened up at the quasi-threefold axes. This process is presumably driven by charge–charge repulsion between acidic residues that are protonated at pH 5 but deprotonated at pH 7.

In the case of BMV, no similar high-resolution data are available, but hydrogen/ deuterium exchange studies have indicated that the pH-induced swelling of BMV is substantially similar to that of CCMV (Wang et al., 2001). Comparison of the global exchange into undigested capsid protein at pH 5 and 7 (after adjusting the timescale by a factor of approximately 75 to correct the increase in chemical exchange rate) indicated much more rapid exchange at pH 7, suggesting a more open and flexible structure.

Dissociation and proteolytic digestion of the BMV capsid gave a family of peptides spanning the protein sequence. As would be expected from the global exchange data, many of the peptides showed an increase in deuterium incorporation at pH 7 relative to pH 5. The extent of these changes suggested that the structure of most of the protein subunits was opening up. The alterations were grouped into three classes. When the peptides showing the most change were mapped back onto the presumed structure of BMV, they tended to localize at the trimer interface, suggesting that pH-induced expansion in BMV was similar to that in CCMV.

6.5 HYDROGEN/DEUTERIUM EXCHANGE STUDIES OF VIRAL PROTEIN DYNAMICS

6.5.1 Bacteriophage Phi-29 Scaffolding Protein

Assembly of dsDNA bacteriophages is generally a two-step process in which the immediate product of assembly is a procapsid composed of an outer shell of coat protein and an inner core of scaffolding protein. In the second step, the DNA is packaged into the procapsid resulting in the formation of an infectious phage. To package the entire genome within the capsid, the scaffolding protein must be removed. While in many cases this is done by proteolysis in both bacteriophage P22 and Phi-29, the scaffolding protein is released intact and participates in multiple rounds of assembly and release (King and Casjens, 1974; Nelson et al., 1976). As a class, scaffolding proteins are highly flexible and helical, and this has complicated crystallographic efforts (Fane and Prevelige, 2003). Nevertheless, the crystal structure of the Phi-29 scaffolding protein has been solved (Morais et al., 2003) (Fig. 6.4a). In solution and crystal structure, the protein is a parallel coiled-coil dimer with a helix–loop–helix motif at the N-terminus. The helix–loop–helix motif is similar in structure to the helix–loop–helix motif at the C-terminus of the P22 scaffolding protein as solved by NMR (Sun et al., 2000). The P22 helix–loop–helix has been shown to be involved in binding to the capsid shell (Tuma et al., 1998b). The structural similarity between the two motifs suggested that the N-terminal

FIGURE 6.4 Hydrogen/deuterium exchange of bacteriophage Phi-29 scaffolding protein. (a) The three-dimensional structure of the Phi-29 scaffolding protein determined crystallographically. The protein consists of an N-terminal helix–loop–helix motif followed by two long helices in a parallel coiled-coil arrangement. The C-terminal residues are unstructured. (b) The hydrogen/deuterium exchange profile of a peptide spanning residues 20–31 (left) and the intensity histogram fit as the sum of Gaussian curves (right). (c) (Left) The fraction of completely exchanged species as a function of time for the three peptides showing bimodal exchange as well as the intact protein. Note the congruence of the curves. (Right) The bimodal regions forming a cooperative structural unit mapped on the structure of the monomer. Part (a) of this figure reproduced, with permission from Nature Publishing Group, from Morais et al., 2003. Parts (b) and (c) reproduced, with permission from Cold Spring Harbor Laboratory Press, from Fu and Prevelige, 2006. (See the color version of this figure in the Color Plates section.)

(a)

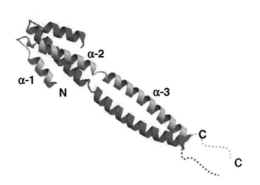

(b)   R 20–31

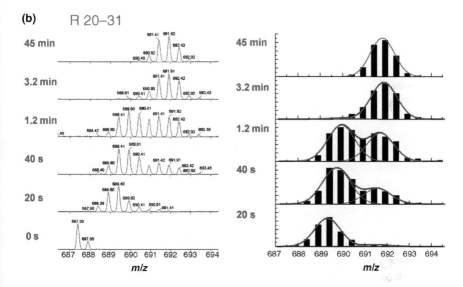

(c)

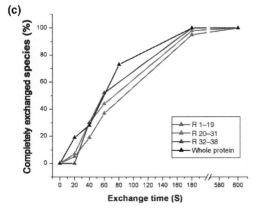

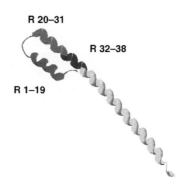

helix–loop–helix motif in Phi-29 scaffolding might likewise fulfill the role of capsid binding. Because it has been difficult to crystallize and solve the structures of procapsids containing scaffolding protein and because cyro-electron microscopic reconstructions lack sufficient resolution to identify the regions of the scaffolding protein interacting with the capsid, hydrogen/deuterium exchange experiments were performed to measure and localize the changes in dynamics in the Phi-29 scaffolding protein upon capsid binding (Fu and Prevelige, 2006).

As expected, owing to its flexible nature, the Phi-29 scaffolding protein exchanged relatively quickly in solution, reaching full exchange within 10 min. Surprisingly, the exchange pattern of the scaffolding protein in both free and bound forms was bimodal. Bimodal exchange represents EX1-type exchange in which the closing rate for an opening motion is slow relative to the chemical exchange rate. In the case of the HIV CA described above, CA existed in two forms, which essentially did not interconvert, always open (no NTD:CTD interaction), and always closed (the mature core). The EX1-type exchange seen for the scaffolding protein in solution could arise from dissociation of the dimer or from an opening motion within a single scaffolding molecule. In the former case, the expectation was that the long helix-mediating dimerization might show bimodal behavior, while in the latter case a more localized bimodal region might be expected.

Proteolytic digestion of the exchanged protein quenched at various time points yielded three peptides that displayed a bimodal exchange pattern (Fig. 6.4b). These peptides were all derived from the helix–loop–helix region. The exchange patterns of each peptide were analyzed by measuring the area under each isotopic peak and fitting the pattern as the sum of two Gaussian distributions. Under EX1 conditions, the exchange rate is proportional to the opening rate of the peptide and therefore the opening rate can be monitored by following the evolution of the more exchanged species over time. The rate of evolution of this species for all three peptides, as well as the undigested protein, were indentical suggesting that the helix–loop–helix opened as a single cooperative unit (Fig. 6.4c). This hypothesis was tested by lowering the temperature at which the exchange experiment was performed in an effort to "freeze" out the cooperative opening motions, which were presumed to be high energy. Because observing EX1 exchange requires that the chemical exchange rate be fast relative to the closing rate, and lowering the temperature will lower the chemical exchange rate, the pH of the buffer was adjusted to maintain a constant chemical exchange rate. At 10°C, the exchange was no longer bimodal, and the protein exchanged by an alternative local EX2 mechanism. This result suggested that the helix–loop–helix motif opened as a single cooperative unit via global opening at room temperature but became stable at 10°C.

While both free and bound scaffolding proteins displayed EX1 exchange when the opening rates of the two forms were compared, it was found that the bound protein exchanged more rapidly than the free protein. The observation that the dynamics of opening was altered in the capsid was consistent with the hypothesis that this region is involved in capsid binding and furthermore led to the hypothesis that the opening of the N-terminal helix–loop–helix motif might be involved in the switch that promotes exit of the scaffolding. Mutants in which the first eight residues of the scaffolding protein were deleted were prepared. These mutants were incapable of promoting assembly (Fu et al., 2007). When analyzed by circular dichroism and analytical ultracentrifugation, these mutants were found to be unfolded monomers. Thus, it is possible that opening of this helix–loop–helix motif results in dimer dissociation and denaturation and thereby provides a switch for scaffolding exit.

6.5.2   Packaging Motor P4 from dsRNA Bacteriophages Phi-8 and Phi-12

Bacteriophages Phi-8 and Phi-12 are dsRNA-containing viruses of the family Cystoviridae. Their life cycle includes a step during which single-stranded viral RNA precursors are packaged into a preformed capsid. Hexameric complexes of the protein P4 located at each of the icosahedral vertices act as a motor protein that uses the energy of nucleotide triphosphate hydrolysis to package the RNA. The crystal structure of the motor protein apoenzyme as well as substrate- and product-bound forms has been solved (Mancini et al., 2004). The hexamer is a dome-shaped structure with a funnel-shaped central channel with a diameter of approximately 25 Å. Flexible loops protrude from each subunit into the central channel. These loops are locked in an either up or down position depending upon the liganded state of the subunit. The loops are thought to make contact with the RNA, and a coordinated wavelike motion of the loops is thought to act to translocate the RNA into the capsid.

While the crystallographic analysis provided detailed structures of the protein complex in different liganded states, it did not include the RNA. Translocation intermediates that include RNA have proven refractory to crystallography. However, it has proven possible to obtain hydrogen/deuterium exchange data for the P4 protein hexamer from a related phage in complex with poly(rC), the nonhydrolyzable ATP analogue AMPPNP, AMPPNP plus poly(rC), and ADP as well as during the actual process of translocation in the presence of ATP (Lisal et al., 2005). Approximately, 85% coverage of the primary sequence was obtained. Peptides that were derived from the nucleotide binding cavity displayed increased protection upon nucleotide binding but no change upon RNA binding, whereas peptides corresponding the protruding loops showed only slight protection upon RNA binding. To detect conformational changes upon ATP binding and unravel the energy transduction path from the ATP-binding site to the RNA translocation site, Lisal et al. searched for peptides that displayed altered exchange kinetics upon both RNA and ATP binding. The region that connects the ATP-binding site to the translocating loop in the central channel displayed faster exchange in the presence of both ATP and RNA, RNA binding altered the exchange kinetics of the arginine fingers, thought to coordinate sequential ATP hydrolysis between the subunits.

Exchange experiments also provided insight into one possible mechanism of RNA loading. The RNA translocates through the central channel of a topologically closed ring of P4 subunits. To detect ring opening, Lisal et al. looked for exchange in the protected interface between the subunits. In the absence of RNA, the interface remained protected. Upon addition of RNA, a bimodal exchange pattern occurred immediately and did not increase over time, suggesting that RNA loading was responsible for the exchange and that once loaded the complexes were stable.

One hexameric P4 complex resides at each of the 12 pentameric icosahedral vertices of the capsid. This local symmetry mismatch has made it difficult to achieve a sufficiently high-resolution EM structure of the complex to resolve the question of which face of the P4 complex is in contact with the capsid. By comparing the exchange profiles of the free P4 complex and the P4 complex bound to the capsid, it was possible to determine that the C-terminal dome of the complex is in contact with the capsid (Lisal et al., 2006). Peptides not in direct contact with the surface of the capsid underwent minimal alterations in exchange rate. Interestingly, peptides at the intersubunit interface underwent exchange in solution suggesting frequent ring opening, but these motions were damped upon docking of the ring to the capsid surface, suggesting that the capsid controls the ring opening rate. During RNA packaging, the RNA first docks onto the capsid, which recognizes a specific RNA sequence. The RNA is then fed into the central channel of the P4 ring. It is tempting to speculate that docking of the RNA on the surface of the capsid promotes capsid-controlled ring opening and that this control mechanism provides a means for ensuring that only viral-specific RNA is packaged.

6.6   TECHNICAL ASPECTS OF PERFORMING HYDROGEN/DEUTERIUM EXCHANGE EXPERIMENTS ON VIRUSES

6.6.1   Dissociation of Structures

Viral capsids function as protective coats for the viral genome and have evolved to provide protection as the virus transits through space and time. As a result, they can be surprisingly stable. For example, bacteriophage P22 is stable under acidic quench conditions. Therefore, it is frequently necessary to include a chaotrope such as 8 M urea in the quench buffer to ensure complete dissociation of the particles. The presence of this concentration of denaturant reduces but does not fully inhibit the activity of pepsin. The urea can readily be removed from the digest during the desalting step. (Guanidine HCl is difficult to remove.) It is advisable to configure the LC system with a shunt to avoid the introduction of 8 M urea into the source.

6.6.2   Presence of Nucleic Acid

Contained within the viral capsid is the genomic nucleic acids, either RNA or DNA. We have observed suppression of the signal due to the presence of these polyelectrolytes. A strategy that we have found to be effective is to rapidly precipitate the nucleic acid by including protamine sulfate to a final concentration of $1 \, mg \, ml^{-1}$ and fractionate the mixture using a 1-min centrifugation in a microcentrifuge.

6.6.3 Potential for Strain Variation

It should be noted that viruses undergo mutation and that there may be many published sequences for any given virus. It should also be pointed out that given the mutation rates, particularly of RNA viruses, the sequence may not actually correspond to the one the investigator believes it to be. For this reason, we find it useful to verify that the mass of the intact protein corresponds to that expected from the sequence. Knowing the protein sequence accurately greatly simplifies the assignment of the peptides following digestion.

6.6.4 Presence of BSA

For virus derived from eukaryotic hosts grown in tissue culture, the presence of serum in the media can pose a significant obstacle. Given the high concentration of serum in typical growth media, even stringent purification protocols often fail to completely remove the contaminating serum albumin. In our experience, the best policy is to produce the virus in cells grown in serum-free media. It is often possible to wash the cells and resuspend them in serum-free media approximately one generation time in advance of harvesting the virus.

6.6.5 Complexity and Size

While viral capsids have high molecular weight, the fact that they are composed of hundreds of copies of chemically identical protein subunits greatly reduces spectral complexity. However, even a typically sized protein (80 kDa) can, in principle, produce a large number of isobaric peptides. For example, digestion of the 80 kDa bacteriophage P22 portal motor protein resulted in the production of 43 analyzable peptides. Thirteen of these peptides had two theoretical candidates within 2 ppm of each other, and seven of the peptides had identical elemental compositions (though not identical protein composition). For this reason, it is highly advisable to identify the peptides by sequencing rather than by mass matching, even when using high-resolution instrumentation.

REFERENCES

Berthet-Colominas, C., Monaco, S., Novelli, A., Sibai, G., Mallet, F., and Cusack, S., 1999. Head-to-tail dimers and interdomain flexibility revealed by the crystal structure of HIV-1 capsid protein (p24) complexed with a monoclonal antibody Fab. *Embo J* 18, 1124–1136.

Bothner, B., Dong, X. F., Bibbs, L., Johnson, J. E., and Siuzdak, G., 1998. Evidence of viral capsid dynamics using limited proteolysis and mass spectrometry. *J Biol Chem* 273, 673–676.

Bothner, B., Schneemann, A., Marshall, D., Reddy, V., Johnson, J. E., and Siuzdak, G., 1999. Crystallographically identical virus capsids display different properties in solution. *Nat Struct Biol* 6, 114–116.

Caspar, D. L. D. and Klug, A., 1962. Physical principles in the construction of regular viruses. *Cold Spring Harb Symp Quant Biol* 27, 1–24.

Fane, B. A. and Prevelige, P. E., Jr., 2003. Mechanism of scaffolding-assisted viral assembly. *Adv Protein Chem* 64, 259–299.

Freed, E. O., 1998. HIV-1 gag proteins: diverse functions in the virus life cycle. *Virology* 251, 1–15.

Fu, C. Y. and Prevelige, P. E., Jr., 2006. Dynamic motions of free and bound O29 scaffolding protein identified by hydrogen deuterium exchange mass spectrometry. *Protein Sci* 15, 731–743.

Fu, C. Y., Morais, M. C., Battisti, A. J., Rossmann, M. G., and Prevelige, P. E., Jr., 2007. Molecular dissection of o29 scaffolding protein function in an in vitro assembly system. *J Mol Biol* 366, 1161–1173.

Gamble, T. R., Yoo, S., Vajdos, F. F., von Schwedler, U. K., Worthylake, D. K., Wang, H., McCutcheon, J. P., Sundquist, W. I., and Hill, C. P., 1997. Structure of the carboxyl-terminal dimerization domain of the HIV-1 capsid protein. *Science* 278, 849–853.

Gitti, R. K., Lee, B. M., Walker, J., Summers, M. F., Yoo, S., and Sundquist, W. I., 1996. Structure of the amino-terminal core domain of the HIV-1 capsid protein. *Science* 273, 231–235.

Kang, S. and Prevelige, P. E., Jr., 2005. Domain study of bacteriophage p22 coat protein and characterization of the capsid lattice transformation by hydrogen/deuterium exchange. *J Mol Biol* 347, 935–948.

Kang, S., Hawkridge, A. M., Johnson, K. L., Muddiman, D. C., and Prevelige, P. E., Jr., 2006. Identification of subunit–subunit interactions in bacteriophage P22 procapsids by chemical cross-linking and mass spectrometry. *J Proteome Res* 5, 370–377.

King, J. and Casjens, S., 1974. Catalytic head assembling protein in virus morphogenesis. *Nature* 251, 112–119.

Lanman, J., Tuma, R., and Prevelige, P. E., Jr., 1999. Identification and characterization of the domain structure of bacteriophage P22 coat protein. *Biochemistry* 38, 14614–14623.

Lanman, J., Lam, T. T., Barnes, S., Sakalian, M., Emmett, M. R., Marshall, A. G., and Prevelige, P. E., Jr., 2003. Identification of novel interactions in HIV-1 capsid protein assembly by high-resolution mass spectrometry. *J Mol Biol* 325, 759–772.

Lanman, J., Lam, T. T., Emmett, M. R., Marshall, A. G., Sakalian, M., and Prevelige, P. E., Jr., 2004. Key interactions in HIV-1 maturation identified by hydrogen–deuterium exchange. *Nat Struct Mol Biol* 11, 676–677.

Li, Q., Yafal, A. G., Lee, Y. M., Hogle, J., and Chow, M., 1994. Poliovirus neutralization by antibodies to internal epitopes of VP4 and VP1 results from reversible exposure of these sequences at physiological temperature. *J Virol* 68, 3965–3970.

Li, S., Hill, C. P., Sundquist, W. I., and Finch, J. T., 2000. Image reconstructions of helical assemblies of the HIV-1 CA protein. *Nature* 407, 409–413.

Lisal, J., Lam, T. T., Kainov, D. E., Emmett, M. R., Marshall, A. G., and Tuma, R., 2005. Functional visualization of viral molecular motor by hydrogen–deuterium exchange reveals transient states. *Nat Struct Mol Biol* 12, 460–466.

Lisal, J., Kainov, D. E., Lam, T. T., Emmett, M. R., Wei, H., Gottlieb, P., Marshall, A. G., and Tuma, R., 2006. Interaction of packaging motor with the polymerase complex of dsRNA bacteriophage. *Virology* 351, 73–79.

Mancini, E. J., Kainov, D. E., Grimes, J. M., Tuma, R., Bamford, D. H., and Stuart, D. I., 2004. Atomic snapshots of an RNA packaging motor reveal conformational changes linking ATP hydrolysis to RNA translocation. *Cell* 118, 743–755.

Momany, C., Kovari, L. C., Prongay, A. J., Keller, W., Gitti, R. K., Lee, B. M., Gorbalenya, A. E., Tong, L., McClure, J., Ehrlich, L. S., Summers, M. F., Carter, C., and Rossmann, M. G., 1996. Crystal structure of dimeric HIV-1 capsid protein. *Nat Struct Biol* 3, 763–770.

Morais, M. C., Kanamaru, S., Badasso, M. O., Koti, J. S., Owen, B. A., McMurray, C. T., Anderson, D. L., and Rossmann, M. G., 2003. Bacteriophage phi29 scaffolding protein gp7 before and after prohead assembly. *Nat Struct Biol* 10, 572–576.

Nelson, R. A., Reilly, B. E., and Anderson, D. L., 1976. Morphogenesis of bacteriophage phi 29 of Bacillus subtilis: preliminary isolation and characterization of intermediate particles of the assembly pathway. *J Virol* 19, 518–532.

Prevelige, P. E., Jr. and King, J., 1993. Assembly of bacteriophage P22: a model for ds-DNA virus assembly. *Prog Med Virol* 40, 206–221.

Speir, J. A., Munshi, S., Wang, G., Baker, T. S., and Johnson, J. E., 1995. Structures of the native and swollen forms of cowpea chlorotic mottle virus determined by X-ray crystallography and cryo-electron microscopy. *Structure* 3, 63–78.

Sun, Y., Parker, M. H., Weigele, P., Casjens, S., Prevelige, P. E., Jr., and Krishna, N. R., 2000. Structure of the coat protein-binding domain of the scaffolding protein from a double-stranded DNA virus. *J Mol Biol* 297, 1195–1202.

Tang, C., Loeliger, E., Kinde, I., Kyere, S., Mayo, K., Barklis, E., Sun, Y., Huang, M., and Summers, M. F., 2003. Antiviral inhibition of the HIV-1 capsid protein. *J Mol Biol* 327, 1013–1020.

Tang, S., Murakami, T., Agresta, B. E., Campbell, S., Freed, E. O., and Levin, J. G., 2001. Human immunodeficiency virus type 1 N-terminal capsid mutants that exhibit aberrant core morphology and are blocked in initiation of reverse transcription in infected cells. *J Virol* 75, 9357–9366.

Tuma, R., Prevelige, P. E., Jr., and Thomas, G. J., Jr., 1998a. Mechanism of capsid maturation in a double-stranded DNA virus. *Proc Natl Acad Sci USA* 95, 9885–9890.

Tuma, R., Parker, M. H., Weigele, P., Sampson, L., Sun, Y., Krishna, N. R., Casjens, S., Thomas, G. J., Jr., and Prevelige, P. E., Jr., 1998b. A helical coat protein recognition domain of the bacteriophage P22 scaffolding protein. *J Mol Biol* 281, 81–94.

von Schwedler, U. K., Stray, K. M., Garrus, J. E., and Sundquist, W. I., 2003. Functional surfaces of the human immunodeficiency virus type 1 capsid protein. *J Virol* 77, 5439–5450.

Wang, L., Lane, L. C., and Smith, D. L., 2001. Detecting structural changes in viral capsids by hydrogen exchange and mass spectrometry. *Protein Sci* 10, 1234–1243.

Worthylake, D. K., Wang, H., Yoo, S., Sundquist, W. I., Hill, C. P., 1999. Structures of the HIV-1 capsid protein dimerization domain at 2.6 A resolution. *Acta Crystallogr D Biol Crystallogr* 55 (Pt 1), 85–92.

CHAPTER 7

Use of Enhanced Peptide Amide Hydrogen/Deuterium Exchange-Mass Spectrometry (DXMS) in the Examination of Protein–Protein Interactions

YOSHITOMO HAMURO and STEPHEN J. COALES

ExSAR Corporation, Monmouth Junction, NJ, USA

LORA L. HAMURO

Provid Pharmaceutical, North Brunswick, NJ, USA

VIRGIL L. WOODS JR.

Department of Medicine and Biomedical Sciences Graduate Program, University of California San Diego, La Jolla, CA, USA

7.1 INTRODUCTION

Protein–protein interactions play a critical role in many biological functions. Proteins may interact with each other for a long time to form a stable complex, such as the hemoglobin tetramer and holoenzyme complexes of kinases, or proteins can interact with each other transiently, such as in protease–substrate interactions. Protein–protein interactions can be detected by various biophysical methods, such as coimmunoprecipitation, yeast two-hybrid screen, and tandem affinity purification. Once detected, interaction can be further characterized by additional methods, such as analytical untracentrifugation, light scattering, fluorescence spectroscopy, and surface plasmon resonance.

To study protein–protein interactions at the submolecular level, X-ray crystallography and mutagenesis can be employed. X-ray crystallographic data can provide

Mass Spectrometry Analysis for Protein–Protein Interactions and Dynamics, Edited by Mark Chance
Copyright © 2008 John Wiley & Sons, Inc.

information regarding the contact residues that are involved in some protein–protein interactions (Davies et al., 1990; Janin and Chothia, 1990; Vos et al., 1992; Stanfield et al., 1993). Mutational studies can reveal the critical sequence determinants, "hot spots," of protein–protein interactions (Clackson and Wells, 1995; Arkin and Wells, 2004). Although the utilities of these approaches are well recognized, each has problems. The crystallization step is a big hurdle for X-ray crystallography, and conformational/dynamic changes induced by mutation decrease the robustness of the mutagenesis approach (Arkin et al., 2003; Whitty and Kumaravel, 2006).

In this chapter, we describe a new approach to study protein–protein interactions, enhanced peptide amide hydrogen/deuterium exchange coupled with proteolysis, liquid chromatography, and mass spectrometry (DXMS). This approach is applicable to the study of most protein–protein interactions and is free from the complications caused by mutagenesis. Below we present four examples of the use of DXMS to probe protein–protein interactions. We first describe how DXMS can be used to complement thermodynamic information of a protein–protein interaction by comparing the binding of wild-type human growth hormone (hGHwt) and that of a high-affinity variant (hGHv) to human growth hormone binding protein (hGHbp). Second, we illustrate how DXMS can be used to monitor conformational changes associated with cAMP activation of the multisubunit kinase, cAMP-dependent protein kinase A (PKA). Third, we describe how DXMS can be used to compare isoform-specific differences in binding of a common ligand by comparing the differences in binding of regulatory subunit isoforms of PKA to the binding domain of an A-kinase anchoring protein (AKAP). Finally, we show that DXMS can be used to map the epitope of the horse cytochrome *c*-reactive monoclonal antibody E8 and we compare the results with those derived from X-ray crystallography.

7.2 THEORY OF H/D EXCHANGE

7.2.1 Amide H/D Exchange

The exchange rate of a backbone amide hydrogen reflects its unique environment within the protein's three-dimensional structure. There is one such hydrogen for each amino acid in the protein, except for proline. A backbone amide hydrogen can exhibit highly variable exchange rates with solvent hydrogen, with rates ranging over eight orders of magnitude in folded proteins (Engen and Smith, 2001). Amide hydrogen exchange rates in peptides lacking secondary and tertiary structure vary only about 100-fold, depending primarily on neighboring amino acid side chains (Bai et al., 1993).

The exchange kinetics of amide hydrogens can be followed by deuterium or tritium isotope labeling with exchange times ranging from seconds to days. The exchange rates of hydrogens on $-OH$, $-SH$, $-NH_2$, $-COOH$, and $-CONH_2$ groups and the amino and carboxy termini are much faster. Carbon-centered hydrogens do not exchange under normal conditions, and undergo isotope substitution only following activation by chemical treatment, such as reaction with hydroxyl radicals (Goshe and Anderson, 1999).

Amide hydrogens can be exchanged with solvent hydrogen through acid, base, or water-catalyzed reactions (Bai et al., 1993).

$$k_{ch} = k_H[H^+] + k_{OH}[OH^-] + k_{H_2O} \qquad (7.1)$$

At low pH, the acid-catalyzed reaction dominates, while the rates of the base-catalyzed reactions increase at higher pH. The water-catalyzed reaction is independent of pH. Given the temperature dependence of exchange rates, the slowest exchange rates at room temperature are observed at about pH 2.7. Much higher exchange rates are observed near neutral pH at which the amide hydrogen exchange reaction is mostly base catalyzed.

7.2.2 Protection Factor

Several features affect the rate of amide hydrogen exchange, including an amide's participation in hydrogen bonding (Hilser and Freire, 1996), its distance from the protein surface (Resing et al., 1999), and the flexibility of the peptide chain (Zhang et al., 1996). The degree of retardation in the amide hydrogen exchange rate that results from the amide's physical environment is termed its "protection factor (pf)"

$$pf = k_{ch}/k_{ex} \qquad (7.2)$$

where k_{ex} is the observed exchange rate and k_{ch} is the "intrinsic" exchange rate calculated at a given pH and temperature in unstructured peptide chain (Bai et al., 1993).

7.2.3 Backbone Amide Hydrogens as Thermodynamic Sensors

Formalisms to relate the observed rates of amide hydrogen exchange to thermodynamic stabilization of proteins have been developed (Englander and Kallenbach, 1984). Amide hydrogens of proteins in the native, folded state are proposed to exchange according to the following equation.

$$\text{closed} \underset{k_{cl}}{\overset{k_{op}}{\rightleftarrows}} \text{open} \overset{k_{ch}}{\longrightarrow} \text{exchanged} \qquad (7.3)$$

$$k_{ex} = k_{op} \times k_{ch}/(k_{cl} + k_{ch}) \qquad (7.4)$$

where k_{op} is the rate at which amide hydrogen converts from closed state to open state and k_{cl} is the rate the amide hydrogen converts from the open state to the closed state. For most proteins at or below neutral pH, amide H/D exchange occurs by an EX2 mechanism (Sivaraman et al., 2001), where $k_{cl} \gg k_{ch}$. In Ex2 condition, Equation 7.4 can be simplified as

$$k_{ex} = k_{op} \times k_{ch}/k_{cl} = k_{ch}/K_{cl} \qquad (7.5)$$

The closing equilibrium constant at each amide ($K_{cl} = k_{cl}/k_{op}$) is equal to the protection factor (pf) and can be translated into the stabilization free energy of the closed state (ΔG_{cl}) by Equation 7.6.

$$\Delta G_{cl} = -RT \ln(K_{cl}) = -RT \ln(\text{pf}) = -RT \ln(k_{ch}/k_{ex}) \qquad (7.6)$$

Therefore, the ratio of measured H/D exchange rates in the folded protein (k_{ex}) and the calculated "intrinsic" rates (k_{ch}) can be converted into the free energy of amide hydrogen at a given condition. Thus, the measurement of exchange rates of backbone amide hydrogen serves as a precise thermodynamic sensor of the local environment.

7.2.4   H/D Exchange for Protein–Protein Interactions

Frequently, the hydrogen exchange rates of two or more physical states of a protein, such as with and without a protein binding partner (here represented by k_{ex+} and k_{ex-}), are measured to locate stabilization free energy changes upon the perturbation ($\Delta G_{- \rightarrow +}$).

$$\Delta G_{- \rightarrow +} = \Delta G_{+} - \Delta G_{-} = -RT \ln(k_{ex-}/k_{ex+}) \qquad (7.7)$$

The change in free energy upon protein binding ($\Delta G_{- \rightarrow +}$) can be monitored by H/D exchange rates.

7.3   OVERVIEW OF DXMS TECHNOLOGY FOR PROTEIN–PROTEIN INTERACTIONS

Dramatic advances in mass spectrometry and improvements in the various steps within the experimental peptide amide hydrogen exchange procedures have resulted in the development of automated systems for high-throughput, high-resolution H/D exchange analysis, collectively termed "DXMS" (Woods, 1997, 2001a, b, 2003; Woods and Hamuro, 2001; Hamuro et al., 2002a, 2002b, 2003a, 2003b, 2004; Englander et al., 2003; Zawadzki et al., 2003; Black et al., 2004, 2007; Pantazatos et al., 2004; Spraggon et al., 2004; Wong et al., 2004, 2005; Burns-Hamuro et al., 2005; Del Mar et al., 2005; Derunes et al., 2005, 2006; Garcia et al., 2005; Iyer et al., 2005; Yang et al., 2005; Begley et al., 2006; Brudler et al., 2006; Melnyk et al., 2006; Golynskiy et al., 2007). The system (Fig. 7.1), described in this section, incorporates the latest of these enhancements, including solid phase proteolysis, automated liquid handling, and streamlined data reduction software (Hamuro et al., 2003a).

7.3.1   On-Exchange Reaction

An H/D exchange reaction can be initiated by a simple mixing of a protein sample, initially in nondeuterated buffer, and a deuterated buffer. There are almost no restrictions on reaction conditions. To follow the deuterium buildup of individual amide hydrogen or sets of hydrogens, several on-exchange time points are sampled for each condition employed.

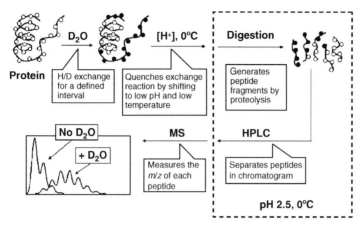

FIGURE 7.1 Overall DXMS experiment. After quenching the exchange reaction, proteolytic digestion and HPLC separation are carried out in low pH and low temperature condition to prevent back exchange.

7.3.2 Quench of Exchange Reaction

The addition of a cold, acidic solution (pH ~ 2.5 and $0°C$) to the exchange solution quenches the exchange reaction, following incubation in a deuterated environment for a defined interval. The quench conditions significantly slow down the amide exchange reaction (average half-life of amide hydrogen exchange reaction is about 1 h) and limit undesirable back exchange, as the amide hydrogen exchange reaction is primarily a base-catalyzed reaction around neutral pH. Subsequent experimental procedures are conducted near the quench conditions to minimize the loss of incorporated deuterium.

When the exchange reaction is quenched, typically high concentration of denaturant, such as 8 M urea or 4 M guanidine hydrochloride (GuHCl), is also introduced to facilitate proteolytic digestion in the following step.

7.3.3 Protein Fragmentation by Proteolysis

To localize the rate of deuterium buildup to specific amides, the analyte protein is fragmented into a collection of peptides using acid active protease(s) such as pepsin. Owing the low pH of the quench conditions in which the protein and peptide samples are maintained after deuterium labeling, standard proteases, such as trypsin, cannot be used (Woods and Hamuro, 2001; Englander et al., 2003).

To ensure highest sequence coverage possible for the analyte protein, the optimization of digestion conditions is usually performed prior to conducting the DXMS experiments. Also, in general, the more peptides, the better. Calculation of the differences in deuterium content between overlapping peptides is the preferred method to improve the resolution (Anand et al., 2002; Hamuro et al., 2002a). Variable digestion parameters are optimized, including the type and bed volume of the protease columns, the transit time of the protein over the protease columns, the type and concentration of denaturant in quenching buffer (Hamuro et al., 2002a), and inclusion of reducing reagents such as Tris(2-carboxyethyl) phosphine hydrochloride (TCEP) (Yan et al., 2002).

7.3.4 HPLC Separation

The peptides generated by proteolysis are typically separated by reverse-phase HPLC. The optimization of HPLC condition may also be performed to minimize mass overlap and ionization suppression caused by ion competition in the electrospray source (Woods, 1997). Increased sensitivity can be achieved by using capillary HPLC columns and nanoelectrospray methods (Wang and Smith, 2003).

7.3.5 Mass Analysis

The majority of H/D studies that have been reported employ quadrupole ion-trap (QIT) instruments due to their ease of use, excellent sensitivity, ability to perform MS/MS experiments, compact size, and low cost. Other reports discuss the use of instruments with higher mass resolving power such as the hybrid QqTOF instruments (Wang and Smith, 2003) and FT-ICR mass spectrometry (Akashi and Takio, 2002; Lanman et al., 2003).

7.3.6 DXMS of a Protein with or without Protein Binding Partner

When a protein–protein interaction is studied by DXMS technique, at least two sets of experiments are carried out: the DXMS of the protein alone and the DXMS of the protein in complex with its binding partner (Fig. 7.2). The DXMS behavior of the

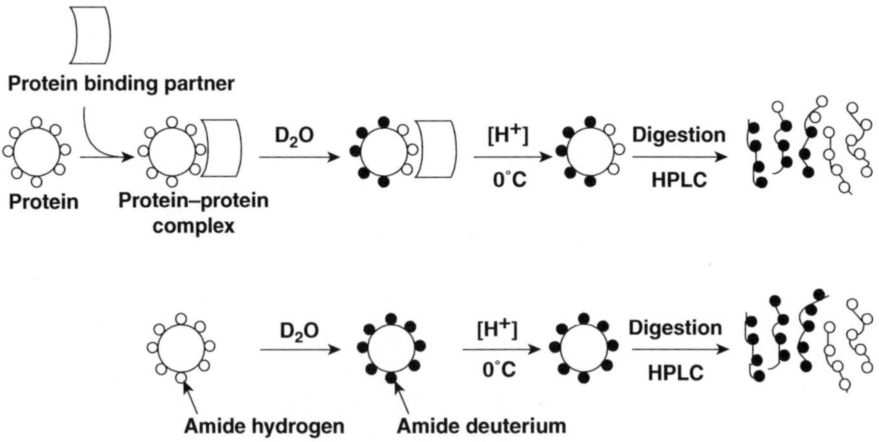

FIGURE 7.2 The concept of DXMS experiments for a protein with or without protein binding partner. An analyte protein is mixed with deuterated buffer in the presence (top) or the absence (bottom) of its binding partner. The protein–protein interface should be protected from H/D exchange in the presence of the protein binding partner (top).

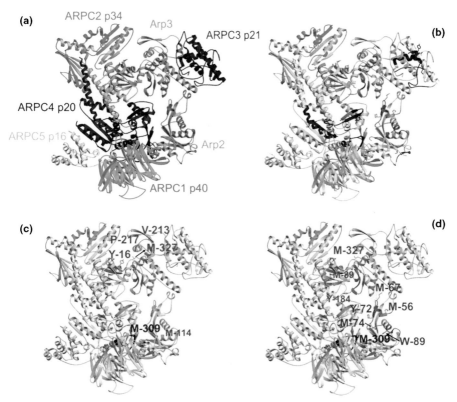

FIGURE 3.5 (a) Schematic representation of the structure of apo-form of the Arp2/3 complex with the modeled subdomains 1 and 2 of Arp2. Color codes for subunits are as follow: Arp3, orange; Arp2, red for subdomains 3 and 4, pink for the actin backbone model of subdomains 1 and 2; ARPC1 p40, green; ARPC2 p34, cyan; ARPC3 p21, magenta; ARPC4 p20, dark blue; and ARPC5 p16, yellow. (b) Peptides that were reactive in the absence of ATP are color coded as per (a); (c) Reactive peptides whose oxidation rate decreased in the presence of 1 mM ATP including 5–18, 212–225 and 318–329 (Arp3), and 300–316 (Arp2) are color coded as per (a); peptide 107–118 (Arp2) whose oxidation rate increased is also shown. (d) Peptides identified within Arp2/3 complex whose oxidation rate decreased on binding of WASp and 1 mM ATP including 80–91, 162–191 and 318–329 (Arp3), and 54–65, 66–80, 87–97, 300–316 (Arp2). Side chains of the oxidized peptide are shown. This figure was modified from Kiselar et al., (2007).

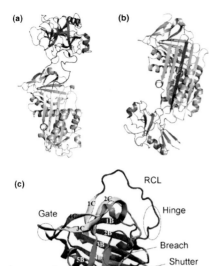

FIGURE 4.2 The structural and inhibitory properties of α_1-antitrypsin. (a) The encounter complex between a serpin and a target protease (PDBID 1K9O) (Ye et al., 2001). The protease is shown in orange. (b) The inhibitory complex of a serpin and a translocated protease (PDBID 1EZX) (Huntington et al., 2000). The inserted RCL is colored red. (c) The crystal structure of active α_1AT (PDB ID 1QLP) (Elliott et al., 2000). Functionally important regions as described by Whisstock et al. are indicated in circles. β-sheets A, B, and C are in green, purple, and yellow, respectively. All α-helices are in blue. The RCL is shown in red. Reproduced, with permission from Elsevier Ltd., from Zheng et al., 2008.

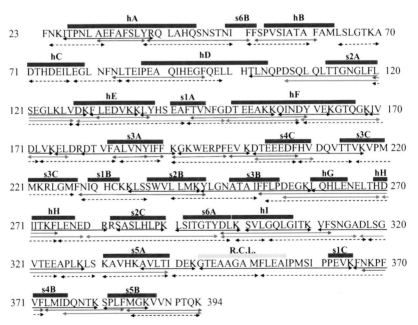

FIGURE 4.3 Peptide mapping for footprinting and H/D exchange mass analysis. (*See text for full caption.*)

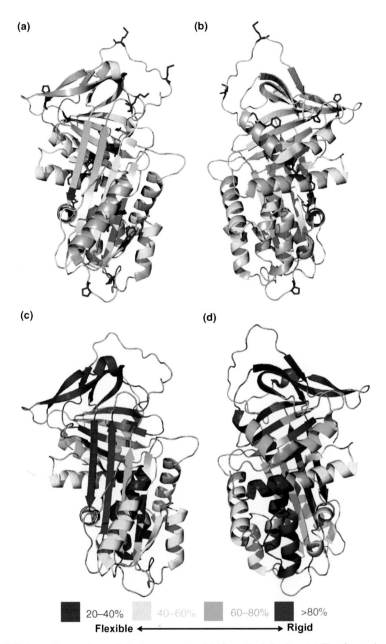

| ■ 20–40% | ▨ 40–60% | ▨ 60–80% | ■ >80% |

Flexible ◄——————————► Rigid

FIGURE 4.6 Comparison of the two structural MS methodologies. (a) The front view and (b) the back view of the three-dimensional structure of active α_1AT (PDB ID 1QLP) with α-helices and β-strands colored according to the ratio of the number of slow-exchanging hydrogens/the number of protected hydrogens as described in *Results*. Loops are colored gray in both figures. (c) The front view and (d) the back view of the structure of active α_1AT. Amino acid side chains identified to be oxidized are shown in red. Peptides identified to be unmodified are colored yellow.

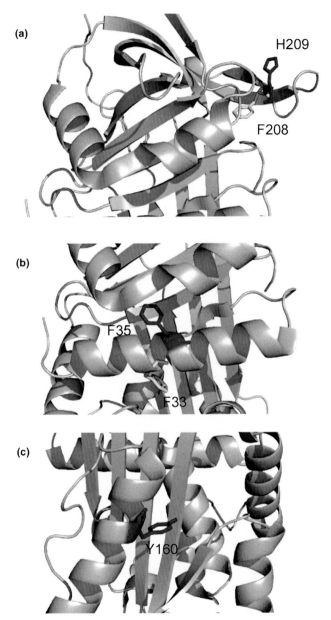

FIGURE 4.7 Solvent accessibility and dynamics information revealed by footprinting study. (a), (b) Oxidized probe residues are colored red and unmodified residues are colored yellow. (c) Oxidized probe residue Tyr 160 (red) is located in the C-terminus of helix F. Reproduced, with permission from Elsevier Ltd., from Zheng et al., 2008.

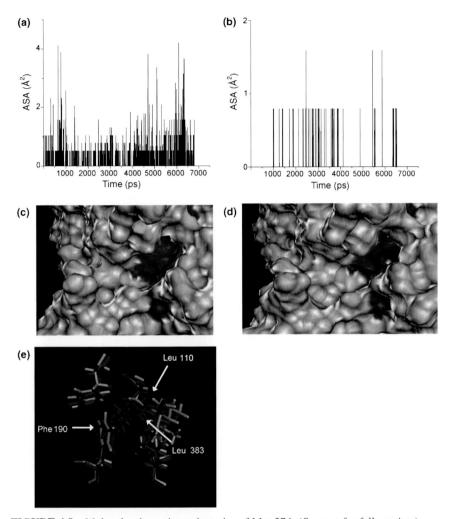

FIGURE 4.8 Molecular dynamics trajectories of Met 374. (*See text for full caption.*)

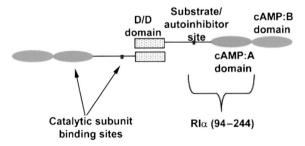

FIGURE 5.1 A schematic showing the domain organization of intact RIα. The dimerization/docking domain (10–61) is hatched, the pseudosubstrate region (94–97) is indicated in red, and the cAMP-binding domain A is in cyan and B is in blue. The region encompassing the two binding sites, the pseudosubstrate and the unknown site in the cAMP-binding A domain, is marked with arrows, and the truncated RIα(94–244) is indicated with a bracket.

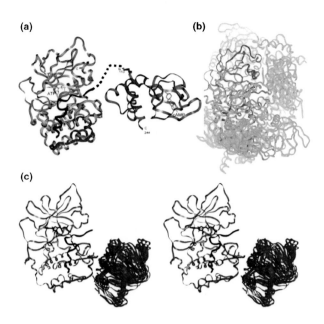

FIGURE 5.3 (a) Structure of the protein kinase A catalytic subunit (gray) showing the inhibitor/pseudosubstrate peptide (black). (*See text for full caption.*)

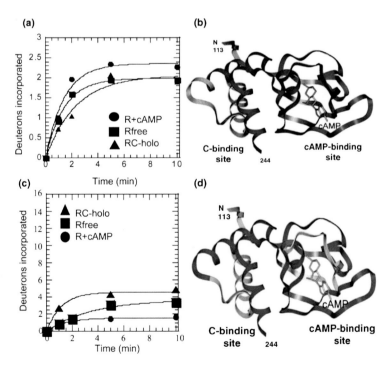

FIGURE 5.4 (a) The time course of deuterium incorporation into the peptide $m/z = 1594.73$, corresponding to residues 136–148 of the A-helix, shows that the most deuterium is incorporated in the R + cAMP state (●), while somewhat less is incorporated into the free R-subunit (■), and the least is incorporated when the C-subunit is bound (▲). (*See text for full caption.*)

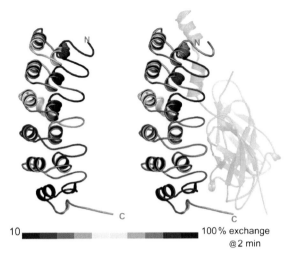

FIGURE 5.7 Summary of the H/D exchange results on free versus bound IκBα. (*See text for full caption.*)

FIGURE 6.1 Hydrogen/deuterium exchange of bacteriophage P22 capsid. (*See text for full caption.*)

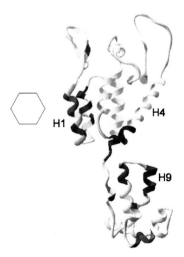

FIGURE 6.2 Hydrogen/deuterium exchange of HIV CA. (*See text for full caption.*)

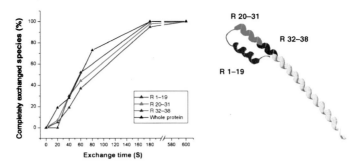

FIGURE 6.4 Hydrogen/deuterium exchange of bacteriophage Phi-29 scaffolding protein. (*See text for full caption.*)

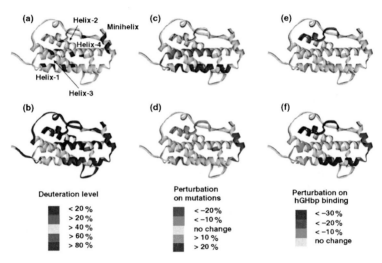

FIGURE 7.3 The DXMS results of hGHwt and hGHv in the presence or the absence of hGHbp. (*See text for full caption.*)

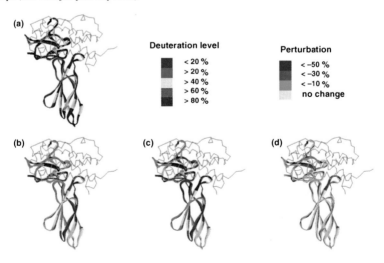

FIGURE 7.4 The DXMS results of hGHbp in the presence or absence of hGH. (*See text for full caption.*)

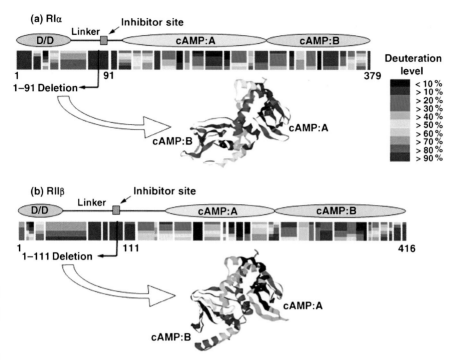

FIGURE 7.5 Domain organization and DXMS results in cAMP-bound state of RIα (a) and RIIβ (b). (*See text for full caption.*)

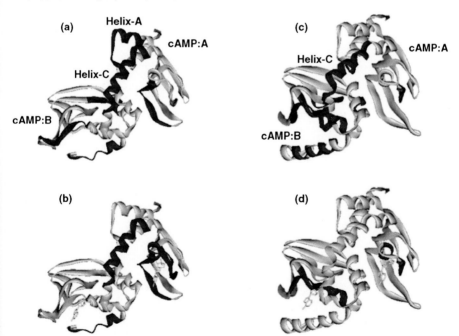

FIGURE 7.7 Average differences in deuteration of R-subunit upon C-subunit or cAMP binding overlaid on the crystal structure of an R-subunit deletion mutant (1RGS for RIα and 1CX4 for RIIβ). (*See text for full caption.*)

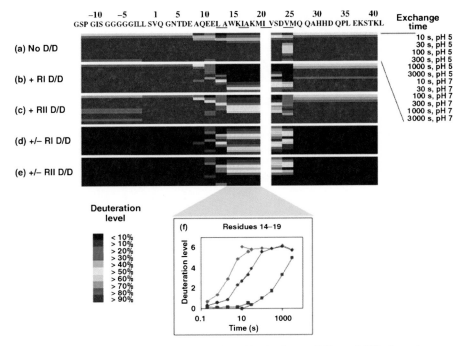

FIGURE 7.8 DXMS results of the AKB ligand in the free and RIα and RIIα-bound states. (*See text for full caption.*)

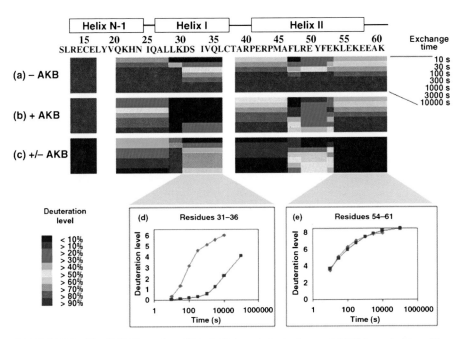

FIGURE 7.9 The DXMS results of RIα D/D domain in the free and AKB-bound states. (*See text for full caption.*)

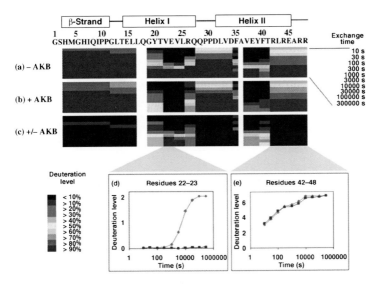

FIGURE 7.10 The DXMS results of RIIα D/D domain in the free and AKB-bound states. (*See text for full caption.*)

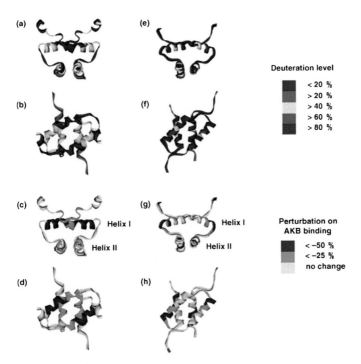

FIGURE 7.11 Deuteration levels of AKB-free D/D domains and average difference in deuteration levels of D/D domains with or without AKB ligand mapped onto the structures of RIα and RIIα D/D domain. (*See text for full caption.*)

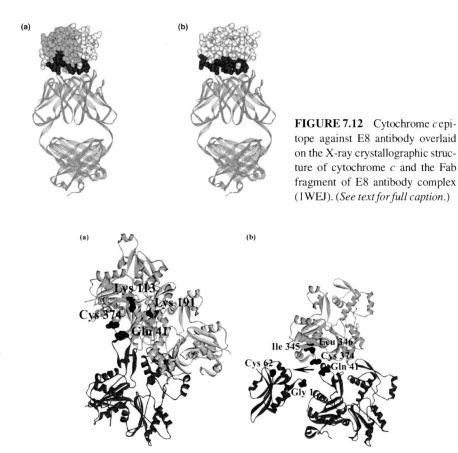

FIGURE 7.12 Cytochrome *c* epitope against E8 antibody overlaid on the X-ray crystallographic structure of cytochrome *c* and the Fab fragment of E8 antibody complex (1WEJ). (*See text for full caption.*)

(a)

(b)

Lys 113

Lys 191

Cys 374

Gln 41

Ile 345　Leu 346

Cys 374

Cys 62

Gln 41

Gly 1

FIGURE 9.2 (a) Three protomers in the molecular model of F-actin (Holmes et al., 1999). (b) Schematic representation of cofilin-F-actin binding. (*See text for full caption.*)

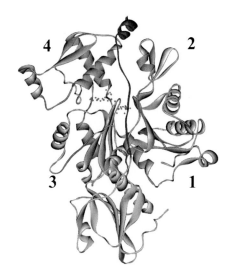

4

2

3

1

FIGURE 9.4 The structure of gelsolin segment 1–thymosin β4–actin complex. The actin monomer is depicted in grey, with the bound ATP in purple and the four subdomains labeled 1–4. The N-terminus of the chimera protein composed of the gelsolin segment 1 half (shown in cyan) caps actin at the barbed end between subdomains 1 and 3, while the thymosin β4 half (shown in pink) wraps around actin, intercalating the C-terminus between subdomains 2 and 4.

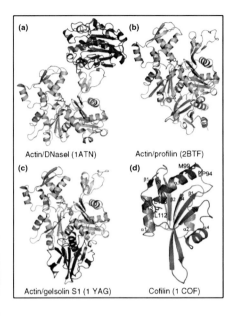

Actin/DNaseI (1ATN)   Actin/profilin (2BTF)

Actin/gelsolin S1 (1 YAG)   Cofilin (1 COF)

FIGURE 10.1 Crystal structures of G-actin/actin binding protein complexes and cofilin. (a) Structures of G-actin/DNaseI (PDB code 1ATN), (b) G-actin/GS1 (1YAG), (c) G-actin/profilin (2BTF), and (d) cofilin (1COF). The red-colored region in the cofilin structure is the G-actin binding site (G/F site) established by footprinting (Guan et al., 2002), mutagenesis (Lappalainen et al., 1997), and NMR (Pope et al., 2004). The specific footprinting probe residues that form the G-actin binding site are indicated with "stick models." Helices and β-strands are indicated with α and β symbols, including their serial numbers.

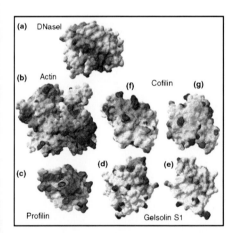

FIGURE 10.2 Electrostatic potential surfaces of actin and actin binding proteins. (a) DNaseI in the actin binding orientation, (b) actin, (c) profilin in the actin binding orientation, (d) gelsolin S1 in the actin binding orientation, (e) gelsolin S1 rotated 180° from the actin binding orientation, (f) cofilin in the actin binding orientation, (g) cofilin rotated 90° to right from the actin binding orientation. Red color is for negative electrostatics, blue color is for positive electrostatics, and white is for neutral.

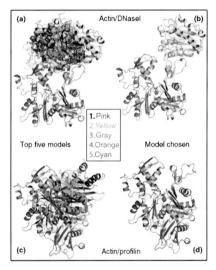

FIGURE 10.3 Modeling of G-actin/DNaseI and G-actin/profilin. (*See text for full caption.*)

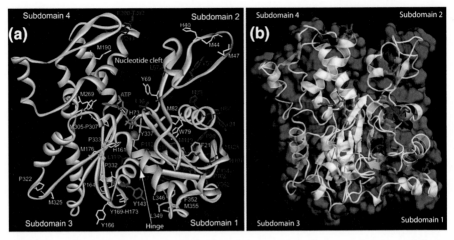

FIGURE 10.4 G-actin structure indicating the protection sites revealed by radiolytic footprinting. (*See text for full caption.*)

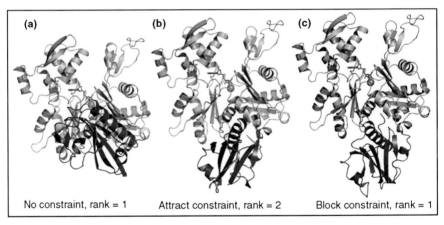

No constraint, rank = 1   Attract constraint, rank = 2   Block constraint, rank = 1

FIGURE 10.5 Modeling of G-actin/GS1. (*See text for full caption.*)

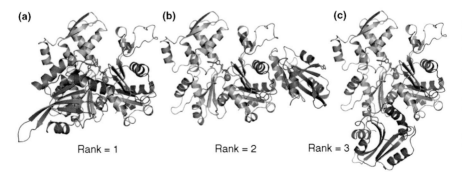

Rank = 1   Rank = 2   Rank = 3

FIGURE 10.6 Three top-ranked G-actin/cofilin models from the computational modeling strategy without incorporating experimental constraints. (*See text for full caption.*)

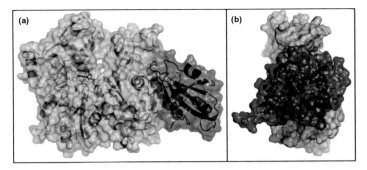

FIGURE 10.7 Three-dimensional model of G-actin/cofilin complex. (*See text for full caption.*)

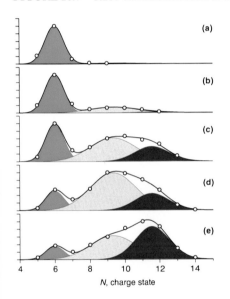

(a)

(b)

(c)

(d)

(e)

4   6   8   10   12   14

N, charge state

FIGURE 11.2 Charge state distributions of ubiquitin (Ub) in ESI mass spectra acquired in 10 mM ammonium acetate, pH 7.0 (a); 10 mM ammonium acetate, pH 7.0, and 60% methanol by volume (b); 10 mM ammonium acetate, pH adjusted to 2.0 with acetic acid (c); 10 mM ammonium acetate, pH adjusted to 2.0 with acetic acid, and 60% methanol by volume (d); H$_2$O/methanol (40:60, *v:v*), pH adjusted to 2.0 with acetic acid (e). The charge state distributions were deconvoluted by using a chemometric approach described in (Mohimen et al., 2003). The three conformers contributing to the overall ionic signal are assigned as the native state (blue), the (a) state (yellow), and the random coil (red). (*See text for full caption.*)

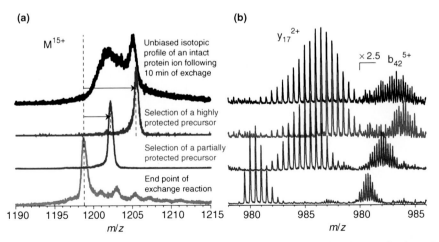

FIGURE 11.7 Application of direct HDX MS/MS measurements for characterization of local dynamics in *wt**-CRABP I in a conformer-specific fashion. (*See text for full caption.*)

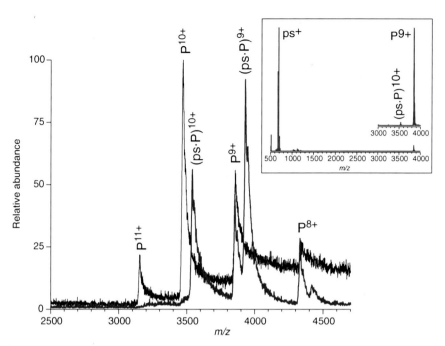

FIGURE 11.9 ESI mass spectrum of pepsin (P) acquired at pH 2.3 (black trace). The brown trace shows a mass spectrum of pepsin incubated with excess of pepstatin (ps). Inset: CAD mass spectrum of (ps·P)$^{10+}$ ion suggests that enzyme–inhibitor complex dissociation in the gas phase proceeds via charge separation.

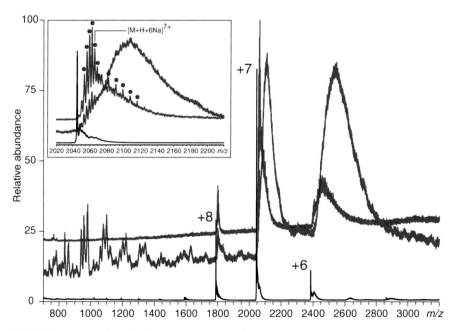

FIGURE 11.11 ESI MS of 10 μM solution of chicken egg white lysozyme acquired in 100 mM ammonium bicarbonate in the absence of nonvolatile electrolytes (black trace) and in the presence of 50 mM NaCl (blue and red traces). (*See text for full caption.*)

protein in the absence or presence of its binding partner can describe the dynamic properties of the protein in each condition at the submolecular level. The difference in DXMS patterns can shed light on how the protein dynamics changes upon the complex formation.

The DXMS experiment of a protein–protein complex can be carried out by three different methods. (i) When the protein–protein complex is stable enough to isolate, the complex may be treated as a normal protein sample. The H/D exchange reaction can be initiated by mixing the complex solution with a deuterated buffer. (ii) When the complex is not stable enough to isolate, the complex may be formed by mixing protein A with a slight excess of its binding partner protein B to ensure more than 95% of protein A is in the bound form. Then, the H/D exchange for protein A may be initiated by mixing the complex solution with a deuterated buffer. Conversely, when the information on DXMS behavior of protein B is necessary, an excess amount of protein A has to be used to ensure 95% of protein B is in the bound form. (iii) When the protein binding partner is large, such as an antibody, the peptide fragments from the binding partner may suppress the signal of the peptides from the analyte protein. In such a case, the binding partner may be immobilized by the solid phase before the initiation of the H/D exchange reaction, to facilitate the separation of the two proteins.

7.4   DXMS OF HUMAN GROWTH HORMONE AND ITS BINDING PROTEIN

7.4.1   Human Growth Hormone High Affinity Variant

The human growth hormone high affinity variant (hGHv) was discovered by phage display mutagenesis. The hGHv binds through Site-1 to the extracellular domain of the receptor (hGHbp) with a Kd < 10 pM, which is significantly tighter than the Kd (1 nM) for human growth hormone wild type (hGHwt). This variant contains 15 mutations in the Site-1 binding interface of helix-1, helix-4, and the minihelix and retains full biological activity (Fig. 7.3a) (Pearce et al., 1999).

While the mutagenesis studies showed that the functional epitope of hGHwt is characterized by a discrete, well defined hot spot, where much of the binding energy comes from seven to eight residues (Cunningham and Wells, 1993; Clackson and Wells, 1995; Wells, 1996), hGHv does not have a similar hot spot as hGHwt (Pal et al., 2003). The new set of hydrogen bonds and van der Waals interactions introduced in hGHv do not appear to add much binding energy to the interaction (Pal et al., 2003).

Another interesting characteristic of the hGHv–hGHbp interaction is the distribution of enthalpic (ΔH) and entropic ($-T\Delta S$) contributions (Kouadio et al., 2005). While the binding of hGHwt to hGHbp has a favorable ΔH ($-9\,\text{kcal}\,\text{mol}^{-1}$) and $-T\Delta S$ ($-3\,\text{kcal}\,\text{mol}^{-1}$), that of hGHv binding to hGHbp has an even more favorable ΔH ($-36\,\text{kcal}\,\text{mol}^{-1}$) and very unfavorable entropy $-T\Delta S$ ($21\,\text{kcal}\,\text{mol}^{-1}$). The X-ray crystallographic data showed that the extent of reorganization upon

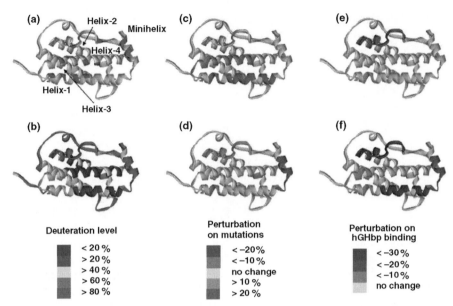

FIGURE 7.3 The DXMS results of hGHwt and hGHv in the presence or the absence of hGHbp. (a) The mutation sites (indicated by red) in hGHv overlaid onto X-ray crystal structure (1HWH). (b) The average deuteration level of hGHwt overlaid onto X-ray crystal structure. The four-helix bundle has lower deuterium content, indicating less dynamic nature compared to loop regions. (c) The average perturbations in deuteration levels upon mutations in unbound state. The helices 1 and 4 exchange faster in unbound hGHv than in unbound hGHwt. (d) The average perturbations in deuteration levels upon mutations in the presence of hGHbp. Miniloop regions exchange slower in hGHv than hGHwt in the presence of hGHbp. (e) The average perturbations in deuteration levels of hGHwt upon hGHbp binding. Very little perturbation was observed. (f) The average perturbations in deuteration levels of hGHv upon hGHbp binding. The exchange rates of helix-1 and minihelix are retarded significantly. (See the color version of this figure in the Color Plates section.)

complex formation is most likely similar for both hGHwt and hGHv (Schiffer et al., 2002). The huge discrepancy in the thermodynamic parameters upon binding and the structural similarity between the hGHwt–hGHbp and hGHv–hGHbp complexes suggest that the time-averaged structures of the unbound and bound molecules cannot explain the underlying mechanisms of the two binding interactions. We used DXMS to determine if changes in backbone dynamics of this variant versus wild type in the free and hGHbp complexed state could contribute to the enhanced binding affinity of hGHv to its receptor (Horn et al., 2006).

7.4.2 DXMS Experiments of Human Growth Hormone and Its Binding Protein

H/D exchange reactions of hGHwt or hGHv were initiated by mixing $10\,\mu l$ of hGH solution ($33\,\mu M$ hormone with or without 20% molar excess of hGHbp) and $10\,\mu l$ of deuterated buffer at 25°C. Each exchange reaction was quenched by 30 μl of 8 M urea, 1 M tris(2-carboxyethyl)phosphine hydrochloride (TCEP) at 1°C after predetermined exchange period (30, 100, 300, 1000, 3000, and 10,000 s). The quenched solution was digested by pepsin, separated by reverse-phase HPLC, and analyzed by mass spectrometer (Horn et al., 2006). This condition gave 93% sequence coverage for hGHwt and 88% sequence coverage for hGHv.

Analogous procedures were employed to monitor the deuterium buildup of each segment in hGHbp. Again, for the measurement of deuterium contents in hGHbp with hGH, 20% molar excess hGH was used to ensure more than 99% of hGHbp is in the bound form. The sequence coverage for hGHbp was 100%.

7.4.3   DXMS of hGHwt and hGHv without hGHbp

The deuterium content in each segment of hGHwt was consistent with the X-ray crystallographic structure: The four helix bundle regions of the protein exchanged slowly whereas the loop regions exchanged fast (Fig. 7.3b). The general deuterium buildup pattern of hGHv was similar to that of hGHwt, indicating the overall structure of the hGHv is similar to hGHwt as X-ray crystal structure revealed. However, the exchange rates of helix-1 of hGHv (segments 12–15, 18–25, and 28–31) were significantly faster than those of hGHwt and those of helix-4 (segments 166–176 and 179–191) were marginally so (Fig. 7.3c), implying these protein–protein interface regions are more dynamic in the mutant.

7.4.4   DXMS of hGHwt and hGHv with hGHbp

The deuterium content in each segment of hGHwt and hGHv was also determined in the presence of hGHbp. The deuterium content of hGHv is somewhat similar to hGHwt in the presence of hGHbp, unlike in the absence of hGHbp as described above. The only significant difference in perturbation between hGHwt and hGHv in the presence of hGHbp was the retardation of exchange rates for the minihelix (residues 33–44; Fig. 7.4d). Helix-1 and helix-4, which exhibited a faster exchange in uncomplexed hGHv compared to uncomplexed hGHwt, showed no significant difference when bound to hGHbp, suggesting that the bound complexes for mutant and wild type had similar backbone energetics.

The deuterium incorporation of hGHwt decreased upon binding to hGHbp at three segments, while that to hGHv decreased at seven segments (Fig. 7.3e and f). Four extra segments were perturbed in hGHv upon hGHbp binding. Among the four segments, the exchange rates of helix-1 (segments 18–25 and 28–31) and helix-4 (segment 179–191) were slowed upon hGHbp binding only in hGHv, as these segments of hGHbp-free hGHv exchanged faster than hGHbp-free hGHwt, while these segments of hGHbp-bound hGHv and hGHbp-bound hGHwt exchanged at a similar rate. On the other hand, the deuterium content of the minihelix (segment 33–44) was lower in hGHv in the presence of hGHbp than in hGHwt, while the deuterium contents of this segment in hGHbp-free hGHwt and hGHbp-free hGHv were about the same.

7.4.5 DXMS of hGHbp with or without hGH

The deuterium content in each segment of unbound hGHbp and that in the presence of hGHwt or hGHv was determined. On average, the amide hydrogens of unbound hGHbp exchanged a lot faster than those in unbound hGHwt or unbound hGHv (Fig. 7.4a), indicating hGHbp is more dynamic than hGHwt or hGHv in the respective free state.

The presence of hGHwt or hGHv retarded the amide hydrogen exchange rates of hGHbp significantly throughout the molecule (Fig. 7.4b and c). The retardation was observed not only at the protein–protein interface but also at the segments far away from the interface. This is probably caused by diminished breathing motion of the β–strand core of the hGHbp upon hormone binding. The H/D exchange rates of three segments (103–108, 123–126, and 167–172) were slowed significantly more in the presence of hGHv than in the presence of hGHwt (Fig. 7.4d).

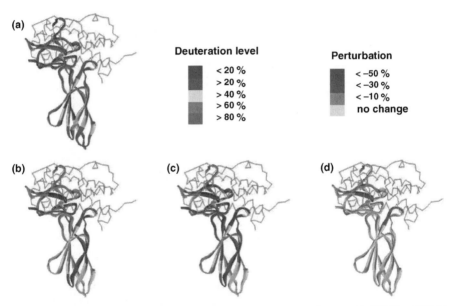

FIGURE 7.4 The DXMS results of hGHbp in the presence or absence of hGH. The DXMS results of hGHwt and hGHv in the presence or absence of hGHbp. (a) The average deuteration level of hGHbp overlaid onto X-ray crystal structure (1HWH). The hGHbp appears very red, indicating this is a very dynamic protein. (b) The average perturbations in deuteration levels of hGHbp upon hGHwt binding. Almost entire molecule looks blue, indicating most of the molecules tightened upon hGHwt binding. (c) The average perturbations in deuteration levels of hGHbp upon hGHv binding. (d) The average difference in deuteration levels between the hGHbp with hGHv and the hGHbp with hGHwt. Blue indicates the regions where hGHbp exchanges slower in the presence of hGHv. (See the color version of this figure in the Color Plates section.)

7.4.6 Enhanced Affinity by Increasing the Free Energy of the Unbound State

The H/D exchange rates of unbound hGHv was significantly faster than that of unbound hGHwt in many regions (Fig. 7.3c). Also, chemical denaturation studies using circular dichroism showed that hGHv is about $3\,kcal\,mol^{-1}$ less stable than hGHwt. These two facts showed that unbound hGHv is higher in energy than unbound hGHwt. On the other hand, the DXMS patterns of hGHbp-bound hGHwt and hGHbp-bound hGHv are similar (Fig. 7.4d), indicating these two hormones have similar backbone energetics in hGHbp-bound state. It is clear from this analysis that, hGHv has enhanced affinity to hGHbp compared to hGHwt, at least partially, by increasing the free energy of the unbound state.

The lowered free energy in the bound state also contributes to the enhanced affinity of hGHv. The hGHv-bound hGHbp exchanged slower than the hGHwt-bound hGHbp, suggesting a lower free energy. The minihelix of hGHbp-bound hGHv exchanged slower than that of hGHbp-bound hGHwt (Fig. 7.4d), also indicating the lower free energy of the hGHv–hGHbp complex.

From the DXMS studies, we have a clearer picture of the backbone dynamics of wild type and variant binding to hGHbp. The backbone of free hGHv is more dynamic than that of free hGHwt with little difference in the backbone dynamics of the complexes, implying that hGHv–hGHbp interaction gains more favorable hydrogen bonding and/or electrostatic interactions and loses more mobility upon complexation than the hGHwt–hGHbp interaction. This is consistent with the more favorable ΔH and less favorable $-T\Delta S$ for the hGHv–hGHbp interaction compared to the hGHwt–hGHbp interaction.

7.5 DXMS OF PKA REGULATORY SUBUNITS

7.5.1 Protein Kinase A (PKA) Regulatory Subunits

PKA is a key component of cAMP signaling pathways in the cell and phosphorylates diverse protein substrates in response to cAMP binding. The inactive holoenzyme contains two regulatory subunits (R) and two catalytic (C) subunits. In the absence of cAMP, the R-subunit binds to the active site cleft of the C-subunit, inhibiting activity. cAMP binding to the R-subunit releases the inhibition, leading to an active C-subunit (reviewed in Taylor et al., 2005). DXMS has proven useful to characterize the conformational changes in the R-subunit that are associated with cAMP binding and release of the C-subunit (Anand et al., 2002). We measured the DXMS patterns of three states of the R-subunit: cAMP-free state, cAMP-bound state, and C-subunit-bound state. By comparing the different states, we can localize the backbone perturbations specific to each ligand, allowing us to begin to understand how the small molecule, cAMP, can be coupled to changes in the R:C interface, leading to C-subunit activation.

There are two types of R-subunits (type I and II) with each type further classified into α and β subtypes. Although the two major isoforms of the R-subunit (RI and RII) differ in their subcellular localization and physiological function, they have similar domain organizations (Skalhegg and Tasken, 2000).

The R-subunits are highly modular with ordered and disordered regions, consisting of an N-terminal dimerization/docking (D/D) domain, which dimerizes the R-subunit and provides a docking surface for A-kinase anchoring proteins; a variable linker region, which includes an inhibitory site that binds to the active site of the C-subunit; and two tandem, C-terminal cAMP binding domains (cAMP:A and cAMP:B) (Fig. 7.5).

7.5.2   DXMS Experiments of PKA Regulatory Subunits

7.5.2.1   Expression and Purification   Proteins were expressed in *E. coli* cells and purified as described previously using cAMP-agarose resin (Diller et al., 2001). Following cell lysis and precipitation, the pellets were resuspended, incubated overnight with the cAMP resin, and eluted at room temperature. For cAMP-bound R-subunit, the protein was eluted with buffer containing 25 mM cAMP. For cAMP-free R-subunit, the protein was eluted with buffer containing 25 mM cGMP. The protein eluates were then dialyzed overnight at 4°C against 20 mM MOPS buffer.

7.5.2.2   Holoenzyme Formation   The cAMP-free R-subunit was added to wild-type C-subunit in a 1:1.2 molar ratio and dialyzed overnight. The complex was purified by elution through a S200 gel filtration column to remove excess C-subunit.

7.5.2.3   DXMS Experiments   Deuterated samples were prepared at $23 \pm 1°C$ by diluting 2 µl of protein stock solution with 18 µl of deuterated buffer, followed by "on-exchange" incubation for varying times (10, 30, 100, 300, 1000, and 3000 s) prior to quenching in 30 µl of 3.2 M guanidine hydrochloride (GuHCl), 0.8% formic acid, at 0°C. The quenched solution was immediately passed over a pepsin column with contemporaneous collection of proteolytic products by a C18 column. Subsequently, the C18 column was eluted with a linear gradient of 8%–32% acetonitrile over 10 min. This condition gave 100% sequence coverage for RIα and 99% sequence coverage for RIIβ.

7.5.3   DXMS of cAMP-Bound PKA R-Subunits

The DXMS behaviors of RIα and RIIβ are consistent with a modular domain structure containing both ordered and disordered regions (Fig. 7.5) (Hamuro et al.,2003b, 2004). The slowest exchanging regions of both RIα and RIIβ were in the cAMP binding sites (121–379 for RIα and 158–414 for RIIβ). RIIβ is somewhat more deuterated compared to RIα, especially in cAMP:B, but overall they had similar DXMS maps. The N-terminal D/D domain (1–61 for RIα and 1–45 for RIIβ) showed intermediate protection, followed by the linker region (62–120 for RIα and 46–157 for RIIβ) with a little to no protection, suggesting a very disordered region in the

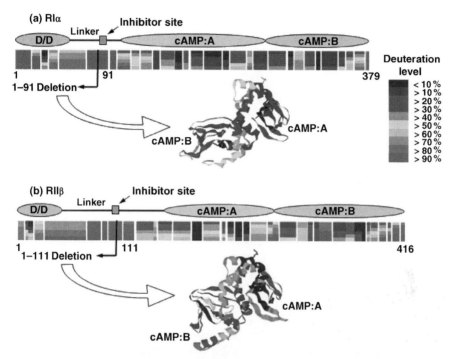

FIGURE 7.5   Domain organization and DXMS results in cAMP-bound state of RIα (a) and RIIβ (b). Each R-subunit is composed of three structured regions, an N-terminal dimerization/docking domain (D/D), and two cAMP binding domains (cAMP:A and cAMP:B) connected by an unstructured linker region. Each block in the DXMS results indicates a peptide analyzed. Each block contains six time points: (from top) 10, 30, 100, 300, 1000, and 3000 s. The deuteration levels of each segment at each time point are shown by different colors from blue (<10% deuteration) through red (>90% deuteration) as indicated on the right of the figure. A mutant of each R-subunit devoid of D/D domain and linker region produced X-ray crystallographic structure (1RGS for RIα and 1CX4 for RIIβ). The deuteration level of each segment was overlaid on to the respective crystallographic structure. (See the color version of this figure in the Color Plates section.)

absence of the C-subunit. It is noteworthy that X-ray structures are available only for the deletion mutants (Δ1–91 for RIα and Δ1–111 for RIIβ) in which most of the linker region is deleted (Su et al., 1995; Diller et al., 2001). Presumably, the very flexible (= fast exchanging) linker region was detrimental to the crystallization process. About 20 residues in the deletion mutants are fast-exchanging residues and they were not seen in the respective crystal structures (Fig. 7.5). The DXMS results and the crystallization compatible constructs imply that identification of flexible regions by DXMS can optimize the constructs for crystallization. Proof of concept was shown in previous work on crystallization optimization (Pantazatos et al., 2004; Spraggon et al., 2004).

7.5.4   Interaction between R-Subunits and C-Subunit

The comparison between the DXMS behaviors of the R-subunit in the cAMP-free state and the C-subunit-bound (holoenzyme) state can shed light on conformation flexibility of the R-subunit in the absence of ligand and map conformational changes associated with ligand binding. The deuteration levels of five segments in RIα (80–101, 138–148, 232–247, 348–353, and 377–379) (Fig. 7.6a) and four segments (102–115, 150–152, 253–268, and 271–275) in RIIβ (Fig. 7.6b) were significantly lower in the presence of C-subunit, suggesting that these regions were either directly or indirectly involved in the R:C interaction. Residues 80–101 for RIα and 102–115 for RIIβ include the inhibitor site (94–99 for RIα and 109–113 for RIIβ), which binds to the active site cleft of the C-subunit (Taylor et al., 2005). Therefore, DXMS picked up these direct interaction site for the R- and C-subunits.

In addition to the inhibitor site protection, we saw protection of the A helix (138–148) and C helix (232–247) in cAMP:A domain of RIα upon C-subunit binding. Most of the protected areas map to cAMP:A and the inhibitor segment with only minor protection in cAMP:B. Isoform RIIβ also showed protection in the C helix of cAMP:A (253–268 and 271–275) upon C-subunit binding, but there was less protection in the A helix, which predicts that the R:C interface will be somewhat different in these two isoforms (Fig. 7.7a and c).

Previous deletion mapping studies and H/D exchange studies of a deletion mutant of RIα (94–244) bound to the C-subunit indicated that the A and C helices of cAMP:A, located at the N- and C-terminal regions of the cAMP binding domain, are necessary for high affinity binding to the C-subunit (Fig. 7.7a and c) (Huang and Taylor, 1998; Anand et al., 2002, 2003). A recent crystal structure of RIα (91–244) bound to the C-subunit has confirmed these interface contacts (Kim et al., 2006). The cAMP binding domain is made up of helical segments that punctuate each side of an eight-stranded β-barrel (153–199 and 212–225) that forms the binding pocket for cAMP. There was very little structural change in the β-barrel itself upon C-subunit binding. This was consistent with the DXMS experiments described here (Figs. 7.6a and 7.7a).

Notably, three segments in RIα (190–201, 312–321, and 348–353) and eight segments (228–233, 303–312, 321–325, 341–353, 354–363, 377–379, 390–396, and 399–401) in RIIβ, all of them in cAMP binding domains, exchanged faster in the presence of C-subunit. Binding of cAMP is cooperative with cAMP first binding to cAMP:B, which increases cAMP binding to cAMP:A. The fact that there was an increase in exchange in these segments of cAMP:B in the R:C complex suggested that these regions were more disordered in the complex and perhaps primed to recognize cAMP. This region of disorder was even greater for the RIIβ isoform, suggesting a conformational difference in the cAMP:B binding pockets for the isoforms (Fig. 7.7a and c). Based on RIα studies, the cAMP:B domain is believed to act solely as a gatekeeper to control cAMP binding to the cAMP:A domain and it was presumed that there was little communication between the cAMP:B domain and the C-subunit (Herberg et al., 1996). The current studies, however, have demonstrated that C-subunit binding enhanced the dynamics within the cAMP:B domain (312–321, 348–353 for RIα; 321–325, 341–353, 354–363,

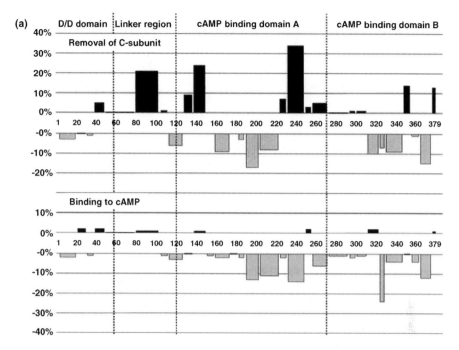

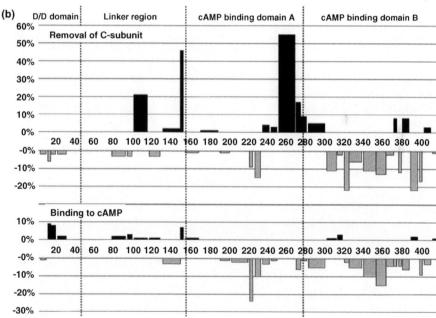

FIGURE 7.6 Average deuterium incorporation changes upon activation of RIα (a) and RIIβ (b). The deuteration level change upon the removal of C-subunit is shown on the top of the figure (comparing cAMP-free RIα and holoenzyme). A positive value represents increased deuteration and a negative value represents decreased deuteration upon C-subunit removal. The deuteration level change upon the binding to cAMP is shown at the bottom of the figure (comparing cAMP-bound R-subunit and cAMP-free R-subunit). A positive value represents increased deuteration and a negative value represents decreased deuteration upon cAMP binding.

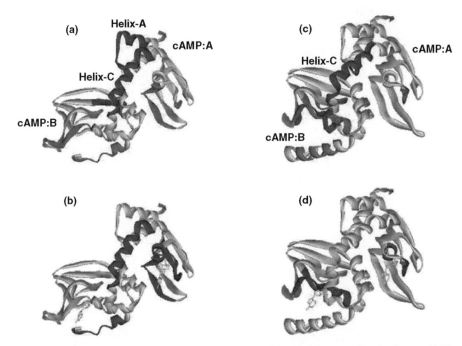

FIGURE 7.7 Average differences in deuteration of R-subunit upon C-subunit or cAMP binding overlaid on the crystal structure of an R-subunit deletion mutant (1RGS for RIα and 1CX4 for RIIβ). (a) RIα with/without C-subunit, (b) RIα with/without cAMP, (c) RIIβ with/without C-subunit, and (d) RIIβ with/without cAMP. Blue indicates protected regions in the presence of C-subunit or cAMP, and red indicates regions exhibiting increased exchange in the presence of C-subunit or cAMP. (See the color version of this figure in the Color Plates section.)

377–379, 390–396, and 399–401 for RIIβ) and suggests that the cAMP:B domain is malleable upon holoenzyme formation.

7.5.5 Interaction between R-Subunit and cAMP

As discussed in the previous section, the binding of C-subunit causes not only the retardation of H/D exchange rates at direct contact sites between the two subunits, but also global effects on cAMP:A and cAMP:B domains. In contrast, the majority of the effects upon cAMP binding to the R-subunit appeared very local. Five segments (190–201, 204–221, 232–247, 324–327, and 365–374) in RIα and four segments (222–224, 228–233, 341–355, and 354–363) in RIIβ were protected upon cAMP binding (Figs. 7.6, 7.7b and d). All of these protected segments, except the C helix of cAMP:A in RIα (232–247), were located near one of the two cAMP binding sites, indicating these protections were the result of direct contact with cAMP.

While the segment spanning the C helix of the cAMP:A domain (residues 253–268) did not show any change upon cAMP binding in the RIIβ (Hamuro et al., 2003b), the equivalent region in RIα (residues 232–247) showed a decreased deuteration level upon cAMP binding. The C helix was strongly protected when C-subunit was bound to RIα, and the fact that we see changes in the C helix upon cAMP binding suggests a communication link between cAMP binding and the R:C interface. The different effects of cAMP binding on RIα compared to RIIβ are interesting and suggest isoform-specific differences in the mechanism of cAMP-dependent activation. Indeed, earlier studies characterizing the type I and II isoforms of the PKA holoenzymes have shown that the type I holoenzyme is activated *in vitro* at lower concentrations of cAMP compared to that of the type II holoenzyme (Zawadzki and Taylor, 2004).

DXMS has provided us with a conformational model for cAMP activation of PKA. In this model, cAMP binding to RIα induces conformational changes in the C helix that propagate to the R:C interface, relieving inhibition at the active site and releasing the active C-subunit. The DXMS experiments also suggest differences in the contact surface between R:C and the mechanism of cAMP activation.

7.5.6 Lack of Significant Effects on the D/D Domain upon Binding to cAMP or C-Subunit

The D/D domain serves to dimerize the R-subunit, which once dimerized serves as a binding surface for A-kinase anchoring proteins (AKAPs), a family of proteins that bind and tether PKA to various intracellular sites. The D/D domain functions independently from the rest of the molecule as evidenced by similar binding affinities for full-length versus the isolated D/D domain, and the DXMS behaviors of the isolated D/D domain (described below) are similar to those of the intact protein (Burns et al., 2003). There was no significant change in deuteration levels (significance defined as >10% difference in deuteration levels) of the D/D domain upon binding to the C-subunit or cAMP for either R-subunit isoform. The lack of deuteration change implied the lack of communication between the D/D domain and the rest of the molecule.

7.6 DXMS OF PKA R-SUBUNIT D/D DOMAINS AND D-AKAP2 AKB DOMAIN

7.6.1 PKA R-Subunit D/D Domains and D-AKAP2 AKB Domain

As described above, the dimerization/docking (D/D) domain of the PKA R-subunit appears to function independent of the rest of the molecule. This binding module is critical for subcellular localization of the kinase via a family of scaffold proteins known as A-Kinase Anchoring Proteins (AKAPs) (Colledge and Scott, 1999). AKAPs tether PKA through a critical interaction motif, the A-kinase binding (AKB) domain. For most AKAPs, this consists of a 15–20 amino acid amphipathic, helical motif that binds to the D/D domain of the dimerized R-subunit. DXMS was used to address how structural changes in the D/D domain of the type I and type II isoforms contribute to the differences in affinity and specificity of AKAP binding to each isoforms (Burns-Hamuro et al., 2005).

The structure of the D/D domain for the type I and type II isoforms have been solved by NMR. While they have the same overall X-type, four-helix bundle fold, the docking surfaces have some significant differences (Newlon et al., 2001; Banky et al., 2003). The RIα D/D domain contains an extended N-terminal helix, helix N-1, which overlies the four-helix bundle, lending a hindered and rugged surface to the AKAP-binding groove. A small hinge segment allows for significant variation in the positioning of the N-1 helix. The RIIα D/D domain contains a shorter, β-strand-like N-terminus, which creates a more accessible AKAP-binding surface (Newlon et al., 1999). The charge distribution is also very different for these isoforms. The RIα D/D domain contains more acidic and basic residues that line the surface of the AKAP-binding groove, whereas the AKAP-binding surface of RIIα contains primarily hydrophobic residues (Newlon et al., 1999; Banky et al., 2000; Burns et al., 2003). These structural differences also translate into functional differences. The RI subunit binds AKAPs, in general, weaker than the RII subunit. Localization of PKA-RI is more dynamic with more partitioning between the cytoplasm and organelles. In contrast, PKA-RII is usually always localized to specific structures in the cell and binds tighter *in vitro* to most AKAPs (Feliciello et al., 2001).

D-AKAP2 is a multisubunit protein containing two putative regulators of G-protein signaling (RGS) domains and a 40-amino acid C-terminal domain containing the AKB helix and a PDZ binding motif. It is one of the unusual AKAPs in that it can bind to both RI and RII, with a 25-fold increased affinity for RII (Burns et al., 2003). We evaluated the DXMS properties of a peptide corresponding to the AKB helix of D-AKAP2 bound to both RI and RII D/D in the free and complexed form.

7.6.2   DXMS Experiments of PKA R-Subunit D/D Domains and D-AKAP2 AKB Domain

7.6.2.1 Protein Expression and Purification   The AKB domain from mouse D-AKAP2 was subcloned, expressed, and purified as previously described (Burns, 2003). This protein fragment contains 40 amino acids from the C-terminus of mouse D-AKAP2 containing the AKB domain and an additional 15 N-terminal amino acids introduced by the vector (GSPGISGGGGILLS). These additional residues have little effect on binding to the R-subunit D/D domains.

7.6.2.2 DXMS Experiments   Deuterated samples were prepared by diluting 5 μl of protein solution (either analyte protein alone or with 50 mole% excess of binding partner) with 15 μl of deuterated buffer, followed by "on-exchange" incubation at room temperature ($23 \pm 1°C$) for varying times (10–300,000 s) prior to quenching in 30 μl of 0.8% formic acid with GuHCl at 0°C. At 0°C, the quenched solution was immediately passed over a pepsin column with contemporaneous collection of proteolytic products using a C18 column. Subsequently, the C18 column was eluted with a linear gradient of 8%–40% solvent acetonitrile over 10 min. Mass spectrometric analysis was carried out with a Finnigan LCQ mass spectrometer.

7.6.3   DXMS of D-AKAP2 AKB Domain with or without PKA R-Subunit D/D Domains

7.6.3.1 DXMS of Free D-AKAP2 AKB Domain   The DXMS behavior of the D-AKAP2 AKB domain in the absence of the D/D domain was measured at pH 7. This domain was so dynamic that all the amide hydrogens were fully exchanged even at the shortest exchange time employed (10 s) (Fig. 7.8). This was not surprising for a peptide of this size (53aa) and was consistent with circular dichroism measurements,

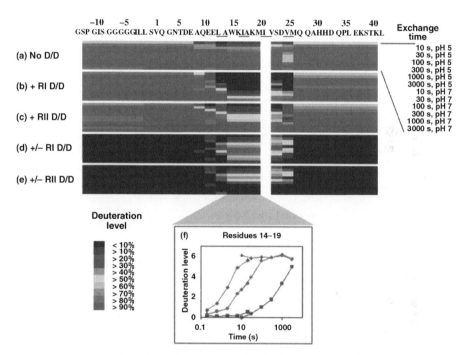

FIGURE 7.8   DXMS results of the AKB ligand in the free and RIα and RIIα-bound states. (a) is the deuterium content of each segment in free state AKB ligand. Each block represents a peptide analyzed. Each block contains 11 time points as indicated on the right. The deuteration levels of each segment at each time point are shown by different colors from blue (<10% deuteration) through red (>90% deuteration) as indicated on the bottom left of the figure. The free AKB state is nearly fully deuterated except very short time points at pH 5. (b) and (c) are the deuteration of AKB when bound to RIα or RIIα, respectively. (d) and (e) are the differences in deuteration levels with or without a D/D domain. The difference in deuteration levels of each segment at each time point are shown by different colors from blue (<10% difference) through red (>90% difference). When AKB is bound to D/D domain, the deuteration levels lower at residues 12–25 and the magnitude of the protection is larger with RIα D/D domain. The bottom insert (f) displays deuteration versus time plot for segment 14–19 in the free state (green), RIIα-bound state (red), or RIα-bound state (blue). As the amide H/D exchange is primarily base-catalyzed reaction at neutral pH, the intrinsic exchange rate is about 50 times faster at pH 7 than at pH 5 (see Section 5.2). The pH 5 time points were converted to pH 7 equivalent (e.g., 10 s at pH 5 is equivalent to 0.2 s at pH 7). (See the color version of this figure in the Color Plates section.)

which showed a lack of a helical signature at pH 7.0 (Burns et al., 2003). As the amide hydrogen exchange reaction is primarily a base-catalyzed reaction at neutral pH, the experiments were repeated at pH 5 to slow the intrinsic amide hydrogen exchange rates and observe the behaviors of fast-exchanging amide hydrogens. The intrinsic amide hydrogen exchange rate at pH 5 is about 1/50 of that at pH 7. At pH 5, the exchange behavior of the central part of the AKB peptide was detected. The fact that the deuterium buildup curve of pH 5 and that of pH 7 overlaid onto each other after taking the intrinsic exchange rate change into account indicated the dynamic properties of the protein were not altered by the pH shift.

7.6.3.2 *RIαD/D Domain Protects a Wider Region of the AKB Domain*
The DXMS behaviors of the AKB:D/D complex was also measured at pH 5 and 7. Fifty mole percent excess of the D/D domain was used to ensure at least 99% of the AKB domain was in the complexed form in the exchange solution.

Binding of both RIα D/D and RIIα D/D domains to the AKB ligand resulted in a significant slowing of deuteron incorporation into the backbone amides of the AKB ligand (Fig. 7.8). Residues 12–25 showed varying degrees of protection from exchange in RIα, while RIIα complexation protected residues 14–25. These regions of AKB protection when bound to RIα and RIIα were consistent with recent peptide array analysis (Burns-Hamuro, 2003). The critical contact areas mapped out by the peptide substitution array extended the length of the AKB helix for RIα binding (residues 12–25), but were more localized to the C-terminal residues of the AKB helix for RIIα binding (residues 16–25) (Burns-Hamuro et al., 2005). Therefore, substitutional analysis and DXMS were consistent with the idea that a broader region of the AKB helix is stabilized when bound to the lower affinity RI interaction surface.

7.6.3.3 *RIα D/D Domain Protects the AKB Domain More Strongly than RIIα*   Interestingly, the magnitude of protection for the AKAP peptide was different when bound to each isoform, again suggesting differences in helical stabilization of the docked AKAP ligand. The binding affinity for the AKAP peptide is 25-fold weaker for RIα than RIIα (Burns et al., 2003), yet was protected to a greater extent when docked to the RIα isoform, suggesting an increase in helical stabilization of the AKB ligand. Deuteration versus log-time plots can provide a more quantitative picture of the individual peptide segments (Fig. 7.8f). A first-order reaction appeared sigmoidal in plots of logarithm of time versus the reaction. The retardation of reaction rates appears as a shift toward the right. The presence of RIIα resulted in less than one order of magnitude shift to the right, indicating this segment is about 1 kcal mol^{-1} more stable in the presence of RIIα than the free state. On the other hand, the presence of RIα resulted in about a two orders of magnitude shift to the right, indicating this segment is close to 3 kcal mol^{-1} more stable in the presence of RIα for segment 14–19 (Fig. 7.8f).

Stronger protection of the AKB ligand when bound to RIα suggested that the ligand backbone was less dynamic when bound to RIα than when bound to RIIα. There are more charged residues on the docking surface of RIα compared to that of RIIα, which has a primarily, hydrophobic surface. The stronger protection by RIα may be a consequence of a tighter hydrogen bonding network that is stabilized by electrostatic and/or hydrogen bonding upon AKB-RIα complex formation. On the contrary, the interaction between the AKB ligand and the predominately hydrophobic binding surface of RIIα is most likely driven by solvent exclusion and does not alter the dynamics of the AKB backbone to the same extent.

7.6.4   DXMS of PKA R-Subunit D/D Domains with or without D-AKAP2 AKB Domain

7.6.4.1 DXMS of Free PKA R-Subunit D/D Domains   The DXMS behaviors of the R-subunit D/D domains, RIα and RIIα, in the absence of the AKB ligand were measured at pH 7 (Figs. 7.9a and 7.10a). The DXMS patterns for the isolated D/D domains are similar for the two isoforms, reflecting the fact that both

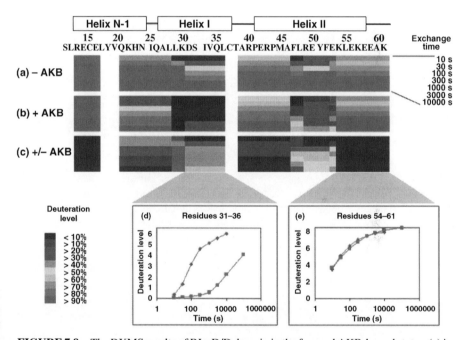

FIGURE 7.9   The DXMS results of RIα D/D domain in the free and AKB-bound states. (a) is the deuterium content of each segment in free state RIα D/D domain. Each block represents a peptide analyzed. Each block contains 11 time points as indicated on the right. The deuteration levels of each segment at each time point are shown by different colors from blue (<10% deuteration) through red (>90% deuteration) as indicated on the bottom left of the figure. (b) is the deuteration of RIα D/D domain when bound to AKB ligand. (c) is the difference in deuteration levels with or without AKB ligand. The difference in deuteration levels of each segment at each time point are shown by different colors from blue (<10% difference) through red (>90% difference). The bottom inserts (d) and (e) display deuteration versus time plots for segments 31–36 and 54–61 in the free state (green) and AKB-bound state (red). (See the color version of this figure in the Color Plates section.)

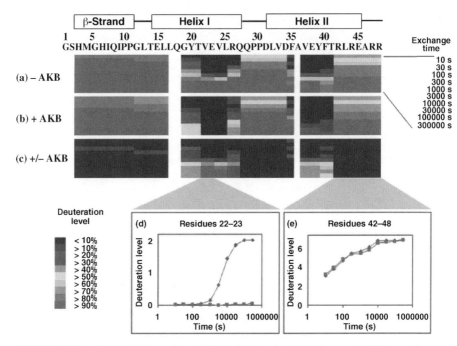

FIGURE 7.10   The DXMS results of RIIα D/D domain in the free and AKB-bound states. (a) is the deuterium content of each segment in free state RIIα D/D domain. Each block represents a peptide analyzed. Each block contains 11 time points as indicated on the right. The deuteration levels of each segment at each time point are shown by different colors from blue (<10% deuteration) through red (>90% deuteration) as indicated on the bottom left of the figure. (b) is the deuteration of RIIα D/D domain when bound to AKB ligand. (c) is the difference in deuteration levels with or without AKB ligand. The differences in deuteration levels of each segment at each time point are shown by different colors from blue (<10% difference) through red (>90% difference). The bottom inserts (d) and (e) display deuteration versus time plots for segments 22–23 and 42–48 in the free state (green) and AKB-bound state (red). (See the color version of this figure in the Color Plates section.)

domains have similar four-helix bundle structures. Concentrated regions of protection map to helix I and helix II for both of these isoforms. The slow exchanged regions of the AKB-free state were consistent with previous H/D exchange experiments by NMR, which map slowly exchanging amide hydrogens to similar regions (Newlon et al., 1999; Banky et al., 2000). The difference between the two isoforms was the stability. While helices I and II of RIα were about 50% deuterated after 100 s of exchange reaction, those of RIIα were not deuterated at all, suggesting that the RIIα helical bundle was more stable than RIα (Fig. 7.11a, b, e, and f).

7.6.4.2 *AKB Binding Perturbs a Wider Range of the RIα Surface than the RIIα Surface*   Helix I and helix I′ form the AKAP docking surface and helix II and helix II′ form the primary dimerization contacts for this domain (Newlon et al.,

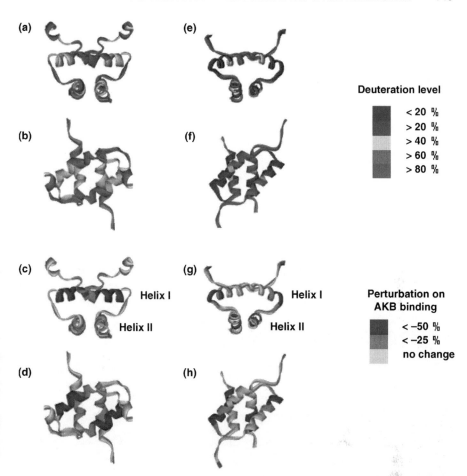

FIGURE 7.11 Deuteration levels of AKB-free D/D domains and average difference in deuteration levels of D/D domains with or without AKB ligand mapped onto the structures of RIα and RIIα D/D domain. The deuteration levels of RIα D/D at 100 s of on-exchange at pH 7 are shown in (a) for side view and (b) for top view. The deuteration levels of RIIα D/D at 100 s of on-exchange at pH 7 are shown in (e) for side view and (f) for top view. The average differences in deuteration levels of RIα D/D domains with or without AKB ligand are shown in (c) for side view and (d) for top view. The average differences in deuteration levels of RIIα D/D domains with or without AKB ligand are shown in (g) for side view and (h) for top view. The structures were provided by Dr. Jennings (Newlon et al., 2001; Banky et al., 2003). (See the color version of this figure in the Color Plates section.)

1999; Banky et al., 2000). R binding to the AKB ligand slowed the deuterium incorporation over a wider segment of RIα (segments 21–36, 16 residues near helix I) compared to RIIα (segments 19–27, 9 residues in helix I). This corresponded well with the earlier observation that H/D exchange of the AKB ligand was protected over a wider range when bound to RIα than when bound to RIIα. This wider binding surface of RI may be a major reason why the RIα surface is more discriminating and has only been shown to bind a few AKAPs (Feliciello et al., 2001). On the other hand, a relatively narrow and more hydrophobic binding surface for RII may enable binding to a wider range of AKAPs.

7.6.4.3 Intradomain Communication within the X-Type Four-Helix Bundle Motif   The binding to the AKB ligand slowed the deuterium incorporation in not only helix I, which maps to the AKB ligand interface, but also helix II of both isoforms. This area of protection was consistent with recent H/D exchange experiments by NMR, which showed a similar protection pattern for helix II and helix II' when the AKAP peptide, HT31, was bound to RIIα, indicating that protection in helix II is not specific to a given AKAP ligand (Fayos et al., 2003). Protection in these helices was presumably due to longer range effects that were propagated through the helices I and II interface, as there was no evidence to suggest from mutagenesis or structural experiments that helix II interacts directly with the AKAP. These longer range effects potentially highlight communication networks within the domain.

DXMS has proven useful to dissect the backbone contributions to ligand binding to the D/D isoforms. Both interacting partners were mapped by DXMS for each complex, which allowed us to propose a structural model as to why AKAP ligands prefer the more hydrophobic and more dynamic RII binding surface.

7.7 EPITOPE MAPPING BY DXMS

7.7.1 Epitope Mapping

The monoclonal antibody market was over $10 billion in 2005 and is expected to exceed $30 billion by 2010. As such, it is perhaps the fastest growing sector in the pharmaceutical industry. The major targets of monoclonal antibodies are in oncology, rheumatology, dermatology, and inflammatory bowel diseases, and many other disease targets are on the horizon.

The epitope mapping of the relevant antigen–antibody interaction is a key step in characterizing the interaction of the therapeutic agent with its target. Although a few antigen–antibody complexes have been solved by X-ray crystallography (Amit et al., 1986; Davies et al., 1988), this method is not applicable to all complexes due to crystallization difficulties. Mutagenesis (Benjamin et al., 1984) and the protection of the antigenic surface against proteolysis (Jemmerson and Paterson, 1986) or chemical modification (Burnens, 1987) by antibody binding have been used to map antibody-binding epitopes. False positives due to the alteration of antigen structure/dynamics can be a problem for mutagenesis, and low resolution is an issue for proteolysis and chemical modification. A popular strategy is the binding assay of antigen fragments with antibody (Scott and Smith, 1990). However, this strategy works well only when the epitope is linear and continuous.

Here, we review how DXMS can be applied to map the epitope of the interaction between cytochrome c and its anti-cytochrome c antibody E8, a classical binding system developed and characterized by Paterson's group (Paterson et al., 1990). The DXMS behaviors of cytochrome c in the presence and absence of the monoclonal antibody were monitored and the segments protected from amide hydrogen exchange reactions by antibody binding were determined.

7.7.2 DXMS Experiments of Cytochrome c in the Presence and Absence of Antibody

7.7.2.1 Immobilization of Antibody
To determine the epitope, DXMS experiments with the antigen must be carried out in the presence and absence of the antibody. However, as the antibody is much larger than cytochrome c ($\sim$150 kDa) and is digested very well by pepsin under the conditions employed, the numerous peptic fragments from the antibody suppress the signal of the cytochrome c originated peptides. To avoid this complication, it is desirable to have the antibody removed quickly from antigen prior to the digestion step. To facilitate the separation from the antigen, the antibody was immobilized onto a perfusive solid phase by Schiff's base chemistry and packed into a column prior to the exchange experiments. The DXMS experiments of antigen in the presence of antibody were carried out in the antibody column.

7.7.2.2 DXMS Experiments of Cytochrome c
On-exchange (H/D exchange) followed by off-exchange (D/H exchange) experiments were performed to map the cytochrome c epitope. First, cytochrome c was incubated in deuterated buffer for 300 s in solution (on-solution). The partially deuterated cytochrome c was then introduced into the antibody column with deuterated buffer and the unbound protein was washed away. The off-exchange reaction was initiated by washing the column with copious aqueous nondeuterated buffer. After various incubation times, the off-exchange reaction was quenched and simultaneously the antigen was eluted out by a chilled acidic buffer. The quenched solution was immediately passed over a pepsin column followed by LC-MS analysis to determine the deuterium content in each peptic fragment. Analogous on–off exchange experiments of cytochrome c were carried out in the absence of monoclonal antibody column as control experiments.

7.7.3 Antibody Binding Site on Cytochrome c by DXMS

In the on–off exchange experiments, two segments of cytochrome c, residues 50–64 and 97–104, retained significantly more deuteriums in the presence of E8 antibody compared to in the absence of the antibody (unpublished results, Fig. 7.12). Three segments, residues 3–21, 24–36, and 39–47, also showed weak perturbations upon the antibody binding. The other segments showed no perturbations.

7.7.4 Comparison with X-ray Crystallographic Structure

The DXMS defined epitope is, in general, in good agreement with the X-ray crystallographic structure defined epitope (Fig. 7.12). A major advantage of

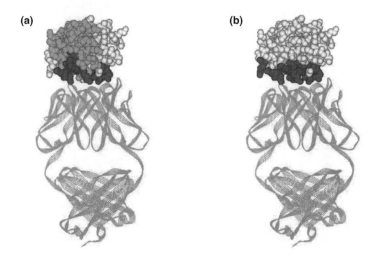

(a)   (b)

FIGURE 7.12   Cytochrome c epitope against E8 antibody overlaid on the X-ray crystallographic structure of cytochrome c and the Fab fragment of E8 antibody complex (1WEJ). (a) DXMS defined epitope. Dark blue indicates the regions where the H/D exchange rates are strongly retarded in the presence of antibody. Light blue indicates the regions where the H/D exchange rates are mildly retarded in the presence of antibody. Gray indicates the regions unaffected by the antibody. Pink is the antibody light chain and orange is the antibody heavy chain. (b) X-ray crystallographic defined contact residues. (See the color version of this figure in the Color Plates section.)

crystallographic structure over DXMS defined epitope mapping is its resolution; in this case, 1.8 Å for crystallographic structure and about 10 residues long for DXMS approach. A major advantage of DXMS approach is that it should be readily applicable to any antigen–antibody interaction. One intriguing difference between the two approaches is the ability to identify binding-induced conformational changes in the DXMS approach. Our results indicate that there is small conformational and/or dynamic change in cytochrome c distant from the binding site upon antigen binding.

7.8   CONCLUSIONS

In this chapter, we have reviewed how the DXMS patterns of various proteins in the presence and absence of their protein binding partners can be deduced and interpreted: hGHs–hGHbp interactions, PKA R-subunit–C-subunit interactions, PKA R-subunit–DAKAP2 interactions, and the cytochrome c–E8 antibody interaction. In most of the interactions, the perturbation upon binding to each protein-binding partner was larger than what was expected with mere footprinting by the binding protein. In hGH–hGHbp interactions, the amide hydrogen exchange rates of more than 80% of hGHbp residues were slowed down in the presence of hGH (Fig. 7.4). In PKA R-subunit–C-subunit interactions, the cAMP binding domains of PKA R-subunit, which are far away from the C-subunit binding site, were more dynamic in the presence of C-subunit (Fig. 7.7). In PKA R-subunit–DAKAP2 interactions, contacting not only helix I of PKA R-subunit D/D domains but also helix II in the opposite side was protected upon binding to DAKAP2 (Figs. 7.9 and 7.10). In the cytochrome *c*–E8 antibody interaction, amides distant from the binding site were weakly protected from exchange upon binding to E8 antibody in addition to the strongly protected amides in the binding site.

Taken together, these findings demonstrate that DXMS of protein–protein complexes affords information regarding not only the binding surface between proteins, but also the structural/dynamic changes that are induced in other portions of the proteins by the binding event.

ABBREVIATIONS

AKAP	A kinase anchoring protein
AKB	A kinase binding
ATP	adenosine triphosphate
C-subunit	catalytic subunit
cAMP	cyclic adenosine monophosphate
cAMP:A	cAMP binding domain A
cAMP:B	cAMP binding domain B
cGMP	guanosine 3'-5'-cyclic monophosphate
DAKAP	dual specific A kinase anchoring protein
D/D domain	dimerization/docking domain
DTT	dithiothreitol
DXMS	enhanced peptide amide hydrogen/deuterium exchange mass spectrometry
EDTA	ethylenediaminetetraacetic acid
ESI	electrospray ionization
GuHCl	guanidine hydrochloride
H/D exchange	hydrogen/deuterium exchange
hGH	human growth hormone
hGHbp	human growth hormone binding protein (receptor)
hGHv	phage display optimized human growth hormone variant
hGHwt	wild-type human growth hormone
HPLC	high-performance liquid chromatography
MS	mass spectrometry
NMR	nuclear magnetic resonance
PKA	protein kinase A
RGS	regulators of G-protein signaling
R-subunit	regulatory subunit
TCEP	tris(2-carboxyethyl)phosphine hydrochloride
TFA	trifluoroacetic acid

ACKNOWLEDGMENTS

We would like to sincerely thank Mark Fisher, Paul DeStefano, Robert Johnston, David Smith, and Walter Englander for their vision and support of this work, which would have not been possible without their efforts. This work was supported by NIH Grants CA099835 and CA118595 (VLW). VLW has equity interests in ExSAR Corporation.

REFERENCES

Akashi, S. and Takio,K., 2002. Melittin-diacylphosphatidylcholine interaction examined by electrospray ionization Fourier transform ion cyclotron resonance mass spectrometry. *J Mass Spectrom Soc Jpn* 50, 67–71.

Amit, A. G., Mariuzza, R. A., Phillips, S. E., and Poljak, R. J., 1986. Three-dimensional structure of an antigen–antibody complex at 2.8 Å resolution. *Science* 233, 747–753.

Anand, G. S., Hughes, C. A., Jones, J. M., Taylor, S. S., and Komives, E. A., 2002. Amide H/2H exchange reveals communication between the cAMP and catalytic subunit-binding sites in the RIα subunit of protein kinase A. *J Mol Biol* 323, 377–386.

Anand, G. S., Law, D., Mandell, J. G., Snead, A. N., Tsigelny, I., Taylor, S. S., Ten Eyck, L. F., and Komives, E. A., 2003. Identification of the protein kinase A regulatory RIalpha-catalytic subunit interface by amide H/2H exchange and protein docking. *Proc Natl Acad Sci USA* 100, 13264–13269.

Arkin, M. R., Randal, M., DeLano, W. L., Hyde, J., Luong, T. N., Oslob, J. D., Raphael, D. R., Taylor, L., Wang, J., McDowell, R. S., et al., 2003. Binding of small molecules to an adaptive protein–protein interface. *Proc Natl Acad Sci USA* 100, 1603–1608.

Arkin, M. R. and Wells, J. A., 2004. Small-molecule inhibitors of protein–protein interactions: progressing towards the dream. *Nat Rev Drug Discov* 3, 301–317.

Bai, Y., Milne, J. S., Mayne, L. C., and Englander, S. W., 1993. Primary structure effects on peptide group hydrogen exchange. *Proteins Struct Funct Genet* 17, 75–86.

Banky, P., Newlon, M. G., Roy, M., Garrod, S., Taylor, S. S., and Jennings, P. A., 2000. Isoform-specific differences between the type I.alpha and II.alpha cyclic AMP-dependent protein kinase anchoring domains revealed by solution NMR. *J Bio Chem* 275, 35146–35152.

Banky, P., Roy, M., Newlon, M. G., Morikis, D., Haste, N. M., Taylor, S. S., and Jennings, P. A., 2003. Related protein–protein interaction modules present drastically different surface topographies despite a conserved helical platform. *J Mol Biol* 330, 1117–1129.

Begley, M. J., Taylor, G. S., Brock, M. A., Ghosh, P., Woods, V. L., Jr., and Dixon, J. E., 2006. Molecular basis for substrate recognition by MTMR2, a myotubularin family phosphoinositide phosphatase. *Proc Natl Acad Sci USA* 103, 927–932.

Benjamin, D. C., Berzofsky, J. A., East, I. J., Gurd, F. R., Hannum, C., Leach, S. J., Margoliash, E., Michael, J. G., Miller, A., Prager, E. M., et al, 1984. The antigenic structure of proteins: a reappraisal. *Annu Rev Immunol* 2, 67–101.

Black, B. E., Brock, M. A., Bedard, S., Woods, V. L., Jr., and Cleveland, D. W., 2007. An epigenetic mark generated by the incorporation of CENP-A into centromeric nucleosomes. *Proc Natl Acad Sci USA* 104, 5008–5013.

Black, B. E., Foltz, D. R., Chakravarthy, S., Luger, K., Woods, V. L., Jr., and Cleveland, D. W., 2004. Structural determinants for generating centromeric chromatin. *Nature* 430, 578–582.

Brudler, R., Gessner, C. R., Li, S., Tyndall, S., Getzoff, E. D., and Woods, V. L., Jr., 2006. PAS domain allostery and light-induced conformational changes in photoactive yellow protein upon I2 intermediate formation, probed with enhanced hydrogen/deuterium exchange mass spectrometry. *J Mol Biol* 363, 148–160.

Burnens, A., Demotz, S., Corradin, G., Binz, H., and Bosshard, H. R., 1987. Epitope mapping by chemical modification of free and antibody-bound protein antigen. *Science* 235, 780–783.

Burns, L. L., Canaves, J. M., Pennypacker, J. K., Blumenthal, D. K., and Taylor, S. S., 2003. Isoform specific differences in binding of a dual-specificity A-kinase anchoring protein to type I and type II regulatory subunits of PKA. *Biochemistry* 42, 5754–5763.

Burns-Hamuro, L., Hamuro, Y., Kim, J., Sigala, P., Fayos, R., Stranz, D., Jennings, P., Taylor, S., and Woods, V. L., Jr., 2005. Distinct interaction modes of an AKAP bound to two regulatory subunit isoforms of protein kinase A revealed by amide hydrogen/deuterium exchange. *Protein Sci* 14, 2982–2992.

Clackson, T. and Wells, J. A., 1995. A hot spot of binding energy in a hormone–receptor interface. *Science* 267, 383–386.

Colledge, M. and Scott, J. D., 1999. AKAPs: from structure to function. *Trends Cell Biol* 9, 216–221.

Cunningham, B. C. and Wells, J. A., 1993. Comparison of a structural and a functional epitope. *J Mol Biol* 234, 554–563.

Davies, D. R., Sheriff, S., and Padlan, E. A., 1988. Antibody–antigen complexes. *J Biol Chem* 263, 10541–10544.

Davies, D. R., Padlan, E. A., and Sheriff, S., 1990. Antibody–antigen complexes. *Annu Rev Biochem* 59, 439–473.

Del Mar, C., Greenbaum, E. A., Mayne, L., Englander, S. W., and Woods, V. L., Jr., 2005. Structure and properties of α-synuclein and other amyloids determined at the amino acid level. *Proc Natl Acad Sci USA* 102, 15477–15482.

Derunes, C., Briknarova, K., Geng, L., Li, S., Gessner, C. R., Hewitt, K., Wu, S., Huang, S., Woods, V. L., Jr., and Ely, K. R., 2005. Characterization of the PR domain of RIZ1 histone methyltransferase. *Biochem Biophys Res Commun* 333, 925–934.

Derunes, C., Burgess, R., Iraheta, E., Kellerer, R., Becherer, K., Gessner, C. R., Li, S., Hewitt, K., Vuori, K., Pasquale, E. B., Woods, V. L., Jr., and Ely, K. R., 2006. Molecular determinants for interaction of SHEP1 with Cas localize to a highly solvent-protected region in the complex. *FEBS Lett.* 580, 175–178.

Diller, T. C., Madhusudan Xuong, N. H., and Taylor, S. S., 2001. Molecular basis for regulatory subunit diversity in cAMP-dependent protein kinase: crystal structure of the type II beta regulatory subunit. *Structure* 9, 73–82.

Engen, J. R. and Smith, D. L., 2001. Investigating protein structure and dynamics by hydrogen exchange MS. *Anal Chem* 73, 256A–265A.

Englander, S. W. and Kallenbach, N. R., 1984. Hydrogen exchange and structural dynamics of proteins and nucleic acids. *Q Rev Biophys* 16, 521–655.

Englander, J., Del Mar, C., Li, W., Englander, S., Kim, J., Stranz, D., Hamuro, Y., and Woods, Jr. V., 2003. Protein structure change studied by hydrogen–deuterium exchange, functional labeling, and mass spectrometry. *Proc Natl Acad Sci USA* 100, 7057–7062.

Fayos, R., Melacini, G., Newlon, M. G., Burns, L., Scott, J. D., and Jennings, P. A., 2003. Induction of flexibility through protein–protein interactions. *J Biol Chem* 278, 18581–18587.

Feliciello, A., Gottesman, M. E., and Avvedimento, E. V., 2001. The biological functions of A-kinase anchor proteins. *J Mol Biol* 308, 99–114.

Garcia, R. A., Pantazatos, D. P., Gessner, C. R., Go, K. V., Woods, V. L., Jr., and Villarreal, F. J., 2005. Molecular interactions between matrilysin and the matrix metalloproteinase inhibitor doxycycline investigated by deuterium exchange mass spectrometry. *Mol Pharmacol* 67, 1128–1136.

Golynskiy, M., Li, S., Woods, V. L., Jr., and Cohen, S. M., 2007. Conformational studies of the manganese transport regulator (MntR) from *Bacillus subtilis* using deuterium exchange mass spectrometry. *J Biol Inorg Chem* 12, 699–709.

Goshe, M. B. and Anderson, V. E., 1999. Hydroxyl radical-induced hydrogen/deuterium exchange in amino acid carbon–hydrogen bonds. *Radiat Res* 151, 50–58.

Hamuro, Y., Burns, L. L., Canaves, J. M., Hoffman, R. C., Taylor, S. S., and Woods, V. L., Jr., 2002a. Domain organization of D-AKAP2 revealed by enhanced deuterium exchange-mass spectrometry (DXMS). *J Mol Biol* 321, 703–714.

Hamuro, Y., Wong, L., Shaffer, J., Kim, J. S., Jennings, P. A., Adams, J. A., and Woods, V. L., Jr., 2002. Phosphorylation-driven motion in the COOH-terminal Src kinase, Csk, revealed through enhanced hydrogen–deuterium exchange and mass spectrometry (DXMS). *J Mol Biol* 323, 871–881.

Hamuro, Y., Coales, S. J., Southern, M. R., Nemeth-Cawley, J. F., Stranz, D. D., and Griffin, P. R., 2003. Rapid analysis of protein structure and dynamics by hydrogen/deuterium exchange mass spectrometry. *J Biomol Techn* 14, 171–182.

Hamuro, Y., Zawadzki, K. M., Kim, J. S., Stranz, D., Taylor, S. S., and Woods, V. L., Jr., 2003b. Dynamics of cAPK type IIb activation revealed by enhanced amide H/2H exchange mass spectrometry (DXMS). *J Mol Biol* 327, 1065–1076.

Hamuro, Y., Anand, G., Kim, J., Juliano, C., Stranz, D., Taylor, S., and Woods, V. L., Jr., 2004. Mapping intersubunit interactions of the regulatory subunit (RIα) in the type I holoenzyme of protein kinase A by amide hydrogen/deuterium exchange mass spectrometry (DXMS). *J Mol Biol* 340, 1185–1196.

Herberg, F. W., Taylor, S. S., and Dostmann, W. R., 1996. Active site mutations define the pathway for the cooperative activation of cAMP-dependent protein kinase. *Biochemistry* 35, 2934–2942.

Hilser, V. J. and Freire, E., 1996. Structure-based calculation of the equilibrium folding pathway of proteins: correlation with hydrogen exchange protection factors. *J Mol Biol* 262, 756–772.

Horn, J. R., Kraybill, B., Petro, E. J., Coales, S. J., Morrow, J. A., Hamuro, Y., and Kossiakoff, A. A., 2006. The role of protein dynamics in increasing binding affinity for an engineered protein–protein interaction established by H/D exchange mass spectrometry. *Biochemistry* 45, 8488–8498.

Huang, L. J. and Taylor, S. S., 1998. Dissecting cAMP binding domain A in the RIalpha subunit of cAMP-dependent protein kinase: distinct subsites for recognition of cAMP and the catalytic subunit. *J Biol Chem* 273, 26739–26746.

Iyer, G. H., Garrod, S., Woods, V. L., Jr., and Taylor, S. S., 2005. Catalytic independent functions of a protein kinase as revealed by a kinase-dead mutant: study of the Lys72His mutant of cAMP-dependent kinase. *J Mol Biol* 351, 1110–1122.

Janin, J. and Chothia, C., 1990. The structure of protein–protein recognition sites. *J Biol Chem* 265, 16027–16030.

Jemmerson, R. and Paterson, Y., 1986. Mapping epitopes on a protein antigen by the proteolysis of antigen–antibody complexes. *Science* 232, 1001–1004.

Kim, C., Vigil, D., Anand, G., and Taylor, S. S., 2006. Structure and dynamics of PKA signaling proteins. *Eur J Cell Biol* 85, 651–654.

Kouadio, J. L., Horn, J. R., Pal, G., and Kossiakoff, A. A., 2005. Shotgun alanine scanning shows that growth hormone can bind productively to its receptor through a drastically minimized interface. *J Biol Chem* 280, 25524–25532.

Lanman, J., Lam, T. T., Barnes, S., Sakalian, M., Emmett, M. R., Marshall, A. G., and Prevelige, P. E., 2003. Identification of novel interactions in HIV-1 capsid protein assembly by high-resolution mass spectrometry. *J Mol Biol* 325, 759–772.

Melnyk, R. A., Hewitt, K. M., Lacy, D. B., Lin, H. C., Gessner, C. R., Li, S., Woods, V. L., Jr., and Collier, R. J., 2006. Structural determinants for the binding of anthrax lethal factor to oligomeric protective antigen. *J Biol Chem* 281, 1630–1635.

Newlon, M. G., Roy, M., Morikis, D., Carr, D. W., Westphal, R., Scott, J. D., and Jennings, P. A., 2001. A novel mechanism of PKA anchoring revealed by solution structures of anchoring complexes. *EMBO J* 20, 1651–1662.

Newlon, M. G., Roy, M., Morikis, D., Hausken, Z. E., Coghlan, V., Scott, J. D., and Jennings, P. A., 1999. The molecular basis for protein kinase A anchoring revealed by solution NMR. *Nat Struct Biol* 6, 222–227.

Pal, G., Kossiakoff, A. A., and Sidhu, S. S., 2003. The functional binding epitope of a high affinity variant of human growth hormone mapped by shotgun alanine-scanning mutagenesis: insights into the mechanisms responsible for improved affinity. *J Mol Biol* 332, 195–204.

Pantazatos, D., Kim, J. S., Klock, H. E., Stevens, R. C., Wilson, I. A., Lesley, S. A., and Woods, V. L., Jr., 2004. Rapid refinement of crystallographic protein construct definition employing enhanced hydrogen/deuterium exchange MS. *Proc Natl Acad Sci USA* 101, 751–756.

Paterson, Y., Englander, S. W., and Roder, H., 1990. An antibody binding site on cytochrome *c* defined by hydrogen exchange and two-dimensional NMR. *Science* 249, 755–759.

Pearce, K. H., Cunningham, B. C., Fuh, G., Teeri, T., and Wells, J. A., 1999. Growth hormone binding affinity for its receptor surpasses the requirements for cellular activity. *Biochemistry* 38, 81–89.

Resing, K. A., Hoofnagle, A. N., and Ahn, N. G., 1999. Modeling deuterium exchange behavior of ERK2 using pepsin mapping to probe secondary structure. *J Am Soc Mass Spectrom* 10, 685–702.

Schiffer, C., Ultsch, M., Walsh, S., Somers, W., de Vos, A. M., and Kossiakoff, A., 2002. Structure of a phage display-derived variant of human growth hormone complexed to two copies of the extracellular domain of its receptor: evidence for strong structural coupling between receptor binding sites. *J Mol Biol* 316, 277–289.

Scott, J. K. and Smith, G. P., 1990. Searching for peptide ligands with an epitope library. *Science* 249, 386–390.

Sivaraman, T., Arrington, C. B., and Robertson, A. D., 2001. Kinetics of unfolding and folding from amide hydrogen exchange in native ubiquitin. *Nat Struct Biol* 8, 331–333.

Skalhegg, B. S. and Tasken, K., 2000. Specificity in the cAMP/PKA signaling pathway: differential expression, regulation, and subcellular localization of subunits of PKA. *Front Biosci* 5, D678–D693.

Spraggon, G., Pantazatos, D., Klock, H. E., Wilson, I. A., Woods, V. L., Jr., and Lesley, S. A., 2004. On the use of DXMS to produce more crystallizable proteins: structures of the *T. maritima* proteins TM0160 and TM1171. *Protein Sci* 13, 3187–3199.

Stanfield, R. L., Takimoto-Kamimura, M., Rini, J. M., Profy, A. T., and Wilson, I. A., 1993. Major antigen-induced domain rearrangements in an antibody. *Structure* 1, 83–93.

Su, Y., Dostmann, W. R. G., Herberg, F. W., Durick, K., Xuong, N. H., Ten Eyck, L., Taylor, S. S., and Varughese, K. I., 1995. Regulatory subunit of protein kinase A: structure of deletion mutant with cAMP binding domains. *Science* 269, 807–813.

Taylor, S. S., Kim, C., Vigil, D., Haste, N. M., Yang, J., Wu, J., and Anand, G. S., 2005. Dynamics of signaling by PKA. *Biochim Biophys Acta* 1754, 25–37.

Vos, A. M. d, Ultsch, M., and Kossiakoff, A. A., 1992. Human growth hormone and extracellular domain of its receptor: crystal structure of the complex. *Science* 255, 306–312.

Wang, L. and Smith, D. L., 2003. Downsizing improves sensitivity 100-fold for hydrogen exchange-mass spectrometry. *Anal Biochem* 314, 46–53.

Wells, J. A., 1996. Binding in the growth hormone receptor complex. *Proc Natl Acad Sci USA* 93, 1–6.

Whitty, A. and Kumaravel, G., 2006. Between a rock and a hard place? *Nat Chem Biol* 2, 112–118.

Wong, L., Lieser, S., Chie-Leon, B., Miyashita, O., Aubol, B., Shaffer, J., Onuchic, J. N., Jennings, P., Woods, V. L., Jr., and Adams, J. A., 2004. Dynamic coupling between the SH2 domain and active site of the COOH terminal Src kinase, Csk. *J Mol Biol* 341, 93–106.

Wong, L., Lieser, S. A., Miyashita, O., Miller, M., Tasken, K., Onuchic, J. N., Adams, J. A., Woods, V. L., Jr., and Jennings, P. A., 2005. Coupled motions in the SH2 and kinase domains of Csk control Src phosphorylation. *J Mol Biol* 351, 131–143.

Woods, V. L., Jr., 1997. Method for characterization of the fine structure of protein binding sites employing amide hydrogen exchange. US Patent No. 5,658,739. ExSAR Corporation, Monmouth Junction, NJ.

Woods, V. L., Jr., 2001a. Method for characterization of the fine structure of protein binding sites using amide hydrogen exchange. US Patent No. 6,331,400. ExSAR Corporation, Monmouth Junction, NJ.

Woods, V. L., Jr., 2001b. Methods for the high-resolution identification of solvent-accessible amide hydrogens in polypeptides or proteins and for the characterization of the fine structure of protein binding sites. US Patent No. 6,291,189. ExSAR Corporation, Monmouth Junction, NJ.

Woods, V. L., Jr., 2003. Methods for identifying hot-spot residues of binding proteins and small compounds that bind to the same. US Patent No. 6,599,707. ExSAR Corporation, Monmouth Junction, NJ.

Woods, V. L., Jr. and Hamuro, Y., 2001. High resolution, high-throughput amide deuterium exchange-mass spectrometry (DXMS) determination of protein binding site structure and dynamics: utility in pharmaceutical design. *J Cell Biochem* S37, 89–98.

Yan, X., Zhang, H., Watson, J., Schimerlik, M. I., and Deinzer, M. L., 2002. Hydrogen/deuterium exchange and mass spectrometric analysis of a protein containing multiple disulfide bonds: solution structure of recombinant macrophage colony stimulating factor-beta (rhM-CSFb). *Protein Science* 11, 2113–2124.

Yang, J., Garrod, S. M., Deal, M. S., Anand, G. S., Woods, V. L., Jr., and Taylor, S., 2005. Allosteric network of cAMP-dependent protein kinase revealed by mutation of Tyr204 in the P + 1 loop. *J Mol Biol* 346, 191–201.

Zawadzki, K. M. and Taylor, S. S., 2004. cAMP-dependent protein kinase regulatory subunit type IIbeta: active site mutations define an isoform-specific network for allosteric signaling by cAMP. *J Biol Chem* 279, 7029–7036.

Zawadzki, K., Hamuro, Y., Kim, J., Garrod, S., Stranz, D., Taylor, S., and Woods, V. L., Jr., 2003. Dissecting interdomain communication within cAPK regulatory subunit type IIbeta using enhanced amide hydrogen/deuterium exchange mass spectrometry (DXMS). *Protein Science* 12, 1980–1990.

Zhang, Z., Post, C. B., and Smith, D. L., 1996. Amide hydrogen exchange determined by mass spectrometry: application to rabbit muscle aldolase. *Biochemistry* 35, 779–791.

CHAPTER 8

Cross-linking as a Tool to Examine Protein Complexes: Examples of Cross-linking Strategies and Computational Modeling

EVGENIY V. PETROTCHENKO and CHRISTOPH H. BORCHERS

University of Victoria-Genome BC Proteomics Center, Department of Biochemistry and Microbiology, University of Victoria, Victoria, BC, Canada

8.1 INTRODUCTION

Protein–protein interactions play an important role in many biological processes. Interactions may vary from relatively simple transient bimolecular interactions to highly stable multisubunit protein complexes. Together with canonical methods for studying protein structure, such as X-ray crystallography and nuclear magnetic resonance (NMR) spectroscopy, cross-linking combined with mass spectrometry (MS) can provide considerable structural information regarding the organization and function of protein complexes, information that may be otherwise impossible to obtain. Experimentally sound quaternary and ternary protein complex structures can be determined by incorporating cross-linking distance constraints into computational modeling processes.

8.2 CROSS-LINKING STRATEGIES

Protein cross-linking came into practice several decades ago. The underlying principle of chemical protein cross-linking is the formation of a covalent bond between functional groups of adjacent protein molecules. The formation of a protein complex through chemical cross-linking provides researchers with information pertaining to the spatial proximity of reacting groups, and the identity of protein subunits composing the

Mass Spectrometry Analysis for Protein–Protein Interactions and Dynamics, Edited by Mark Chance
Copyright © 2008 John Wiley & Sons, Inc.

complex. Information useful for constructing binary interaction maps of cross-linked sites may also be obtained.

In traditional chemical cross-linking experiments, complexed proteins are identified using one of the many biochemical methods, such as specific detection with antibodies, enzymatic assays, or Edman sequencing. Recent advances in the use of mass spectrometry for the analysis of proteins and peptides, however, have greatly expanded the power of cross-linking methodologies. Mass spectrometry not only enables researchers to rapidly and unequivocally determine the identity of individual complexed proteins, it can also be used to identify the site of interaction by determining the amino acid sequence at the cross-link site. This information then can be used to create a quaternary structural topological map of the protein complex, or to elucidate ternary structural details of the protein interaction interfaces.

8.3 CROSS-LINKING METHODOLOGY

In applying mass spectrometry methods to identify sites of cross-linking between interacting proteins, there are two alternative approaches to produce peptides containing the site of cross-linking. In the "top-down" approach, the intact cross-linked protein complex is subjected to mass spectrometric analysis first, followed by fragmentation of the proteins to reduce the complex down to the peptide level for detailed analysis of the interaction site. Although the top-down approach has the advantage of allowing the researcher to carry out the entire experimental analysis of sites of cross-linkage within the mass spectrometer, it is limited by the size and nature of the protein complex, which significantly impacts ionization and fragmentation efficiency.

An alternative that overcomes some of the limitations of top-down analyses is the "bottom-up" approach. The bottom-up method involves enzymatic or chemical digestion of the cross-linked protein complex down to the peptide level prior to mass spectrometry. The identity of cross-linked peptides, also called cross-links, can then be determined by MS analysis of the resulting peptide mixture. This approach uses well-established and reliable techniques for protein digestion, peptide separation, and mass spectrometric analysis. However, as with the top-down approach, bottom-up strategies have several potential drawbacks. These include the challenge of adequately separating the complex peptide mixture that typically results from enzymatic digestion, as well as sensitivity issues stemming from the characteristic low specific content of cross-linked versus free peptides.

A theoretically very attractive strategy for overcoming some or all of the drawbacks of both top-down and bottom-up approaches would be a method combining features of both. In a combined top-down/bottom-up approach, the intact cross-linked protein complex could be subjected to limited proteolysis to produce a smaller pool of peptide fragments of reasonable size that would be suitable for subsequent top-down analysis.

8.4   CHALLENGES ASSOCIATED WITH COMBINING CROSS-LINKING WITH MASS SPECTROMETRY

Combining chemical cross-linking with mass spectrometry to identify sites of protein–protein interaction would appear to be straightforward. Unfortunately, there are several challenges and difficulties that must be overcome to maximize the potential of this strategy.

A number of factors must be taken into account when designing and interpreting cross-linking experiments involving MS. For example, determining cross-linking sites by mass is hampered by the combinatorial nature of the interpeptide cross-links. As the size and number of subunits within cross-linked protein complexes increase, the number of possible binary combinations of linked peptides increases in factorial manner. As a result, a single measured mass can satisfy a number of possible theoretical cross-linked peptide combinations. Nonperfect specificity of the cross-linking reagent, non-specific and missed sites of proteolytic cleavage, and posttranslational modification of the proteins must be taken into account as well, thereby further complicating mass spectrometric analyses.

Proteins within complexes can also be cross-linked in more than one way. Cross-linking reactions may result in large amounts of structurally less informative cross-linked products, such as dead-end and intrapeptide cross-links. In the case of dead-end cross-links, the cross-linker is incorporated into the protein through one of the two reactive groups of the cross-linker; however, the other reactive group does not react with another amino acid residue and is instead deactivated (e.g., by hydrolysis). Intrapeptide cross-links are peptides that have cross-linked two amino acid residues within the same peptide. Thus, the structurally informative interpeptide cross-links (cross-linker connects two amino acid residues that are in close spatial proximity) may constitute a small fraction of the total sample. Moreover, interpeptide cross-links are larger than cognate noncross-linked peptides. This is due to the fact that in general (1) cross-linking is a sum of two peptides, and (2) occasional destruction of enzymatic cleavage sites by cross-linking modification results in larger peptides (e.g., modification of lysine residues by amine-reactive cross-linking reagents may eliminate tryptic cleavage sites). All of these factors increase the difficulty of detecting the signal of interpeptide cross-links within the prevalent pool of free peptides because they may exacerbate ion suppression effects and diminish the ionization efficiency of cross-linked peptides.

Complex, insufficient, and difficult to interpret MS/MS fragmentation poses another obstacle to successful analysis of cross-linked peptides. Interlinked peptide chains produce far more complicated fragment ion series than linear peptides during MS/MS. In addition, fragmentation of interpeptide cross-links usually does not go to completion. Incomplete fragmentation is most likely due to the mutual effect each chain has upon dissociation of its counterpart.

Another challenge associated with combining chemical cross-linking with mass spectrometry to study protein–protein interactions involves the tremendous number of cross-linked peptide species produced. A single top-down/bottom-up experiment will produce hundreds of peptides requiring analysis. From a practical standpoint, conducting these types of experiments without some degree of automation would not be feasible. Fortunately, the need for automating the analysis of the large amount of MS and MS/MS data involved in a typical experiment has spurred development and integration of specialized software programs, which will be discussed subsequently.

The combinatorial nature and low specific abundance of interpeptide cross-links and the typically complicated MS/MS fragmentation pattern of cross-linked peptides represent the primary challenges that must be overcome to confidently detect and identify interpeptide cross-links. Several recent developments have done much to overcome these challenges. Advancements in mass spectrometry instrumentation, the development of new and more sophisticated cross-linking reagents, and production of software analysis tools to improve automation and data analysis have greatly enhanced the attractiveness of cross-linking combined with mass spectrometry compared to more traditional methods for studying protein–protein interactions.

8.5 ADVANCES IN MASS SPECTROMETRY INSTRUMENTATION AND CAPABILITIES

Perhaps the most crucial advancement in mass spectrometry instrumentation with regard to cross-linking studies was the development of Fourier transform ion cyclotron resonance mass spectrometry (FTICR-MS). One of the defining characteristics of FTICR-MS is its extraordinarily high mass accuracy. The unprecedented accuracy of FTICR-MS drastically narrows down the number of possible assignments for a single measured mass observed in a cross-linking experiment. However, it is not straightforward to develop a general analytical expression describing the relation between the number of possible cross-linked peptide combinations and the mass accuracy of the measurement. Consider the case of the anaphase-promoting complex (APC) with a molecular weight of 1 MDa and consisting of 13 protein subunits. Increasing the mass accuracy of peptide measurements from 20 ppm to 1 ppm reduces the number of possible peptide combinations capable of satisfying a given observed mass from hundreds down to only a few. Extending the mass accuracy further into the sub-ppm range may result in a single solution capable of providing an assignment for the cross-link. It must be noted that the number of possible solutions for a given observed mass greatly depends upon the "popularity" of the mass in peptide mass space. For example, the average abundance of atoms in a peptide dictates that there will typically be a considerably greater number of peptides with a mass around 1500.75 Da that must be taken into consideration than there will be peptides with a mass around 1500.05 Da (Mann, 1995).

Another recent development that has tremendously enhanced the power of mass spectrometry as a tool for elucidating the structure of cross-linked protein complexes is the application of new and improved modes of MS^n fragmentation. In addition to the widely used method of collision-induced dissociation (CID) to fragment peptides,

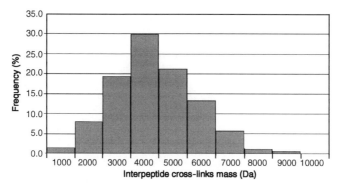

FIGURE 8.1   Mass distribution of theoretically possible interpeptide cross-links for bovine hemoglobin. Tryptic peptides containing one internal lysine residue or N-terminal amino group were taken into account. A majority of the cross-links exceed 3000 Da. (See the color version of this figure in Color Plates section.)

electron-capture dissociation (ECD) and infrared multiple photon dissociation (IRMPD) modes of fragmentation can supply complementary peptide sequence information. Particularly attractive with regard to top-down approaches is the possibility of combining several peptide fragmentation methods into a single cross-linking experiment. Continuing improvements in mass spectrometer sensitivity also make cross-linking combined with MS a promising tool for studying protein interactions. Development of instruments with greater sensitivity, particularly for higher mass/charge (m/z) ranges, is especially beneficial for analyzing the higher mass products that result from the cross-linking of two peptides. The recently introduced Applied Biosystems (ABI) model 4800 matrix-assisted laser desorption ionization-time of flight (MALDI-TOF) instrument, for example, has an improved high mass range sensitivity compared to its progenitor, the ABI model 4700. The importance of extending measurements into higher molecular weight ranges can be illustrated by estimating the distribution of theoretical masses for cross-linked tryptic peptides (Fig. 8.1).

Yet another beneficial advancement in instrumentation involves the ability to more efficiently separate the often complex peptide mixtures that result from proteolytic digestion of multisubunit protein assemblies. This is important because as peptide separation efficiency is increased, so are the chances that the relatively low-abundance cross-linked peptides typically obtained can be detected by mass spectrometry. Sophisticated two-dimensional capillary liquid chromatography systems tailored to the high-efficiency separation of complex protein and peptide mixtures are now in routine use in many proteomics laboratories, and can readily be employed for studying protein interactions. Multidimensional chromatography systems are especially valuable for studying interactions within large protein assemblies because they facilitate separation of higher abundance free peptides that may otherwise totally suppress the ion signal of low-abundance cross-linked peptides.

8.6 NOVEL CROSS-LINKING REAGENTS FOR MASS SPECTROMETRY APPLICATIONS

In addition to improvements that have been and are continuing to be made with respect to instrumentation used in cross-linking experiments, improvements are also being made in relation to the development of cross-linking reagents. Several classes of reagents specifically tailored for use in MS-based cross-linking studies are now available.

One of the most important advances in this regard was the recent introduction of isotopically labeled cross-linking reagents (Müller et al., 2001). Isotopically labeled reagents are based upon substitution of one or more atoms in a cross-linker with heavy stable isotopes, usually deuterium (^{2}H) or carbon isotope ^{13}C, thus producing "heavy" and "light" forms of the reagent that are otherwise chemically indistinguishable. Under experimental conditions, heavy and light isomers of a given cross-linker are mixed in a 1:1 ratio with proteins that are known or believed to form complexes. Following cross-linking of intact proteins, proteolytic digestion, and separation of resulting peptides, peptides cross-linked by isotopically labeled reagents can be distinguished in mass spectra as doublet peaks with 1:1 ion signal ratio separated by a mass corresponding to the difference between heavy and light isomers. This approach allows ready and unequivocal detection of cross-linked peptides in complex mass spectra (Fig. 8.2a). In addition, both heavy and light isotopes can be selected simultaneously for mass spectrometric fragmentation and MS/MS analysis. In this case MS peaks representing fragments of cross-linked peptides will still appear as doublets, which facilitate distinguishing ion series in complicated MS/MS spectra (Fig. 8.2b).

So-called reporter groups represent another means of selectively detecting cross-linked peptides via mass spectrometry. Certain chemical groups can be introduced into the structure of the cross-linkers, and under CID fragmentation conditions these groups will produce specific "reporter" ions. Cross-linked peptides can then be monitored by appearance of the reporter ion signals (Tang et al., 2005).

The use of ^{18}O isotopes during proteolytic digestion represents yet one more possibility for detecting interpeptide cross-links. If digestion takes place in a 1:1 mixture of ^{16}O/^{18}O water, the resulting interpeptide cross-links will have twice the number of incorporated ^{18}O oxygen isotopes than dead-end or interpeptide cross-links resulting in a specific isotopic signature in mass spectrum (Back et al., 2002).

Overcoming the challenge associated with the combinatorial nature of interpeptide cross-links was addressed with the introduction of cleavable isotopically coded cross-linking reagents (Petrotchenko et al., 2005). In that particular case, cleavage of the cross-linker produces two halves, both of which retain their isotopic label. The cleaved peptide halves can be matched to the uncleaved cross-link. Cleavage of the linker allows both peptides to be analyzed separately by MS/MS (Fig. 8.3), thereby reducing the analysis to the well established level of single peptides. The cleavable isotopically coded cross-link approach shows considerable promise for potential proteomic applications, where the identities of

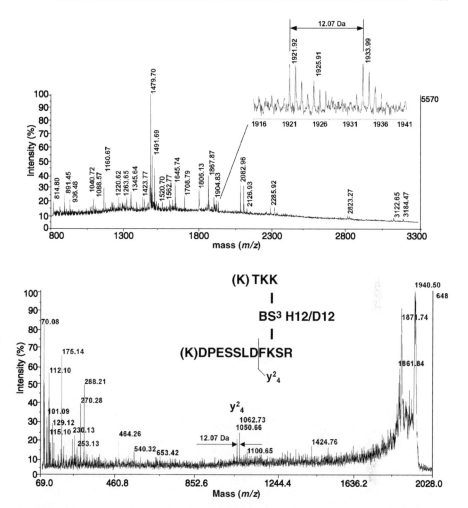

FIGURE 8.2 Use of isotopically coded cross-linkers. (a) MS spectrum of the chromatographic fraction of a tryptic digest of the multisubunit anaphase-promoting complex, crosslinked with a 1:1 molar ratio mixture of BS^3–H12/D12. The peptide containing the cross-linker appears in the spectrum as a doublet of peaks 12.07 Da apart (inset). The substitution of 12 deuterium atoms for hydrogen in the linker region of the heavy form of the cross-linker accounts for the observed mass difference. (b) Tandem MS/MS spectrum of the 1921.92 Da parent ion. The CID fragment containing the isotopically coded cross-linker y_4^2 still appears in the spectrum as a doublet of peaks separated by 12.07 Da. The doublet MS signature of isotopically coded cross-linkers facilitates detection and identification of the cross-link location.

interacting proteins are not known. Isotopically coded cross-linking reagents that are photo-cleavable under MALDI (Petrotchenko and Borchers, 2006) or cleavable at low-energy CID (Chowdhury et al., 2006) represent a recent extension of the concept of cleavable cross-linkers.

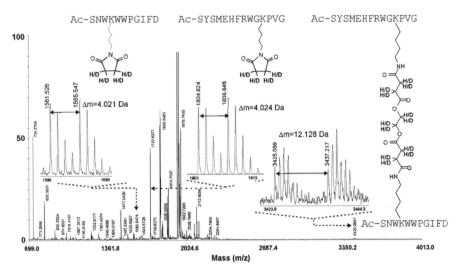

FIGURE 8.3 Detecting cross-linking using the isotopically coded cleavable cross-linking reagent EGS–H12/D12. The interpeptide cross-link with a mass of 3425 Da appears in the MS spectrum as a doublet of signals 12.07 Da apart. Chemical cleavage with ammonia leads to the formation of cleaved individual peptides with masses of 1581 and 1804 Da, which appear in the spectrum as a doublet separated by 4.03 Da. Cleavage of the isotopically coded reagent enables detection and further analysis of the individual cross-linked peptides. Copyright of American Society for Biochemistry and Molecular Biology, Inc. from Petrotchenko et al., 2005.

As discussed earlier, the low specific content of cross-linked peptides in typical digest mixtures makes detecting them via MS analytically challenging. However, the specific content of cross-linked peptides can be increased through affinity enrichment techniques. Conceptually, there are two ways this can be accomplished. First, traditional affinity tags such as biotin can be incorporated into cross-linker molecules. Subsequent affinity purification using an avidin chromatography support results in enrichment of cross-linked peptides. Second, as we recently illustrated, specific antilinker antibodies can be used for affinity enrichment of cross-linker-containing peptides (Petrotchenko et al., 2006). Both approaches eliminate the majority of interfering free peptides, thus reducing ion suppression effects and allowing MS analysis of otherwise undetectable cross-linked peptides (Fig. 8.4).

One area that has as yet not been extensively explored is the possibility of developing cross-linking reagents that have variable chemical reactivity. Such reagents would prove valuable because they would enable researchers to target specific amino acid residues for cross-linking. Overall, considerable progress has been made in developing cross-linking reagents tailored to the specific challenges associated with MS analysis of protein–protein interactions. The development of reagents combining several of the features already described would tremendously expand the potential of mass spectrometry as a tool for studying protein interactions through cross-linking analyses.

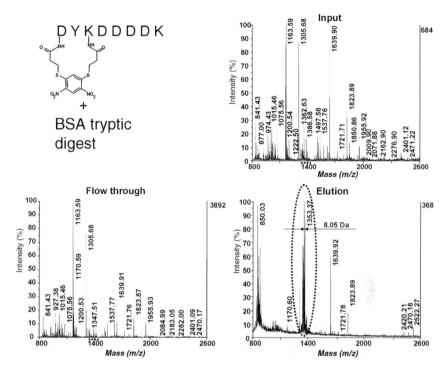

FIGURE 8.4 Cross-link affinity enrichment. An intrapeptide cross-link containing the isotopically coded reagent DNBDPS-H8/D8 was affinity purified using SPE-7 monoclonal antibodies immobilized on agarose beads. The antibodies were directed against the dinitrophenyl moiety of the cross-linker. Regions of spectra corresponding to the cross-link mass are highlighted by a dotted line. Although impossible to detect by MS in the complex peptide mixture (input), the cross-link appears as a dominant signal following affinity enrichment (elution).

8.7 ANALYTICAL SOFTWARE

Mass spectrometry generates extraordinarily large amounts of mass data, making the integration of specialized analysis software a must for cross-linking experiments. Fortunately, cross-linking experiments are readily adaptable to automation. Using the mass signature of isotopically coded cross-linkers, for example, all specific doublet peaks can be automatically detected by screening spectra for the mass difference between peaks corresponding to the isotopic mass increment. Compiled lists of the masses of cross-linker-containing peptides then can be used as a criterion for automatic acquisition of MS/MS spectra.

The most laborious portion of experiments combining cross-linking with mass spectrometry detection is the assignment of the cross-link position. From a practical standpoint, this is not feasible without the assistance of software tools. Our laboratory,

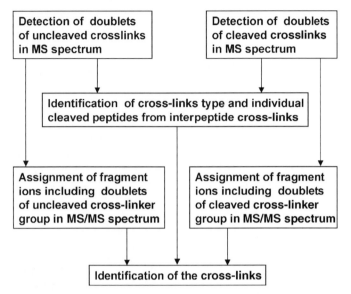

FIGURE 8.5 Data analysis program scheme, which incorporates isotopically coded cross-linker cleavage information. Combining MS and MS/MS data for both uncleaved and cleaved cross-links enables distinguishing between dead-end, intra-, and interpeptide cross-links and allows for ready assignment of fragment ions, thus providing confident cross-link identification.

as well as others, has developed custom programs for assigning isotopically coded cross-links employing MS/MS fragmentation information (Seebacher et al., 2006). The most prominent peaks within the MS/MS spectra can be used as a condition for filtering possible theoretical peptide combinations satisfying a given mass for the parent cross-linked ion.

We have also demonstrated that the principle for isotopically coded cleavable cross-linking data can be adequately incorporated into the powerful identification algorithm (Fig. 8.5). Indeed, symmetrical cleavage of the isotopically coded cross-linker produces halves of the cross-link, which are still coded as in the case of our cross-linker BiPS (Petrotchenko and Borchers, 2006) with half the number of differential isotopes, thus producing another mass signature for the cleaved peptides. Taking into account the chemistry of the cleavage reaction, cleaved peptides can then be matched by the program to their uncleaved ancestors. Based on the specific MS signature for cross-linked peptides after cross-linker cleavage, these peptides can also be automatically selected for MS/MS analysis again, and fragmentation information from both cross-linker cleavage and individual peptides can be used for high-confidence identification of the cross-link. To date, this method is the most advanced and confident algorithm for assignment of interaction sites derived from cross-linking experiments of large protein complexes.

8.8 USING CROSS-LINKING DISTANCE CONSTRAINTS TO BUILD EXPERIMENTAL MODELS OF PROTEIN COMPLEXES

The last, but certainly not the least important, step in experiments combining chemical cross-linking and mass spectrometry is the construction of models to represent the protein complexes under investigation. Several computational methodologies exist for creating three-dimensional ternary and quaternary structures of protein complexes based on peptide sequence data. All of these methodologies are critically dependent upon cross-linking data, since this allows the researcher to discriminate theoretically plausible variants from implausible variants of the final structure. The most valuable data relate to distance constraints between cross-linked sites. Several points related to distance constraints should be considered. First, cross-linking efficiency will mainly depend on both the span of the cross-linker employed and the conformational flexibility of the protein region involved in the cross-linking reaction. In general, longer cross-linking reagents are more efficient (i.e. reagents that can span greater distances), but flexibility of the linker portion of the reagent molecule should also be taken into account. It should be noted that by-products of most cross-linking reactions, such as dead-end and intrapeptide cross-links, may provide valuable information as well. In relation to modeling, it is always a question of how confident the conclusion is and how much biochemical sense can be extracted from the final predicted structure. In the case of cross-linking studies combined with the modeling of protein complexes, at least two aspects deserve consideration: details of the protein interaction interfaces and the general topology of the multisubunit protein complex. Altogether, recent improvements in mass spectrometry instrumentation, cross-linking reagents, and software tools have brought chemical cross-linking combined with mass spectrometry methods closer to the point of routine application for examining protein structure and molecular interactions within protein complexes.

REFERENCES

Back, J. W., Notenboom, V., de Koning, L. J., Muijsers, A. O., Sixma, T. K., de Koster, C. G., and de Jong, L., 2002. Identification of cross-linked peptides for protein interaction studies using mass spectrometry and 18O labeling. *Anal Chem* 74 (17), 4417–4422.

Chowdhury, S. M., Munske, G. R., Tang, X., and Bruce, J. E., 2006. Collisionally activated dissociation and electron-capture dissociation of several mass spectrometry-identifiable chemical cross-linkers. *Anal Chem* 78 (24), 8183–8193.

Mann, M., 1995. Useful tables of possible and probable peptide masses. In: *Proceedings of the 43rd ASMS Conference on Mass Spectrometry and Allied Topics*, May 21–26, Atlanta, GA, p. 639.

Müller, D. R., Schindler, P., Towbin, H., Wirth, U., Voshol, H., Hoving, S., and Steinmetz, M. O., 2001. Isotope-tagged cross-linking reagents. A new tool in mass spectrometric protein interaction analysis. *Anal Chem* 73 (9), 1927–1934.

Petrotchenko, E. V. and Borchers, C. H., 2006. A novel fluorescent isotopically-coded MALDI-cleavable crosslinker, BiPS. In: *Proceedings of the 54th ASMS Conference on Mass Spectrometry and Allied Topics*, May 28–June 1, Seattle, WA.

Petrotchenko, E. V., Olkhovik, V. K., and Borchers, C. H., 2005. Isotopically coded cleavable cross-linker for studying protein–protein interaction and protein complexes. *Mol Cell Proteomics* 4 (8), 1167–1179.

Petrotchenko, E. V., Doant, T., and Borchers, C. H., 2006. A novel chromophoric affinity-tagged isotopically-coded crosslinker, DGDNBS. In: *Proceedings of the 54th ASMS Conference on Mass Spectrometry and Allied Topics*, May 28–June 1, Seattle, WA.

Seebacher, J., Mallick, P., Zhang, N., Eddes, J. S., Aebersold, R., and Gelb, M. H., 2006. Protein cross-linking analysis using mass spectrometry, isotope-coded cross-linkers, and integrated computational data processing. *J Proteome Res* 5 (9), 2270–2282.

Tang, X., Munske, G. R., Siems, W. F., and Bruce, J. E., 2005. Mass spectrometry identifiable cross-linking strategy for studying protein–protein interactions. *Anal Chem* 77 (1), 311–318.

CHAPTER 9

Complex Formation in the Actin Cytoskeleton: Cross-linking Tools to Define Actin Protein Structure and Interactions

SABRINA BENCHAAR and EMIL REISLER

Department of Chemistry and Biochemistry, Molecular Biology Institute, University of California, Los Angeles, CA, USA

9.1 INTRODUCTION

Actin, one of the most abundant proteins in eukaryotic cells, has highly conserved sequence, differing by no more than 5% in taxa as diverse as algae and human. Actin filaments (F-actin) provide mechanical support for cells, drive cell shape changes and diverse motile processes, and participate in cytokinesis, intracellular transport, and force generation processes. The monomer of actin (G-actin) has two major domains, "large" and "small," which, in turn, are divided into two subdomains each, subdomains 1 and 2 in the small domain and 3 and 4 in the large domain. Actin resembles a horseshoe structure with subdomains 2 and 4 located at the opening of the horseshoe, and with the nucleotide cleft in the middle (Fig. 9.1). To carry out its multiple cellular functions, monomeric actin must be polymerized into double-stranded helical filaments. A structural model of these filaments was constructed in 1990 from X-ray diffraction data obtained from oriented actin fibers and the atomic structure of G-actin (Holmes et al., 1990). Recently, the F-actin model has also been constructed on the basis of an atomic structure of a cross-linked dimer (Kudryashov et al., 2005).

To gain insight into the mechanism of actin interactions with its many protein partners in the cell, a detailed structure of actin complexes and its binding interfaces with actin binding proteins is required. This goal has been achieved in several cases: for actin complexes with actin depolymerizing proteins (Kabsch et al., 1990;

Mass Spectrometry Analysis for Protein–Protein Interactions and Dynamics, Edited by Mark Chance
Copyright © 2008 John Wiley & Sons, Inc.

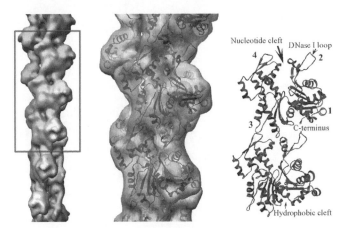

FIGURE 9.1   EM images of frozen-hydrated rabbit muscle F-actin were used in IHRSR procedure (Egelman, 2000) to generate a 3D reconstruction of F-actin at 12 Å resolution (left). A portion of this volume is shown with the G-actin crystal structure fitted into the density map (middle), followed by a more detailed interface between two adjacent protomers along the same long-pitch strand (right). (The figure is courtesy of Vitold E. Galkin, from Galkin et al., 2008). Numbers 1–4 identify subdomains of actin. Black arrows indicate the C-terminus of actin and the hydrophobic and nucleotide binding clefts. The DNase I binding loop is also indicated.

McLaughlin et al., 1993; Schutt et al., 1993), drugs (Morton et al., 2000; Klenchin et al., 2003), and nucleotides (Otterbein et al., 2001; Graceffa and Dominguez, 2003). However, with one exception (Otomo et al., 2005), actin crystals could not be obtained yet with proteins that support filament nucleation, such as cofilin, myosin subfragment 1, gelsolin, and so on. Thus, structural models for several actin complexes have been built using a variety of techniques, and most of all, image reconstruction of electon micrographs of such complexes (Kabsch et al., 1990; Milligan et al., 1990; Hodgkinson et al., 1997; McGough et al., 1997, 1998; Hanein et al., 1998; Sherman et al., 1999; Rosol et al., 2000; Orlova et al., 2001; Volkmann et al., 2001; Holmes et al., 2003; Lilic et al., 2003; Pant et al., 2006). However, the low resolution of EM data precludes in most cases a detailed atomic description of the interacting residues in the protein complex.

Important progress in mapping contact sites in protein complex has been achieved with the development of synchrotron radiation/mass spectrometry tools for probing surface-exposed amino acid residues, as described in Chapter 10. This method has been used successfully for several complexes of actin with actin binding proteins. An alternative and complementary approach to obtaining structural information on actin interaction with ABPs is the use of chemical cross-linking methods and the mapping of the cross-linked amino acids. The cross-linking can provide important distance constraints for structural models of protein complexes. Mass spectrometry is the method of choice for identifying the cross-linked residues in the large mixture of peptides obtained from proteolytic digestion of the cross-linked proteins. This chapter discusses how cross-linking is used as a tool to probe the actin structure and its interactions with actin binding proteins. We will focus on several examples of actin complexes with cytoskeletal proteins and consider some approaches that facilitate the mapping of cross-linked peptides by mass spectrometry.

9.2 MAPPING CROSS-LINKING WITH METHODS OTHER THAN MASS SPECTROMETRY

In the early 1980s, prior to recent developments in mass spectometry methods, Sutoh and coworkers investigated extensively the cross-linking of actin with actin binding proteins by 1-ethyl-3-[3-(dimethylamino) propyl]-carbodiimide (EDC). After the cross-linking, the binding sites of the ABPs (DNAse I, gelsolin, fragmin, depactin, cofilin, myosin) (Sutoh, 1982a, 1982b, 1983, 1984; Sutoh and Mabuchi, 1984, 1986; Sutoh and Hatano, 1986; Sutoh and Yin, 1989) on actin were mapped by partial cleavage of the cross-linked complexes with reagents (hydroxylamine, BNPS-skatole, cyanogen bromide, NTCB) that yielded predictable arrays of peptides (Sutoh, 1984; Sutoh and Hatano, 1986). In all of the above cases, only regions (peptides), but no amino acids involved in the cross-linking, could be assigned. The analysis of the resulting peptide ladder on 1D gels was assisted by either fluorescent labels attached to actin C-terminus (Sutoh and Hatano, 1986) or antibodies to actin's C- and N-termini (Sutoh and Mabuchi, 1984, 1986; Sutoh and Yin, 1989). In a similar fashion, the cross-linked site of actin on depactin was mapped using antibodies to depactin's N- and C-termini. The relative effectivness of these classical protein chemistry methods enabled Onishi et al. (1990) to identify the residues involved in EDC cross-linking of gizzard HMM, when bound to F-actin under rigorous binding conditions. To this end, these authors (Onishi et al., 1990) carried out Edman degradation of HPLC-fractionated proteolytic peptides of the cross-linked actin–HMM. These early mappings of actin cross-linking to several proteins provided the first information on the likely contact sites between actin and the examined proteins and advanced significantly the structural understanding of actin complexes.

9.3 ACTIN-ACTIN CROSS-LINKING

In 2005 Reisler et al. (Kudryashov et al., 2005) reported a 2.5-Å resolution crystal structure for a longitudinal, cross-linked dimer. Up to then, the constraints and the tests of the structural model of F-actin proposed by Holmes et al. (1990) were provided mainly by actin cross-linking data.

9.3.1 Intermolecular Cross-linking in F-Actin by *N,N'-p*-Phenylene-Dimaleimide between Lysine 191 and Cysteine 374

In the mid-1980s, prior to the solution of the 3D structure of monomeric actin, important structural information on F-actin was provided by Elzinga and Phelan (1984). These authors mapped the previously described (Knight and Offer, 1978) cross-linking

(a)   (b)

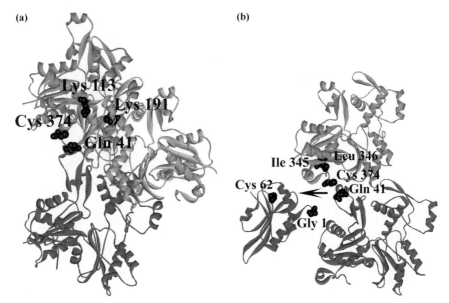

FIGURE 9.2 (a) Three protomers in the molecular model of F-actin (Holmes et al., 2003). The protomers in cyan and pink form a longitudinal contact along the long-pitch helix, while the protomer in yellow belongs to the second strand. ANP, ABP, and pPDM cross-linked residues are Gln41–Cys374, Gln41–Lys113, Cys374–Lys191, respectively. (b) Schematic representation of cofilin-F-actin binding. Yeast cofilin is positioned between two adjacent protomers and then translated out of the interface to visualize the indicated residues (including Gly1 on cofilin and Gln41 on actin that are cross-linked by TGase). (See the color version of this figure in Color Plates section.)

of F-actin by the bifunctional reagent N-N'-p-phenylene-dimaleimide (pPDM) to Cys374 and Lys 191 on an adjacent monomer (Fig. 9.2a, Table 9.1). Although in the absence of actin's atomic structure the cross-linking data could not predict the filament organization, these results were instrumental in modeling the arrangement of protomers in F-actin structure once the G-actin structure was solved (Holmes et al., 1990). To be consistent with the pPDM cross-linked dimer results, the model of F-actin had to position the cross-linked Lys191 and Cys374 across the central axis of the polymer (along the genetic helix), within a maximal distance of ∼21 Å between their Cα atoms. Thus, in the original (Holmes et al., 1990) and subsequently refined (Lorenz et al., 1993) models of F-actin, Cys374 and Lys191 were, respectively, 19 Å (Cα 374–Cα 191) and 16 Å (S 374–N 191) apart, obeying the cross-linking constraint on the filament structure.

9.3.2 Intermolecular Cross-linking in F-Actin by *N*-(4-Azidobenzoyl)-Putrescine between Glutamine 41 and Lysine 113

The publication of the structural model of F-actin stimulated several tests of the predicted interprotomer interfaces and protomer packing in the filament. One of such

TABLE 9.1   List of the cross-linking reagents described herein, their structures and reported lengths.

Cross-linking reagent		Range of S—S distances (Å)
N-(4-azidobenzoyl)-putrescine (ABP)		10.7
N-(4-azido-2-nitrophenyl)-putrescine (ANP)		11.1–12.5
Dithio-bis-maleimdoethane (DTME)		6.68–16.12[1]
N,N'-1,2-phenylene-dimaleimide (pPDM)		9.20–12.29[1]
Benzophenone,4-(N-iodoacetamido)-4'-(N-biotinylamido)		
Benzophenone,4-(N-maleimido)-4'-(N-biotinylamido)		

[1]Average distances according to Green *et al.* [74].

studies was carried out by Hegyi et al. (1992), who synthesized a bifuntional reagent N-(4-azidobenzoyl)-putrescine (ABP) (Fig. 9.2a, Table 9.1), which could be coupled to Gln 41 on G-actin via transglutaminase (TGase) reaction. This approach was based on previous work of Gorman and Folk (1980), who described TGase-mediated coupling of an amino moiety (from aryl azido compounds) to a carboxyl group on an accessible glutamine side chain, and the work of Takashi (1988), who coupled dansyl-cadaverine to Gln 41 on G-actin. Following the protocol of Takashi, Hegyi et al. (1992) first coupled (H^3)ABP through its putrescine moiety to Gln 41 and then polymerized the labeled G-actin and photoactivated the reagent to cross-link actin via the azido group. The cross-linked products were separated and succinylated to limit the subsequent trypsin digestion to arginines only. The resulting tryptic digests were separated chromatographically, with the ABP-containing peptides identified by monitoring radioactive $(H^3)ABP$. Amino acid analysis of the isolated peptides revealed cross-linking between Gln41 and Lys113. The same peptides were analyzed also employing a microcapillary HPLC column and ESI triple quadrupole mass spectrometer. The fragmentation of the cross-linked peptide confirmed sequencing results, which were consistent with the Holmes model of F-actin. In this model, (Holmes et al., 1990) the DNase I binding loop (with its Gln41) contacts subdomain 1 (which contains Lys113) in the protomer above it, and Gln41 and Lys113 are within 10–20 Å from each other along the two-start helices (Fig. 9.2a). The size of the cross-linking reagent, estimated at 10.7 Å, would confine the Cα atoms of Gln41 and Lys113 to within a maximum distance of 22.3 Å, which can be accommodated within the Holmes model.

9.4 INTRASTRAND CROSS-LINKED ACTIN BETWEEN GLN41 AND CYS374

Limitations of reactivity and detection prompted Hegyi et al. (1998) to synthesize an improved version of ABP, the N-(4-azido-2-nitrophenyl)-putrescine (ANP) (Table 9.1), which could be used in the same manner as ABP. The efficiency of photoactivation was improved in ANP, as well as the detection method, which used the chromophore of the nitrophenyl instead of radioactivity. The cross-linked sites in F-actin were mapped using a strategy similar to that employed with ABP. A combination of gel filtration and reverse-phase HPLC allowed the isolation of tryptic peptides. The peaks of interest were N-terminally sequenced. A hybrid tandem mass spectometer was used following electrospray sample introduction. The mass of the peptide was determined, but the cross-linked peptide was not fragmented. The 11.1–12.5 Å length cross-linker was found to link the γ-carboxyl group of Gln41 with the sulfur atom of Cys374 (Fig. 9.2a). In the Holmes and Lorenz models of F-actin, the distance between the Cα atoms of these residues is 12 and 20.7 Å, respectively.

The mapping of Gln41–Cys374 cross-linking in F-actin can serve as a good example of the important value of such information to the overall analysis of filament structure, dynamics, and function. The two structural elements of actin linked by ANP, the DNase I binding loop and the C-terminus, are among the most mobile and flexible parts of the actin molecule (Fig. 9.1). They confer, most likely, dynamic character on the subdomain 1 (upper protomer)–subdomain 2 (lower protomer) interprotomer interface along the filament axis. Thus, the covalent bridging (by ANP) of adjacent protomers paved the way for testing the effect of restricted interprotomer mobility on actin's function. Gerson et al. (2001) used this approach to rule out the hypothesis that motions of subdomain 2 are instrumental in regulation of actin filaments by troponin–tropomyosin. These authors showed similar regulation of uncross-linked and ANP cross-linked F-actin. On the other hand, the same Gln41–Cys374 cross-linking impaired strongly the generation of force and motion by the myosin motor, although most of the kinetic and equilibrium parameters of actomyosin interaction were unchanged by this cross-linking (Kim et al., 1998, 2002). This intriguing finding documented the role of dynamic transitions in F-actin, which were arrested by the cross-linking, in coupling the catalytic events on myosin to the power stroke in actomyosin. However, mechanistic understanding of why the "loss" of mobility in F-actin uncouples mechanical from enzymatic steps in the force/motion generating cycle of myosin is still missing.

Importantly, Gln41–Cys374 cross-linking in F-actin and the resulting stabilization of longitudinal interactions paved the way for another advance in the actin field. This reaction enabled, after F-actin depolymerization, the purification of a cross-linked dimer and its subsequent crystallization and atomic structure determination (Kudryashov et al., 2005). The dimer structure provided the first atomic resolution map of subdomain 3 (upper protomer)–subdomain 4 (lower protomer) contacts and focused attention on their dominant role in the longitudinal dimer. The dynamic and more variable mode of subdomain 1 and subdomain 2 contacts was already deduced from EM studies (Orlova et al., 2001) and the cross-linking studies with yeast actin mutant of Gln41–Cys41 (QC actin). The QC mutant revealed that these residues could be connected even by disulfide bond, thus showing the range of motions of the DNase I binding loop (Orlova et al., 2001).

9.5 REGULATION OF CYTOSKELETON BY ABPs AND MAPPING THEIR INTERFACE WITH ACTIN BY CROSS-LINKING

9.5.1 Actin-Depolymerizing Factor/Cofilin

Extracellular signals often induce the remodeling of cytoskeleton in living cells, which requires rapid changes in the pools of monomeric and polymeric actin. The correct spatial and temporal organization of actin filaments is orchestrated by a number of actin binding proteins, including those involved in severing, capping, and sequestering activities. Among these, the ADF/cofilin family of proteins is a major mediator of the rapid turnover of actin filaments. In a biochemical analysis of ADF, using an *in vitro* reconstructed system of *Listeria* bacterium propulsion, Carlier et al. (Loisel et al., 1999) demonstrated that ADF/cofilin increases the rate of actin treadmilling by increasing the rate of F-actin depolymerization, thereby replenishing the pool of monomeric actin available for new filament formation. The depolymerized actin is then "transfered" to two proteins, profilin and thymosin β4, which maintain the pool of G-actin in the cell.

The severing activity of ADF/cofilin has been linked to structural changes induced by this protein in F-actin. The changes—documented by electron microscopy and image reconstruction methods (McGough et al., 1997; Galkin et al., 2001)—involve the weakening of longitudinal and lateral interprotomer contacts, and the resulting decrease in the filament twist. Attempts to map the actin–cofilin interface by crystallizing this complex have failed so far, most likely because of cofilin's ability to bind two actin molecules (McGough et al., 1997) and nucleate filament formation. In the absence of crystals, the probing of the binding interface between actin and cofilin has been done by molecular dynamics simulations, chemical cross-linking, systematic mutagenesis, and synchrotron radiolysis analysis methods (Yonezawa et al., 1991; Hatanaka et al., 1996; Lappalainen et al., 1997; Wriggers et al., 1998; Guan et al., 2002).

9.5.2 Mapping the Interaction of Cofilin with Subdomain 2 on G-Actin

It was suggested from computational analysis that cofilin binds G-actin between subdomain 1/3 regions, similar to gelsolin segment 1 (Hatanaka et al., 1996; Wriggers et al., 1998). This view was supported later by a hypothesis suggesting a common pattern of G-actin binding for several actin depolymerizing factors, including gelsolin, vitamin D binding protein, and Wiskott–Aldrich syndrome protein (WASP)-homology domain-2-related proteins (Dominguez, 2004). In the filament, as cryo-electron microscopy (McGough et al., 1997) suggested, cofilin appears intercalated between two actin protomers along the filament axis, at the cleft between subdomains 1 and 3 of the upper protomer and at subdomain 2 of the lower protomer. This raised the possibility that a secondary, weak binding site for cofilin might exist in subdomain 2.

We examined this possibility with a transglutaminase-mediated cross-linking reaction, targeting Gln41 on the DNase I binding loop. Actin and cofilin were cross-linked with high efficiency by TGase, forming a heterodimer that could be separated by size exclusion chromatography. LC MS/MS analysis of the proteolytic fragments of this complex mapped the zero-length cross-linking to Gln41 on actin and Gly1 on yeast cofilin (Fig. 9.2b). A unique peptide that was not present in the uncross-linked mixture was spotted. Product ions arising from the peptide backbone cleavage are shown in Fig. 9.3.

The MS/MS fragmentation spectrum was analyzed using the MS2assign automatic structure assignment program (ASAP) (Schilling et al., 2003). The identification of the cross-linked residue on the N-terminus of cofilin was rendered difficult because it does not contain a lysine, which could have provided an ε-amino group for the TGase reaction. To check possible cross-linking targets, the N-terminal peptide of cofilin was purified by HPLC and subjected to ladder sequencing, according to a protocol adapted from Chait et al. (1993). After two cycles of manually performed ladder-generating chemistry, a MALDI-TOF-MS spectrum of the mixture showed two peaks differing by 57 Da, the mass of a glycine residue. The difference between the next two peaks in the ladder sequencing was 87, corresponding to serine. Thus, the amino terminus was a nonacetylated glycine, revealing that Gly1 participated in cross-linking to Gln41 on actin.

In the above case, the cross-linking revealed a secondary, cryptic binding site for cofilin in subdomain 2 of G-actin, in addition to the main, high-affinity binding site. In general, the detection and mapping of weakly bound protein complexes can be facilitated through a progressive accumulation of products of cross-linking reactions. However, caution is needed in data interpretation since nonspecific cross-linking may

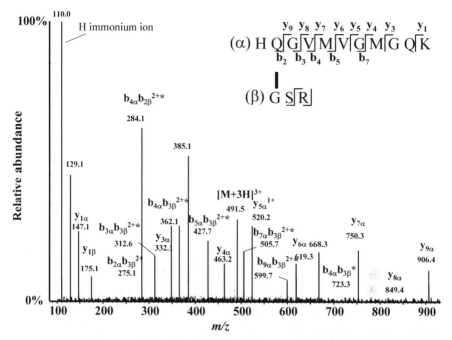

FIGURE 9.3   MS sequencing of the transglutaminase cross-linked actin–cofilin peptides. MS/MS spectrum of the $[M + 3H]^{3+}$ of the transglutaminase cross-linked tryptic peptides at m/z 491.5. Peptide (α) is from G-actin and peptide (β) is from yeast cofilin. Singly and doubly charged y-type product ions were generated from dissociation of the 3^+-charged precursor ion. (Peptide fragments were denoted following the nomenclature for fragmentation of cross-linked oligopeptides (Roepstorff and Fohlman, 1984; Biemann, 1988, 1990). Lack of a superscript denotes a singly charged fragment ion). The asterisk refers to a $+18$-Da product. Reprinted from Benchaar et al. (2007). Copyright 2008 American Chemical Society.

capture collisional complexes of proteins. Thus, cross-linking rates, yield, and product accumulation must be significant to indicate the presence of a specific complex. However, the lack of cross-linking does not prove the complex is not formed since reactivity/availability/orientation of residues and the cross-linker may preclude the coupling reaction.

Our cross-linking of Gln41 on actin to Gly1 on cofilin is consistent with the mode of cofilin binding to F-actin. The N-terminus of cofilin has been previously shown to contribute to actin binding (Morgan et al., 1993; Agnew et al., 1995). This region of cofilin contains a phosphorylation site (Ser3 in yeast cofilin), which serves as an on/off switch of actin binding to cofilin. Moreover, most recently, Mannherz and coworkers (Mannherz et al., 2007) published results in agreement with our findings. These authors used peptide arrays to identify a secondary binding site for cofilin on G-actin, which they mapped to subdomain 2 and the DNase I binding loop, where Gln41 resides.

9.5.3 Cofilin-Induced Switch from Intramolecular to Intermolecular Cross-linking in Skeletal F-Actin

The structural mechanism by which ADF/cofilin dissociates or fragments actin filament remained unknown until electron microscopy studies shed light on one of the unique features of ADF/cofilin. ADF-decorated filaments showed a substantial decrease in their helical twist (5° per subunit) and a reduced filament crossover (although the axial rise per subunit remained constant). Consequently, lateral and longitudinal interfaces in F-actin were changed and weakened by cofilin (McGough and Chiu, 1999; Galkin et al., 2001). The structural elements that form these interfaces, namely, the DNase I binding loop, the C-terminus of actin, and the hydrophobic loop, were all affected by cofilin binding to F-actin (Bobkov et al., 2002, 2004). EM studies suggested that the reorganization of these contacts leads to the formation of tilted filament structures in which protomers are rotated, forming new interprotomer contacts (Orlova et al., 2004). It would be attractive to capture and document such structural shifts via cross-linking methods.

In the mid-1980s Tao et al. (1985) labeled G-actin with the photo-cross-linker benzophenone-4-maleimide (BP) (Table 9.1). BP contains a maleimide group that reacts with Cys374 on actin and a benzophenone moiety, which upon photoactivation can react with a vicinal residue (Dorman and Prestwich, 1994). In G-actin, the BP reaction produced only a small amount of intrasubunit cross-linking. Upon polymerization, BP cross-linking yielded a different result, producing a large amount of internally cross-linked actin and a small amount of cross-linked dimer. The main conclusion drawn from these experiments was that Cys374 environment was changed upon polymerization, indicating the importance of the C-terminus to actin polymerization and intermolecular contacts in F-actin. In our hands, BP maleimide cross-linked both F- and G-actin as described by Tao et al. (1985). Interestingly, the BP reaction with F-actin saturated with cofilin was shifted from intraprotomer to interprotomer cross-linking, reflecting cofilin-induced structural effect on the actin filament (unpublished results).

9.5.4 The Main Cofilin Binding Site on G-Actin

In a recent report, Dominguez (2004) compared the mode of binding of several actin binding proteins and drugs for which crystal structures were solved and proposed a common motif for binding to actin. Kabiramide C, jaspisamide (two marine toxines), gelsolin segment 1, vitamin D binding protein, and ciboulot (a WASP-homology domain-2-related protein) were all shown to bind to actin at the hydrophobic cleft between subdomains 1 and 3. Strikingly, these proteins, although structurally very different from each other, bind actin in a similar fashion through a hydrophobic α-helix. Dominguez hypothesized that, in an analogy to these proteins, cofilin inserts its α3-helix into the hydrophobic cleft between subdomains 1 and 3. This hypothesis offers an attractive mechanism by which cofilin could depolymerize actin: it would displace adjacent actin when competing with it for the binding to the hydrophobic cleft.

An alternative site of cofilin binding to G-actin between subdomains 1 and 2 was identified recently through hydroxyl radical footprinting and computational modeling (Chapter 10). To reexamine cofilin binding to specific region on actin, cofilin and actin mutants need to be generated for cross-linking tests. In a preliminary study, we probed the proximity of cofilin to the hydrophobic cleft between subdomains 1 and 3 using yeast actin mutants with cysteine substitutions for Ile345 and Leu346 (Fig. 9.2b) and Cys374 replaced by alanine. p-Phenylene dimaleimide (pPDM) cross-linked both actin mutants to cofilin with high efficiency. Cross-linked actin cofilin was FPLC purified and digested. The detection of the hydrophobic cross-linked peptide and its sequencing turned out to be challenging but could be simplified with the use of a cleavable cross-linking reagent, dithio-bis-maleimidoethane (DTME). This reagent is a good substitute for pPDM (Table 9.1). After cross-linking and purification, the complex could be reduced and then separated into labeled actin and cofilin. The reduced DTME can be traced to specific tryptic peptides either through an increase in their mass or by attaching a fluorophore and retrieving the labeled peptide by RP-HPLC. LC MS/MS sequencing of TME–cofilin peptide mixture identified the peptide and then the residue on cofilin that is coupled to the reduced cross-linker as Cys 62 (Grintsevich et al., 2008). Top-down approach confirmed the LC MS/MS sequencing results and thus turns out to be an attractive mapping method for cofilin, not requiring prior protein digestion because of the relative small size of cofilin (15 kDa). Clearly, additional cofilin–actin cross-linking sites will need to be mapped to test fully the proposed models of cofilin binding to G-actin.

9.5.4.1 *Profilin*   ADF/cofilin increases the pool of monomeric ADP–actin in the cell, which is not readily used for polymerization at the barbed filament ends. The solution to cellular needs is provided by profilin, which works in synergy with ADF/cofilin to accelerate the conversion of ADP-G-actin to the polymerization-ready ATP-G-actin. When actin filaments are capped, profilin sequesters G-actin and causes depolymerization. Together with β-thymosin and ADF/cofilin, profilin allows cells to maintain a large pool of ATP-G-actin, buffering the high rates of filament elongation.

In the late 1980s, Pollard et al. (Vandekerckhove et al., 1989) used chemical cross-linking to determine contact sites between actin and profilin. Residue Glu364 on actin was readily cross-linked to *Acanthamoeba* profilin Lys115 by zero-length cross-linker, EDC. Unexpectedly, the cross-linked actin–profilin did not cap the barbed ends of filaments but incorporated instead into filaments that retained the same structural parameters and intrinsic properties (phalloidin/S1 binding, nucleotide exchange) as naked F-actin (Gutsche-Perelroizen et al., 1999). A possible explanation for this counterintuitive result would be that the tethered profilin swings out because its affinity to actin decreases upon polymerization. Profilin would then dissociate from the barbed end, allowing for another profilin–actin complex to assemble. This view is not shared by Lindberg et al. (Nyman et al., 2002), who argue that ATP hydrolysis occurs only after profilin comes off because of its ability to inhibit the actin ATPase. These authors cross-linked profilin to β/γ actin with EDC (at residues Lys113 and Glu82 on actin and profilin, respectively) to explore the growth of filament from cross-linked profilin–actin. This cross-linked complex did not polymerize and even interfered with the elongation of filaments. This observation highlights an important fact: ATP and ADP actin structures are sufficiently different to modulate affinity of ABP to actin. Actin growth from profilin–actin is coordinated with the ATP–ADP transition and profilin detachment prior or after ATP hydrolysis.

9.5.4.2 β-Thymosin

Thymosin β4 is a 43-amino acid peptide that together with profilin maintains a pool of G-actin available for rapid filament growth. Thymosin β4 sequesters ATP-G-actin (binding to ATP-actin with a higher affinity) and inhibits actin nucleation or growth at both filament ends, as well as nucleotide exchange. In this way, thymosin β4 provides a reserve of actin that is readily transfered to profilin, since the latter has a higher affinity for actin. However, thymosin β4 buffering system is not essential to all cells since yeast functions without it.

In the late 1990s, the structure of actin–thymosin β4 appeared elusive, although it was of great interest to understand how such a short peptide could block the polymerization and nucleotide exchange on actin. While it was believed that in solution thymosin β4 is disordered, its binding to actin was stipulated to stabilize one of its conformations. To map which sites on actin are in contact with thymosin β4, Reichert et al. (1996) anchored thiol-specific bifunctional reagents of various lengths (alkylene-bis-[5-dithio-(2-nitrobenzoic acid)] and Ellman reagent) to Cys374, Cys10, and ATPγS on actin. Actin was then cross-linked to thymosin β4, in which point mutation to cysteins had been introduced. Upon conjugation, the linker released 2-nitro-5-thiobenzoate group (detectable at 412 nm), which enabled the measurements of the rate and extent of cross-linking. Reichert et al. concluded that the N-terminus of thymosin β4 was within 9.2 Å from Cys374. Similarly, Cys17 and 28 of thymosin β4 were within 9.2 Å from the actin-bound ATPγS nucleotide. In a subsequent study, Safer et al. (1997) cross-linked thymosin β4 to actin with carbodiimide and transglutaminase. Cross-linked residues were mapped with MALDI TOF and electrospray ionization combined with peptide sequencing. Three sites of thymosin β4 cross-linking were indentified: Glu167 in the hydrophobic pocket, the N-terminus of actin, and Gln41 in the DNase I binding loop. On the basis of these cross-linking results, a model of the actin–thymosin β4 complex was built. According to this model, thymosin β4 adopts extended conformation on actin, wrapping around it from the subdomain 3 region, via the N-terminus in subdomain 1 and over to the DNase I binding loop in subdomain 2. The crystal structure of the actin-bound thymosin β4 was solved recently Irobi et al. (2004). To stabilize the complex, the authors prepared a chimera construct composed of the first domain of gelsolin and the C-terminal half of thymosin β4. Gelsolin segment 1 and actin interact through a sequence highly homologous to a WH2 motif that is present in thymosin β4. The 2 Å resolution structure was in agreement with the cross-linking data, with thymosin β4 capping both ends of the monomer by extending and wrapping across the subdomain 1 and 2 regions (Fig. 9.4).

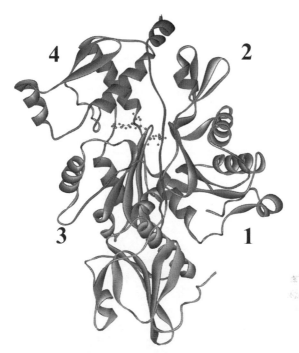

FIGURE 9.4 The structure of gelsolin segment 1–thymosin β4–actin complex Irobi et al. (2004). The actin monomer is depicted in grey, with the bound ATP in purple and the four subdomains labeled 1–4. The N-terminus of the chimera protein composed of the gelsolin segment 1 half (shown in cyan) caps actin at the barbed end between subdomains 1 and 3, while the thymosin β4 half (shown in pink) wraps around actin, intercalating the C-terminus between subdomains 2 and 4. (See the color version of this figure in Color Plates section.)

9.6 CROSS-LINKING OF ACTIN AND MUSCLE PROTEINS—EXAMPLES OF EXPERIMENTAL APPROACHES

A frequent challenge in the application of MS to the analysis of cross-linked peptides is the detection of such peptides in the total mixture of proteolytic fragments of proteins and the enrichment of fractionated samples with the cross-linked products. Examples of three different approaches to this issue are given in the studies of actin and muscle protein interactions.

In a recent report, Luo et al. (2002) have investigated the relationship between troponin inhibitory subunit (TnI), tropomyosin (Tm), and actin. The authors used benzophenone maleimide to cross-link TnI to actin. The product of the reaction was excised from the gel and trypsinized. The cross-linked peptides were separated by reverse-phase HPLC and analyzed by MALDI MS. The authors mapped the cross-linking to Met47 on actin and Cys104 and Cys133 on TnI. In this study, the authors have used the absorbance properties of benzophenone maleimide at 260 nm to identify the cross-linked peptide, but this may not work for all peptides since aromatic amino acids also absorb at this wavelength. To circumvent the detection problem, actin can be labeled with fluorescent probes, but the labeling needs to be done close to the cross-linking site so that the tag remains attached to the peptide after the enzymatic digestion.

Smooth muscle myosins are regulated by the phosphorylation of a regulatory light chain (RLC). In a study aimed at providing structural information about the regulatory domain of smooth muscle heavy meromyosin, Cremo and coworkers (Wahlstrom et al., 2003) focused on cross-linking the RLCs since they contain the serine that is phosphorylated upon regulation. To enrich cross-linked peptides Wahlstrom et al. synthesized a modified version of benzophenone. The sulfhydryl-reactive photo-cross-linkers had a biotin affinity tag attached to the benzophenone maleimide or iodoacetimide (Table 9.1). This allowed the authors to collect the linker-containing peptides with an avidin affinity chromatography column. Because of the high affinity of biotin for avidin, eluting peptides from the column can pose a problem. Cremo and coworkers eluted the biotinylated peptides with formic acid (up to 88%), as the buffers recommended by the manufacturers did not work. Purified, cross-linked peptides were characterized by Edman sequencing and MALDI-MS. The information obtained from the cross-linked peptides and previous cross-linking work done in the same laboratory was used by the authors to build a model of the regulatory domain of myosin. Moreover, mass spectromety analysis of the cross-linked peptides brought new insight into the benzophenone photochemistry: a loss of water was observed upon cross-linking, previously seen only when methionine or glycine was involved in the reaction.

In yet another study, Sinz and Wang (2001) used chemical modification of proteins to generate cross-linking sites for the mapping of molecular interfaces between calmodulin and a fragment of the C-terminus of nebulin. Nebulin, which is regulated by Ca^{2+}-calmodulin, is a very large protein found in the I band of skeletal muscle. Since both proteins are cysteins free, the authors thiolated them with 2-iminothiolane, which targets primary amines to form mercaptobutyramidines. This modification leaves the protein charge unchanged, but opens it to thiol-specific cross-linking by dibromobimane (DBB). The DBB chromophore was then exploited in HPLC fractionation to obtain samples enriched with cross-linked peptides, which were then mapped by MALDI MS and nano-ESI MS/MS. Mapping of the molecular interface between calmodulin and nebulin revealed that actin and calmodulin binding sites on nebulin were overlapping, suggesting that these two proteins might compete with each other, whereas the mapping of intramolecular cross-linking in calmodulin provided information on its conformation in solution.

9.7 CONCLUDING REMARKS

We have shown that chemical cross-linking in conjunction with mass spectrometry provides structural details on molecular interfaces in protein complexes when crystallographic data are not available. Moreover, cross-linking reactions can explore dynamic protein conformations that exist in solution, even when a 3D structure has been solved, since in most cases the crystal structure locks the protein or the complex in a specific conformation. This is illustrated by the various intermolecular cross-linkings documented for F-actin.

Finally, the cross-linking approach can also lead to a detection of weakly bound protein complexes, as well as changes in the structure of a protein due to its binding to various ligands, cofactor and proteins.

ACKNOWLEDGMENT

This work was supported by grants from USPHS (GM-077190) and NSF (MCB-0316269).

REFERENCES

Agnew, B. J., Minamide, L. S., and Bamburg, J. R., 1995. Reactivation of phosphorylated actin depolymerizing factor and identification of the regulatory site. *J Biol Chem* 270, 17582–17587.

Benchaar, S. A., Xie, S., Phillips, M., Ogarzalek Loo, R., Galkin, V. E., Orlova, A., Thevi, M., Muhlrad, A., Almo, S. C., Loo, J. A., Egelman, E. H., and Reisler, E. 2007. Mapping the interaction of cofilin with subdomain 2 on actin. *Biochemistry* 46, 225–233.

Biemann, K., 1988. Contributions of mass spectrometry to peptide and protein structure. *Biol Mass Spectrom* 16, 99–111.

Biemann, K., 1990. Nomenclature for peptide fragment ions (positive ions). *Meth Enzymol* 193, 886–887.

Bobkov, A. A., Muhlrad, A., Kokabi, K., Vorobiev, S., Almo, S. C., and Reisler, E., 2002. Structural effects of cofilin on longitudinal contacts in F-actin. *J Mol Biol* 323, 739–750.

Bobkov, A. A., Muhlrad, A., Shvetsov, A., Benchaar, S., Scoville, D., Almo, S. C., and Reisler, E., 2004. Cofilin (ADF) affects lateral contacts in F-actin. *J Mol Biol* 337, 93–104.

Chait, B. T., Wang, R., Beavis, R. C., and Kent, S. B. H., 1993. Protein ladder sequencing. *Science* 262, 89–92.

Dominguez, R., 2004. Actin-binding proteins—a unifying hypothesis. *Trends Biochem Sci* 29, 572–578.

Dorman, G. and Prestwich, G. D., 1994. Benzophenone photophores in biochemistry. *Biochemistry* 33, 5661–5673.

Egelman, E. H., 2000. A robust algorithm for the reconstruction of helical filaments using single-particle methods. *Ultramicroscopy* 85, 225–234.

Elzinga, M. and Phelan, J. J., 1984. F-actin is intermolecularly crosslinked by N,N′-p-phenylenedimaleimide through lysine-191 and cysteine-374. *Proc Natl Acad Sci USA* 81, 6599–6602.

Galkin, V. E., Orlova, A., Lukoyanova, N., Wriggers, W., and Egelman, E. H., 2001. Actin depolymerizing factor stabilizes an existing state of F-actin and can change the tilt of F-actin subunits. *J Cell Biol* 153, 75–86.

Galkin, V. E., Orlova, A., Cherepanova, O., Lebart, M. C., and Egelman, H. E. 2008. High-resolution cryo-EM structure of the F-actin-fimbrin/plastin ABD2 complex. *Proc Natl Acad Sci USA* 105, 1494–1498.

Gerson, J. H., Kim, E., Muhlrad, A., and Reisler, E., 2001. Tropomyosin–troponin regulation of actin does not involve subdomain 2 motions. *J Biol Chem* 276, 18442–18449.

Gorman, J. J. and Folk, J. E., 1980. Transglutaminase amine substrates for photochemical labeling and cleavable cross-linking of proteins. *J Biol Chem* 255, 1175–1180.

Graceffa, P. and Dominguez, R., 2003. Crystal structure of monomeric actin in the ATP state: structural basis of nucleotide-dependent actin dynamics. *J Biol Chem* 278, 34172–34180.

Green, N. S., Reisler, E., and Houk, K. N., 2001. Quantitative evaluation of the lengths of homobifunctional protein cross-linking reagents used as molecular rulers. *Protein Sci* 10, 1293–1304.

Grintsevich, E. E., Benchaar, S. A. Benchaar S. A., Warshaviak, D., Boontheung, P., Halgand, F., Whitelegge, J. P., Faull, K. F., Loo, R. R., Sept, D., Loo, J. A., and Reisler, E. 2008. Mapping the cofilin binding site on yeast G-actin by chemical cross-linking. *J Mol Biol* 377, 395–409.

Guan, J. Q., Vorobiev, S., Almo, S. C., and Chance, M. R., 2002. Mapping the G-actin binding surface of cofilin using synchrotron protein footprinting. *Biochemistry* 41, 5765–5775.

Gutsche-Perelroizen, I., Lepault, J., Ott, A., and Carlier, M. F., 1999. Filament assembly from profilin–actin. *J Biol Chem* 274, 6234–6243.

Hanein, D., Volkmann, N., Goldsmith, S., Michon, A. M., Lehman, W., Craig, R., DeRosier, D., Almo, S., and Matsudaira, P., 1998. An atomic model of fimbrin binding to F-actin and its implications for filament crosslinking and regulation. *Nat Struct Mol Biol* 5, 787–792.

Hatanaka, H., Ogura, K., Moriyama, K., Ichikawa, S., Yahara, I., and Inagaki, F., 1996. Tertiary structure of destrin and structural similarity between two actin-regulating protein families. *Cell* 85, 1047–1055.

Hegyi, G., Michel, H., Shabanowitz, J., Hunt, D. F., Chatterjie, N., Healy-Louie, G., and Elzinga, M., 1992. Gln-41 is intermolecularly cross-linked to Lys-113 in F-actin by N-(4-azidobenzoyl)-putrescine. *Protein Sci* 1, 132–144.

Hegyi, G., Mák, M., Kim, E., Elzinga, M., Muhlrad, A., and Reisler, E. 1998. Intrastrand cross-linked actin between Gln-41 and Cys-374. I. Mapping of sites cross-linked in F-actin by N-(4-azido-2-nitrophenyl) putrescine. *Biochemistry.* 37, 17784–17792.

Hodgkinson, J. L., el Mezgueldi, M., Craig, R., Vibert, P., Marston, S. B., and Lehman, W., 1997. 3-D image reconstruction of reconstituted smooth muscle thin filaments containing calponin: visualization of interactions between F-actin and calponin. *J Mol Biol* 273, 150–159.

Holmes, K. C., Popp, D., Gebhard, W., and Kabsh, W., 1990. Atomic model of the actin filament. *Nature* 347, 44–49.

Holmes, K. C., Angert, I., Jon Kull, F., Jahn, W., and Schroder, R. R., 2003. Electron cryo-microscopy shows how strong binding of myosin to actin releases nucleotide. *Nature* 425, 423–427.

Irobi, E., Aguda, A. H., Larsson, M., Guerin, C., Yin, H. L., Burtnick, L. D., Blanchoin, L., and Robinson, R. C. 2004. Structural basis of actin sequestretion by thymosin-β4: implications for WH2 proteins. *EMBO J* 23, 3599–3608.

Kabsch, W., Mannherz, H. G., Suck, D., Pai, E. F., and Holmes, K. C., 1990. *Nature* 347, 37–44.

Kim, E., Bobkova, E., Miller, C. J., Orlova, A., Hegyi, G., Egelman, E. H., Muhlrad, A., and Reisler, E., 1998. Intrastrand cross-linked actin between Gln-41 and Cys-374. III. Inhibition of motion and force generation with myosin. *Biochemistry* 37, 17801–17809.

Kim, E., Bobkova, E., Hegyi, G., Muhlrad, A., and Reisler, E., 2002. Actin cross-linking and inhibition of the actomyosin motor. *Biochemistry* 41, 86–93.

Klenchin, V. A., Allingham, J. S., King, R., Tanaka, J., Marriott, G., and Rayment, I., 2003. Trisoxazole macrolide toxins mimic the binding of actin-capping proteins to actin. *Nat Struct Mol Biol* 10, 1058–1063.

Knight, P. and Offer, G., 1978. p-NN′-phenylenebismaleimide, a specific cross-linking agent for F-actin. *Biochem J* 175, 1023–1032.

Kudryashov, D. S., Sawaya, M. R., Adisetiyo, H., Norcross, T., Hegyu, G., Reisler, E., and Yeates, T. O., 2005. The crystal structure of a cross-linked actin dimer suggests a detailed molecular interface in F-actin. *Proc Natl Acad Sci USA* 102, 13105–13110.

Lappalainen, P., Fedorov, E. V., Fedorov, A. A., Almo, S. C., and Drubin, D. G., 1997. Essential functions and actin-binding surfaces of yeast cofilin revealed by systematic mutagenesis. *EMBO J* 16, 5520–5530.

Lilic, M., Galkin, V. E., Orlova, A., VanLoock, M. S., Egelman, E. H., and Stebbins, C. E., 2003. *Salmonella* SipA polymerizes actin by stapling filaments with nonglobular protein arms. *Science* 301, 1918–1921.

Loisel, T. P., Boujemaa, R., Pantaloni, D., and Carlier, M. F., 1999. Reconstitution of actin-based motility of *Listeria* and *Shigella* using pure proteins. *Nature* 401, 613–616.

Lorenz, M., Popp, D., and Holmes, K. C., 1993. Refinement of the F-actin model against X-ray fiber diffraction data by the use of a directed mutation algorithm. *J Mol Biol* 234, 826–836.

Luo, Y., Li, B., Yang, G., Gergely, J., and Tao, T. 2002. Cross-linking between the regulatory regions of troponin-I and troponin-C abolishes the inhibitory function of troponin. *Biochemistry* 41, 12891–12898.

Mannherz, H. G., Ballweber, E., Galla, M., Villard, S., Granier, C., Steegborn, C., Schmidtmann, A., Jaquet, K., Pope, B., and Weeds, A. G., 2007. Mapping the ADF/cofilin binding site on monomeric actin by competitive cross-linking and peptide array: evidence for a second binding site on monomeric actin. *J Mol Biol* 366, 745–755.

McGough, A. and Chiu, W., 1999. ADF/cofilin weakens lateral contacts in the actin filament. *J Mol Biol* 291, 513–519.

McGough, A., Pope, B., Chiu, W., and Weeds, A., 1997. Cofilin changes the twist of F-actin: implications for actin filament dynamics and cellular function. *J Cell Biol* 138, 771–781.

McGough, A., Chiu, W., and Way, M., 1998. Determination of the gelsolin binding site on F-actin: implications for severing and capping. *Biophys J* 74, 764–772.

McLaughlin, P. J., Gooch, J. T., Mannherz, H. G., and Weeds, A. G., 1993. Structure of gelsolin segment 1-actin complex and the mechanism of filament severing. *Nature* 364, 685–692.

Milligan, R. A., Whittaker, M., and Safer, D., 1990. Molecular structure of F-actin and location of surface binding sites. *Nature* 348, 217–221.

Morgan, T. E., Lockerbie, R. O., Minamide, L. S., Browning, M. D., and Bamburg, J. R., 1993. Isolation and characterization of a regulated form of actin depolymerizing factor. *J Cell Biol* 122, 623–633.

Morton, W. M., Ayscough, K. R., and McLaughlin, P. J., 2000. Latrunculin alters the actin-monomer subunit interface to prevent polymerization. *Nat Cell Biol* 2, 376–378.

Nyman, T., Page, R., Schutt, C. E., Karlsson, R., and Lindberg, U., 2002. A cross-linked profilin–actin heterodimer interferes with elongation at the fast-growing end of F-actin. *J Biol Chem* 277, 15828–15833.

Onishi, H., Maita, T., Matsuda, G., and Fujiwara, K., 1990. Lys-65 and Glu-168 are the residues for carbodiimide-catalyzed cross-linking between the two heads of rigor smooth muscle heavy meromyosin. *J Biol Chem* 265, 19362–19368.

Orlova, A., Galkin, V. E., VanLoock, M. S., Kim, E., Shvetsov, A., Reisler, E., and Egelman, E. H., 2001. Probing the structure of F-actin: cross-links constrain atomic models and modify actin dynamics. *J Mol Biol* 312, 95–106.

Orlova, A., Shvetsov, A., Galkin, V. E., Kudryashov, D. S., Rubenstein, P. A., Egelman, E. H., and Reisler, E., 2004. Actin-destabilizing factors disrupt filaments by means of a time reversal of polymerization. *Proc Natl Acad Sci USA* 101, 17664–17668.

Otomo, T., Tomchick, D. R., Otomo, C., Panchal, S. C., Machius, M., and Rosen, M. K., 2005. Structural basis of actin filament nucleation and processive capping by a formin homology 2 domain. *Nature* 433, 488–494.

Otterbein, L. R., Graceffa, P., and Dominguez, R., 2001. The crystal structure of uncomplexed actin in the ADP state. *Science* 293, 708–711.

Pant, K., Chereau, D., Hatch, V., Dominguez, R., and Lehman, W., 2006. Cortactin binding to F-actin revealed by electron microscopy and 3D reconstruction. *J Mol Biol* 359, 840–847.

Reichert, A., Heintz, D., Echner, H., Voelter, W., and Faulstich, H., 1996. Identification of contact sites in the actin–thymosin β4 complex by distance-dependent thiol cross-linking. *J Biol Chem* 271, 1301–1308.

Roepstorff, P. and Fohlman, J., 1984. Proposal for a common nomenclature for sequence ions in mass spectra of peptides. *Biomed Mass Spectrom* 11, 601.

Rosol, M., Lehman, W., Craig, R., Landis, C., Butters, C., and Tobacman, L. S., 2000. Three-dimensional reconstruction of thin filaments containing mutant tropomyosin. *Biophys J* 78, 908–917.

Safer, D., Sosnick, T. R., and Elzinga, M., 1997. Thymosin β4 binds actin in an extended conformation and contacts both the barbed and pointed ends. *Biochemistry* 36, 5806–5816.

Schilling, B., Row, R. H., Gibson, B. W., Guo, X., and Young, M. M., 2003. MS2Assign, automated assignment and nomenclature of tandem mass spectra of chemically crosslinked peptides. *J Am Soc Mass Spectrom* 14, 834–850.

Schutt, C. E., Myslik, J. C., Rozycki, M. D., Goonesekere, N. C. W., and Lindberg, U., 1993. The structure of crystalline profilin-β-actin. *Nature* 365, 810–816.

Sherman, M. B., Jakana, J., Sun, S., Matsudaira, P., Chiu, W., and Schmid, M. F., 1999. The three-dimensional structure of the Limulus acrosomal process: a dynamic actin bundle. *J Mol Biol* 294, 139–149.

Sinz, A. and Wang, K., 2001. Mapping protein interfaces with a fluorogenic cross-linker and mass spectrometry: application to nebulin–calmodulin complexes. *Biochemistry* 40, 7903–7913.

Sutoh, K., 1982a. Identification of myosin-binding sites on the actin sequence. *Biochemistry* 21, 3654–3661.

Sutoh, K., 1982b. An actin-binding site on the 20K fragment of myosin subfragment 1. *Biochemistry* 21, 4800–4804.

Sutoh, K., 1983. Mapping of actin-binding sites on the heavy-chain of myosin subfragment-1. *Biochemistry* 22, 1579–1585.

Sutoh, K., 1984. Actin actin and actin deoxyribonuclease-I contact sites in the actin sequence. *Biochemistry* 23, 1942–1946.

Sutoh, K. and Hatano, S., 1986. Actin–fragmin interactions as revealed by chemical cross-linking. *Biochemistry* 25, 435–440.

Sutoh, K. and Mabuchi, I., 1984. N-terminal and C-terminal segments of actin participate in binding depactin: an actin-depolymerizing protein from starfish oocytes. *Biochemistry* 23, 6757–6761.

Sutoh, K. and Mabuchi, I., 1986. Improved method for mapping the binding site of an actin-binding protein in the actin sequence. Use of A site-directed antibody against the N-terminal region of actin as a probe of its N-terminus. *Biochemistry* 25, 6186–6192.

Sutoh, K. and Yin, H. L., 1989. End-label fingerprintings show that the N-termini and C-termini of actin are in the contact site with gelsolin. *Biochemistry* 28, 5269–5275.

Takashi, R., 1988. A novel actin label: a fluorescent probe at glutamine-41 and its consequences. *Biochemistry* 27, 938–943.

Tao, T., Lamkin, M., and Scheiner, C. J., 1985. The conformation of the C-terminal region of actin: a site-specific photocrosslinking study using benzophenone-4-maleimide. *Arch Biochem Biophy* 240, 627–634.

Vandekerckhove, J. S., Kaiser, D. A., and Pollard, T. D., 1989. Acanthamoeba actin and profilin can be cross-linked between glutamic acid 364 of actin and lysine 115 of profilin. *J Cell Biol* 109, 619–626.

Volkmann, N., Amann, K. J., Stoilova-McPhie, S., Egile, C., Winter, D. C., Hazelwood, L., Heuser, J. E., Li, R., Pollard, T. D., and Hanein, D., 2001. Structure of Arp2/3 complex in its activated state and in actin filament branch junctions. *Science* 293, 2456–2459.

Wahlstrom, J. L., Randall, M. A., Jr., Lawson, J. D., Lyons, D. E., Siems, W. F., Crouch, G. J., Barr, R., Facemyer, K. C., and Cremo, C. R., 2003. Structural model of the regulatory domain of smooth muscle heavy meromyosin. *J Biol Chem* 278, 5123–5131.

Wriggers, W., Tang, J. X., Azuma, T., Marks, P. W., and Janmey, P. A., 1998. Cofilin and gelsolin segment-1: molecular dynamics simulation and biochemical analysis predict a similar actin binding mode. *J Mol Biol* 282, 921–932.

Yonezawa, N., Nishida, E., Iida, K., Kumagai, H., Yahara, I., and Sakai, H., 1991. Inhibition of actin polymerization by a synthetic dodecapeptide patterned on the sequence around the actin-binding site of cofilin. *J Biol Chem* 266, 10485–10489.

Computational Approaches to Examining Protein–Protein Interactions: Combining Experimental and Computational Data in the Era of Structural Genomics

J.K. AMISHA KAMAL

Center for Proteomics, Case Western Reserve University School of Medicine, Cleveland, OH, USA

10.1 INTERACTOME IN STRUCTURAL GENOMICS

Knowledge of the three-dimensional structure of proteins and their structural and functional interactions in protein complexes is critical to understanding a myriad of processes in cell biology. Despite the obvious importance of this structural information, the challenges to obtaining it using current technologies are considerable. It is well known that the number of structural entries for proteins in the protein data bank (PDB) (40,195 protein entries as of May 29, 2007) is 50–100 times lower than the number of open-reading frame sequences in GenBank/TREMBL. When these databases are clustered in terms of homologous sequences, the gap between sequence and structure is not lessened at all. However, structural genomics projects in the United States and worldwide are attempting to bridge this gap by solving structures of a selected set of proteins derived from as many protein families as possible (Bonanno et al., 2005; Chandonia and Brenner, 2005). These template structures, at least one for each family clustered at 30% identity, provide anchor points for high-throughput homology modeling to provide accurate models for the remaining family members (Baker and Sali, 2001; Vitkup et al., 2001). Thus, there is some prospect for providing highly accurate structural models for a large fraction of all known genomes.

For protein–protein complexes, the prospects are daunting by comparison. Protein–protein interactions are one of the most important regulatory mechanisms in

Mass Spectrometry Analysis for Protein–Protein Interactions and Dynamics, Edited by Mark Chance
Copyright © 2008 John Wiley & Sons, Inc.

cells; they underlie intracellular process regulation, intercellular communication, signal transduction, and the regulation of gene expression. Data from high-throughput proteomics projects, including both "pull-down" and yeast 2-hybrid experiments, indicate that the total number of protein interactions in the cell—the interactome—may exceed the total number of proteins (von Mering et al., 2002; Reboul et al., 2003; Tyers and Mann, 2003; Gavin et al., 2006; Ewing et al., 2007). However, our knowledge of the details of these structural interactions, as reflected by the number of protein complexes deposited in PDB, is limited to 3826 crystal structure entries for heterodimers as of May 29, 2007 (http://pqs.ebi.ac.uk/pqs-doc. shtml). Since the number of authentic interactions is at least as large as the number of proteins, the gap between the number of complexes currently solved and the number of interest for structure elucidation likely exceeds 1000-fold, considering that several million open-reading frame sequences currently exist (Chance et al., 2004).

10.2 IMPORTANCE OF COMPUTATIONAL METHODS IN STRUCTURAL GENOMICS

The barriers to determining the structure of complexes include known limitations in crystallography and nuclear magnetic resonance (NMR) technologies: issues such as complex size, crystallizability, solubility, and amounts of materials are well known. In recent years, electron microscopy (EM) and tomography techniques, particularly at low temperatures, have substantially improved and are making important contributions to determining the structure of complexes (Sali et al., 2003; Chiu et al., 2006). These approaches have resolution limitations for many samples and are better for larger complexes or cells due to sample dose issues. This leaves a gap in technological progress for the "medium" size complexes, particularly medium-sized binary complexes (50–200 kDa). This has spurred the development of a host of computational methods that can fill in the gap and contribute to understanding the relationship between protein structure and function.

Progress toward bridging the sequence–structure gap will involve significant advances in experimental and computational methods. As for the structural genomics projects, where a homology-based approach was used to bridge the gap between the number of structures and the number of sequences, it is likely that the structure of carefully selected "template complexes" must be identified and solved; these structures will be leveraged by computational modeling to provide models of the appropriate orthologous "sibling complexes." Concerted efforts to solve these template structures will require a hybrid approach to structure solution that combines both high- and medium-resolution structural data (Sali et al., 2003; Chiu et al., 2006).

10.3 COMBINING COMPUTATIONAL METHOD WITH EXPERIMENTAL DATA IN MODELING THE STRUCTURE OF PROTEIN BINARY COMPLEX

The most reliable computer-based technique for generating three-dimensional models of protein–protein complexes is using docking algorithms that compare surface complementarity. The quality of this comparison highly depends on the quality of the input structures (Halperin et al., 2002; Kontoyianni et al., 2004; Perola et al., 2004). High-accuracy homology modeling (root mean square deviation (RMSD) $\sim$ 1 Å), which can provide high-quality input structures for docking, can be achieved when the target and template proteins have sequence identities of more than 50% (Baker and Sali, 2001); the accuracy drops significantly when the identity is less than 30% (Baker and Sali, 2001). The performance of docking programs is also highly dependent on the nature of the binding site, for example, the driving forces for binding and specificity. Several docking procedures have been recently developed that use different docking algorithms and scoring functions, and these have been evaluated in critical assessment of prediction of interactions (CAPRI) (Janin et al., 2003; Mendez et al., 2003). In the CAPRI competition, the unbound three-dimensional coordinates of the binary partners are given to researchers prior to the release of the coordinates of the cocrystallized complex; this serves as a blind test for evaluating the current status of the field for predicting the structural details of protein–protein interactions. In these competitions, the correct structure can generally be predicted, but cannot be distinguished from a number of equally good possibilities (e.g., many false positives). Recently, a fast algorithm has been developed, namely, ClusPro, which energy filters docked conformations having good surface complementarity and ranks them on the basis of their clustering properties (Comeau et al., 2004a, 2004b; Comeau, 2007). ClusPro is the first automated server that has participated in the CAPRI experiment; it has made several correct predictions within 24 h of the release of the coordinates (Comeau et al., 2005). The free energy filters provided in ClusPro select docked complexes with minimized desolvation or electrostatic energies. Clustering is then used to smooth the local minima and to select the candidates with the broadest energy wells. The robustness of the method was tested for 48 pairs of interacting proteins. The probability of the correct structure being "top scored" is 27%, while the probability of the correct structure being one of the top 10 models is 65%, and within the top 30 models is 81%. Thus, additional experimental information that could reject the false positives in the top 20–30 models would be a valuable addition to the modeling process. In particular, identification of interface residues that could guide the docking step and recognition of the driving forces defining the interface to guide the energy minimization step are crucial to the modeling. With these limitations in mind, docking approaches where ClusPro is constrained with information from mutagenesis (Pons et al., 2006), NMR (Gruschus et al., 2004; Pons et al., 2006), and cryo-EM (Agbulut et al., 2007) experiments have been recently reported in the literature.

10.3.1 General Strategy of the Method

Recently, we used a combination of radiolytic footprinting coupled to mass spectrometry analysis and docking with the ClusPro server to derive a structure for the actin/cofilin binary complex (Kamal et al., 2007). In the following sections, we outline a general approach using multiple examples of binary complexes that include monomeric actin to illustrate the power and limitations of the current method and to provide a guide to other researchers in the field. Specifically, we show how the ClusPro server performs in predicting the structure of the actin/DNaseI and actin/profilin structures in the absence of additional experimental data, how the use of ClusPro in conjunction with previously published footprinting can provide a model of the actin/gelsolin segment-1 (GS1) complex that agrees with crystallographic data, and how the method is used to derive the structure of the actin/cofilin complex that does not have a crystal or NMR structure.

Our general strategy of protein binary complex structure determination involves conducting radiolytic footprinting experiments with mass spectrometry analysis to derive information on the residues participating in the binding interface and/or those involved in conformational reorganization. This is followed by homology modeling if the bound conformation of either protein significantly deviated from the free form, and then protein–protein docking by ClusPro server (Comeau et al., 2004a, 2004b; Comeau, 2007). The top 20,000 docked conformations are filtered either by electrostatics or by desolvation free energy filtering provided in the ClusPro. The choice of energy filter is critical in providing appropriate models. The top 2000 energy-minimized structures are clustered and ranked according to their cluster sizes. The default value of the clustering radius (9 Å) is used if not stated otherwise. The general approach is illustrated in Scheme 10.1 and relevant steps are explained in the examples that follow.

10.3.2 Docking Complexes of Known Crystal Structures without Using Footprinting Constraints

Using ClusPro, we carried out molecular docking of selected actin binding proteins (ABP) to G-actin where the crystal structures are known. This allows us to both develop and judge the effectiveness of the method; the various parameters of the resulting docked models are provided in Table 10.1. The individual components of the indicated crystal structures of actin and its ABP complex were docked to obtain the respective complexes. The resulting models were compared with the respective crystal structures, and their RMSDs were calculated and interface parameters were derived. For docking of actin/DNaseI and actin/profilin, the individual chains from the PDB structures of the two complexes, 1ATN (Fig. 10.1a) and 2BTF (Fig. 10.1b), respectively, were used. Actin and DNaseI exhibit (overall) negative electrostatic potential surface patterns (Fig. 10.2a and b) in accordance with their acidic pI values (~5.2). However, their binding surfaces comprise mostly neutral/hydrophobic residues, suggesting a hydrophobic driven mode of binding. Consistently, the interface of the crystal structure showed a higher percentage for nonpolar atoms over polar atoms for both actin and DNaseI (29% and 41% polar, respectively, Table 10.1). On the basis of this analysis of the interface, we chose desolvation free energy filtering of the docked conformations for actin/DNaseI from the ClusPro energy filtering options (Comeau et al., 2004a, 2004b). Profilin exhibits an overall positively charged electrostatic surface pattern (p$I = 8.4$) (Fig. 10.2c). Its binding site for actin also exhibits many positive and several negatively charged residues. Specifically, the binding interface of actin/profilin complex consists of appropriate positioning of oppositely charged residues from the individual components suggesting that the

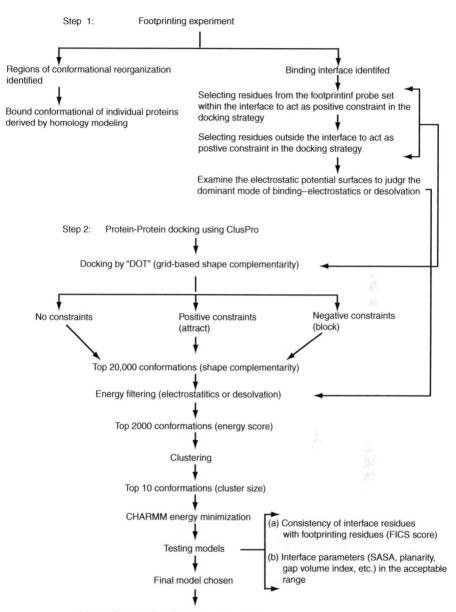

SCHEME 10.1   General strategy of protein binary complex structure determination from footprinting data and computational modeling.

TABLE 10.1 Summary of the Docking and Structural Interface Parameters of the Docked Models of Actin/ABP Complexes

Actin/ABP Complex	Energy Filter	Footprinting Constraints	Rank[a]	CS	RMSD ABP (Å)	RMSD Interface (Å)	Interface SASA (Å²) Actin/ABP	% Polar Atoms[b] Actin/ABP	H-Bonds	Planarity Actin/ABP	GVI	FICS Actin/ABP
Actin/DNaseI	Desolvation	No	2	37	3.2	1.6	887/747 (961/813)	30/44 (29/41)	5 (8)	2.5/2.2 (2.6/2.6)	2.2 (2.1)	–
Actin/profilin	Electrostatic	No	1	44	4.3	2.2	852/866 (1052/1011)	41/35 (44/44)	4 (11)	2.4/2.0 (2.6/2.4)	3.1 (2.6)	–
		No	6	33	7.4	3.8	772/773 (1030/999)	32/35 (36/42)	3 (9)	2.2/2.5 (2.5/2.7)	4.6 (3.0)	0.21/1.0
Actin/GS1	Desolvation	Attract	2	69	6.1	3.7	928/848 (1030/999)	30/41 (36/42)	2 (9)	3.1/2.9 (2.5/2.7)	3.4 (3.0)	0.33/1.0
		Block	1	65	6.7	2.9	869/835 (1030/999)	31/34 (36/42)	2 (9)	2.9/3.0 (2.5/2.7)	3.0 (3.0)	0.29/1.0
Actin/cofilin	Electrostatic	No	2	30	–	–	846/839	53/43	1	2.2/1.9	3.0	0.10/0.20
		Attract	1	57	–	–	1044/1067	54/41	9	2.8/2.3	2.0	0.19/0.20
		Block	1	48	–	–	908/897	56/43	12	2.6/2.2	2.6	0.14/0.20

Values in parenthesis are the respective interface parameters calculated for their crystal structures.
ABP = actin binding protein; CS = cluster size; RMSD = root mean square deviation; SASA = solvent-accessible surface area; H-bonds = hydrogen bonds; GVI = gap volume index; FICS = footprinting interface consistency score.

[a]Rank of the closest model among the 10 top-scoring models to the crystal structure.
[b]Percentage of polar atoms at the interface. Remaining is accounted for nonpolar atoms.

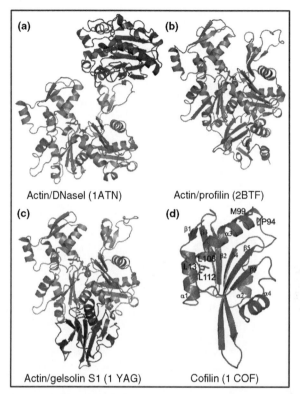

FIGURE 10.1   Crystal structures of G-actin/actin binding protein complexes and cofilin. (a) Structures of G-actin/DNaseI (PDB code 1ATN), (b) G-actin/GS1 (1YAG), (c) G-actin/ profilin (2BTF), and (d) cofilin (1COF). The red-colored region in the cofilin structure is the G-actin binding site (G/F site) established by footprinting (Guan et al., 2002), mutagenesis (Lappalainen et al., 1997), and NMR (Pope et al., 2004). The specific footprinting probe residues that form the G-actin binding site are indicated with "stick models." Helices and β-strands are indicated with α and β symbols, including their serial numbers. (See the color version of this figure in Color Plates section.)

mode of binding is primarily driven by electrostatics. In this case, we chose electro-static free energy filtering for actin/profilin docking.

The five top-scoring models for each docking exercise are shown in Fig. 10.3a and c. The top-ranked model obtained for the actin/DNaseI docking (Fig. 10.3a) does not agree with the crystal structure (Fig. 10.1a); its RMSD compared to the crystal-lographic structure is 29.4 Å. Nevertheless, the second-ranked model (Fig. 10.3b) is very similar to the crystal structure with an RMSD of the DNaseI = 3.2 Å (Table 10.1). To evaluate the geometry of the interface alone, we also calculated the RMSD of the interfacial residues, the residues that showed a minimum of 1 Å^2 solvent-accessible surface area (SASA) decrease upon complex formation. This value for the second-ranked model is 1.6 Å (Table 10.1). The actin segments identified at the interface of the model are 38–49, 60–64, and 202–207, while for DNaseI the interfacial segments are

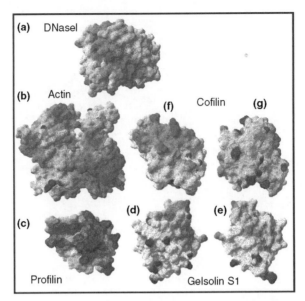

FIGURE 10.2   Electrostatic potential surfaces of actin and actin binding proteins. (a) DNaseI in the actin binding orientation, (b) actin, (c) profilin in the actin binding orientation, (d) gelsolin S1 in the actin binding orientation, (e) gelsolin S1 rotated 180° from the actin binding orientation, (f) cofilin in the actin binding orientation, (g) cofilin rotated 90° to right from the actin binding orientation. Red color is for negative electrostatics, blue color is for positive electrostatics, and white is for neutral. (See the color version of this figure in Color Plates section.)

13–14, 43–69, 91–96, and 112; these overlap closely with those identified in the crystal structure (actin: 38–47, 57–64, and 202–207; DNaseI: 13–14, 43–69, 79, 91–96, and 114). In addition, four of the five predicted hydrogen bonds (H-bonds) at the interface of the model are identical to four of the nine H-bonds seen in the interface of the crystal structure; they are (actin–DNaseI numbers) Arg39–Asp53, Gln41–Tyr65, Val43–Tyr65, and Gly63–His44.

For actin/profilin, the first-ranked model (Fig. 10.3d) is close to the crystal structure (Fig. 10.1b) with RMSD = 4.3 Å and of the interface residues, RMSD = 2.2 Å (Table 10.1). The interface segments identified in the model (actin: 113–116, 166–173, 284–290, 354–355, and 361–375; profilin: 56–60, 71–90, 99, and 119–129) match well with those in the crystal structure (actin: 113–116, 133, 166–173, 283–290, 354–355, and 361–375; profilin: 59–69, 71–91, 97–99, and 117–129). In addition, all the four predicted H-bonds at the interface of the model are identical to 4 of the 11 H-bonds at the interface of the crystal structure; they are (actin–profilin numbers) Lys113–Glu82, Glu361–Lys125, Arg372–Arg74, and Arg372–Thr84. Other parameters such as interface SASA, planarity, and gap volume index (GVI) are also close to the values reported for the crystal structure for both actin/DNaseI and actin/profilin (Table 10.1). Next, we switched the energy filter options between the two complexes, electrostatic free energy filtering for actin/

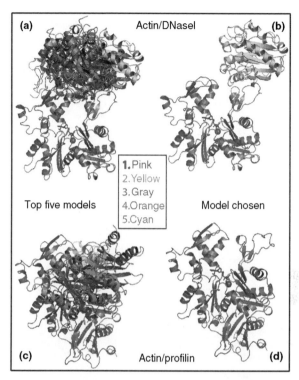

FIGURE 10.3 Modeling of G-actin/DNaseI and G-actin/profilin. (a) Top five models obtained from the docking strategy for G-actin/DNaseI. First-ranked model is colored pink, second-ranked model is colored yellow, third-ranked model is colored gray, fourth-ranked model is colored orange, and fifth-ranked model is colored greenish cyan. (b) The model of G-actin/DNaseI that is closest to the crystal structure. (c) Top five models obtained from the docking for G-actin/profilin. Color codes follow the previous scheme. (d) The model of G-actin/profilin that is closest to the crystal structure. (See the color version of this figure in Color Plates section.)

DNaseI and desolvation free energy filtering for actin/profilin, and no structures close to crystal structure within 10 Å RMSD showed up in the top 10 ranked models.

These data suggest two important conclusions about the ClusPro modeling process. First, ClusPro requires that the mode of interaction for the binary complex (either desolvation or electrostatic) be correctly specified if the program is to provide candidate models consistent with the correct structure (e.g., from crystallography) of complexes with one of these energies being the dominant binding force. For complexes with significant contribution from both binding energies, ClusPro gives reliable models with either of the filters or by pooling conformations from both filters (Comeau et al., 2004a, 2004b; Comeau, 2007). Second, the presumed correct prediction may not be the top-scoring model even when the correct mode of interaction is specified. Thus, for a successful protocol to be developed for docking, the driving forces for binding must be understood and the structural nature of the interface must be defined to a degree. In Section 10.3.4, we describe how footprinting data are used to provide both sets of required information.

10.3.3   Radiolytic Footprinting: G-Actin/GS1 and G-Actin/Cofilin

We conducted radiolytic footprinting experiments for G-actin/GS1 (Goldsmith et al., 2001; Guan et al., 2003) whose crystal structure is known and for G-actin/cofilin (Guan et al., 2002; Kamal et al., 2007) whose crystal or NMR structure is not yet reported. In the footprinting experiment, protein samples in the radiolysis buffer are exposed to a white synchrotron X-ray beam for intervals from 0 to 200 ms. Exposed samples are subjected to enzymatic digestion (trypsin, Asp-N, and Glu-C) followed by quantification of the peptides by liquid chromatography (LC) coupled electrospray ion-source mass spectrometry (ESI-MS). First-order rate constants of modification are derived from dose–response curves (Guan and Chance, 2005; Takamoto and Chance, 2006). These modification rate values are in accordance with the number, type (specific reactivity), and solvent exposure of the reactive amino acids in the respective peptides (Guan et al., 2003, 2004, 2005; Guan and Chance, 2005; Takamoto and Chance, 2006). The details of the method are described in Section 10.5. Footprinting data on the G-actin/GS1 complex (Goldsmith et al., 2001; Guan et al., 2003) revealed five peptides generated from the trypsin digestion of actin (119–147, 157–178, 292–312, 316–326, and 337–359) and the residues Cys374 and Phe375 of the Asp-N peptide 363–375 that are protected as a function of GS1 binding (Table 10.2). These peptides are located in the subdomains 1 and 3, and include the hinge helices (Gly137–Ser145 and Arg335–Ser348) at the cleft; this region is the presumed binding interface based on footprinting alone (McLaughlin et al., 1993). Eleven peptides had rates of oxidation that were unchanged within error (within 20%) upon GS1 binding (1–18, 19–28, 40–50, 63–68, 69–84, 85–95, 96–113, 184–191, 197–206, 329–335, and 360–372) (Guan et al., 2003); the footprinting experiments suggest these are located outside the binding interface. Among the three modifiable Asp-N digested GS1 peptides detected (25–49, 66–83 and 96–109) (Goldsmith et al., 2001), only the peptide 96–109 (probe residue is Phe104) was found protected upon actin/GS1 complex formation (Table 10.2), which corresponds to the peptide 72–84 (probe residue is Phe80) in the crystal structure of actin/GS1 (PDB code 1YAG). Overall, the data indicated specific sites of contact for both actin and GS1 and provided no evidence for any major conformational reorganization of actin or GS1 induced by complex formation.

Table 10.3 lists the oxidation rates of actin and cofilin peptides in free form, and in the bound form in the actin/cofilin complex (Guan et al., 2002; Kamal et al., 2007). Figure 10.4 depicts the locations of various protected residues on the structure of actin monomer. Structural comparisons (Blanchoin and Pollard, 1998; Wriggers et al., 1998; Dominguez, 2004) and competitive binding studies (Blanchoin and Pollard, 1998; Mannherz et al., 2007) suggested that cofilin binds to the cleft formed between subdomains 1 and 3 of actin as precedented by the binding of GS1 and profilin. Since footprinting data did not detect significant protections within the cleft between subdomains 1 and 3 as a function of cofilin binding (Table 10.3, Fig. 10.4),

TABLE 10.2   Modification Rates for Actin and GS1 Peptides that Showed Protections in the Actin/GS1 Complex

Protein	Peptide	Protease	Oxidized Residues	Subdomain	Modification Rate (s^{-1}) Protein	Complex	Degree of Protection[a]
Actin	119–147	Trypsin	**M119, M123, F124, Y143**	1 + 3 (145–)	14 ± 2	10 ± 2	1.4
	292–312		**M305–P307**[b]	3	14 ± 2	10 ± 2	1.4
	316–326		**P322, M325**	3	8.2 ± 0.4	4.3 ± 0.9	1.9
	337–359		**Y337, L346, L349, F352, M355**	1 (338–) +3	33 ± 3	12 ± 2	2.8
	157–178	Asp-N	**H161, P164, Y166, Y169, L171–L173, M176**	3	9.8 ± 2.1	6.1 ± 1.9	1.6
	363–375[c]		**P367, H371, C374, F375**	1	8.5 ± 0.7	6.1 ± 0.4	1.4
GS1	96–109[d]	Asp-N	**F104**	–	0.55 ± 0.16	0	High

The residues shown in bold are identified to be part of the binding interface. Actin data taken from Guan et al. (2003) and GS1 data from Goldsmith et al. (2001).

[a]Defined as ratio of modification rate of protein to complex.
[b]Specific probe sites cannot be determined definitely.
[c]Since another peptide 360–372 showed no change in oxidation (Guan et al., 2003), the probe residues showing protection in peptide 363–375 are C374 and F375.
[d]This peptide corresponds to the GS1 segment 72–84 in the crystal structure of actin/GS1 (PDB code 1YAG) with the probe residue's location being F80.

this hypothetical binding site of cofilin is ruled out in these experiments (Kamal et al., 2007). However, footprinting revealed significant protections for a number of nearby peptides (1–18, 51–61, 69–84, 85–95, 96–113, 118–125, and 360–372) (Table 10.3) within subdomains 1 and 2, indicating the putative binding interface (Kamal et al., 2007). Protections afforded by probe residues within and nearby the nucleotide cleft are accounted for allosteric conformational rearrangement causing the closure of the nucleotide cleft (Kamal et al., 2007). Among the 11 modifiable cofilin peptides detected (Table 10.3) (Guan et al., 2002), only the peptides 4–20, 10–17, 83–96, 91–105, and 106–107 were found protected upon actin binding. The probe residues of these protected peptides (Leu13, Pro94, Met99, Leu108, and Leu112) are shown in Fig 10.1d. This G-actin binding region of cofilin identified by footprinting agreed with the mutational data (Met1–Gly5, Arg96, Lys98, Asp123, and Glu126) (Lappalainen et al., 1997) and was confirmed later by NMR (Ala7, Val95, Lys98, Met99, Tyr101, Ser103, Leu108, Arg109, Gln120, and Asp123) (Pope et al., 2004). The remaining six peptides showed no change in oxidation within the error limit, indicating the absence of any appreciable conformational reorganization of cofilin with actin binding.

TABLE 10.3 Modification Rates for Actin and Cofilin Peptides in Free and in the Actin/Cofilin Complex

Protein	Peptide	Protease	Oxidized Residues	Subdomain	Modification Rate (s^{-1}) Protein	Complex	Degree of Protection[a]
Actin	1–18	Trypsin	**C10, L16**	1	3.1 ± 0.2	0.7 ± 0.1	4.4
	19–28		F21	1	0.7 ± 0.1	0.8 ± 0.1	0.9
	40–50		H40, M44, M47	2	12.9 ± 0.5	5.6 ± 0.6	2.3
	51–61		**Y53**	2	0.9 ± 0.2	0	High
	63–68		L67	2	0.4 ± 0.1	0.06 ± 0.03	6.7
	69–84		**Y69, H73, W79, M82**	1 (70–) +2	5.8 ± 0.6	2.3 ± 0.7	2.5
	85–95		**H87, H88, F90, Y91**	1	2.7 ± 0.3	0.3 ± 0.1	9.0
	96–113		**H101, P102, L110, P112**	1	1.8 ± 0.1	0.16 ± 0.05	11.3
	119–147[b]		M119, M123, F124, Y143	1 + 3 (145–)	4.6 ± 0.1	4.0 ± 0.4	1.2
	148–177		H161, P164, Y166, Y169, L171, P172, H173, M176	3	13.5 ± 0.5	6.3 ± 0.1	2.1
	184–191		M190	4	4.4 ± 0.1	2.3 ± 0.2	1.9
	197–206		F200–T202[c]	4	0.7 ± 0.0	0	High
	239–254		P243	4	2.5 ± 0.1	0.6 ± 0.1	4.2
	292–312		M305–P307[c]	3	3.7 ± 0.1	1.9 ± 0.4	1.9
	316–326		P322, M325	3	4.9 ± 0.1	1.8 ± 0.0	2.7
	329–335		P332, P333	3	1.5 ± 0.3	1.4 ± 0.2	1.1
	337–359		Y337, L346, L349, F352, M355	1 (338–) + 3	2.5 ± 0.3	3.1 ± 0.4	0.8
	360–372		**Y362, P367, H371**	1	2.4 ± 0.2	0.5 ± 0.0	4.8
	118–125	Glu-C	**M119, M123, F124**	1	2.0 ± 0.4	0.5 ± 0.1	4.0
	260–270		M269	4	2.0 ± 0.1	1.2 ± 0.2	1.7
Cofilin	4–20	Trypsin	**L13**	–	0.54 ± 0.02	0.20 ± 0.09	2.7
	27–36		L32	–	0.23 ± 0.03	0.23 ± 0.04	1.0
	43–56		P48, Y50, F53, L54	–	0.53 ± 0.12	0.41 ± 0.06	1.3
	83–96		**P94**	–	0.83 ± 0.16	0.20 ± 0.05	4.2
	10–17	Asp-N	**L13**	–	0.51 ± 0.04	0.10 ± 0.03	5.1
	18–33		L21	–	0.95 ± 0.18	0.92 ± 0.17	1.0
	51–60		F53, L54, L57, P58	–	1.15 ± 0.14	1.05 ± 0.16	1.1

TABLE 10.3 (*Continued*)

Protein	Peptide	Protease	Oxidized Residues	Subdomain	Modification Rate (s^{-1}) Protein	Complex	Degree of Protection[a]
	91–105	**M99**		–	$\approx$44	0.39 ± 0.05	113
	106–117	**L108, L112**		–	0.11 ± 0.02	0.02 ± 0.02	5.5
	123–129	F124		–	0.86 ± 0.06	0.82 ± 0.07	1.0
	130–143	L133, H143		–	0.91 ± 0.05	0.80 ± 0.12	1.1

The residues shown in bold are identified to be part of the binding interface. Actin data taken from Kamal et al. (2007) and cofilin data from Guan et al. (2002).

[a]Defined as ratio of modification rate of protein to complex.

[b]Since peptide 118–125 showed protection and 119–147 showed no protection, the probe residue Y143 shows increased modification.

[c]Specific probe sites cannot be determined definitely.

10.3.4 Docking Complex of Known Crystal Structure Using Footprinting Constraints

Although the overall electrostatic surface pattern of GS1 is negative (pI = 5.5), its actin-binding site suggested by footprinting exhibits predominantly neutral/ hydrophobic and negative electrostatic surface patterns (Fig. 10.2d and e). This suggests a major contribution for desolvation in the binding. Therefore, we used desolvation free energy filtering of the docked conformations. Without using any

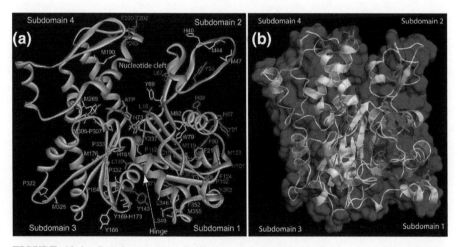

FIGURE 10.4 G-actin structure indicating the protection sites revealed by radiolytic footprinting. (a) Modified amino acids are shown as stick models in the structure of G-actin (1ATN). Residues showing substantial protection are colored in red, moderate protection in yellow, and nearly no protection in cyan. The residue Tyr143 shown in blue-colored stick model shows negative protection (increased modification). (b) Actin structure tilted to left, exposing the substantially protected residues (colored red) clustering in the cleft between subdomains 1 and 2. (See the color version of this figure in Color Plates section.)

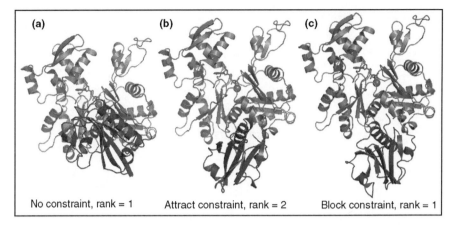

FIGURE 10.5 Modeling of G-actin/GS1. (a) Top-scored model of actin/GS1 without any experimental constraints. (b) Model that is closest to crystal structure with footprinting attract constraint. (c) Model that is closest to crystal structure with footprinting block constraint. (See the color version of this figure in Color Plates section.)

experimental constraints, the top-scored structure obtained (Fig. 10.5a) does not have GS1 located in the cleft between subdomains 1 and 3 and hence is not consistent with the footprinting data or with the crystal structure (the RMSD for GS1 compared to crystallographic data is 22.6 Å). The closest structure among the top 10 ranked models is the 6th-ranked one, which exhibits an RMSD of 7.4 Å for GS1 and 3.8 Å for the crystallographically established interface residues. This model is very different from the crystallographic geometry, especially the geometry of the N-terminus; it does not show interactions with the C-terminus regions of actin (Leu349–Gln353 and Lys370–Phe375), which is one of the crystallographic established contact regions (Fig. 10.1c) and indicated to be a contact point by footprinting data (Table 10.2). The interface SASA value for actin is 772 Å^2 and that for GS1 is 773 Å^2, lower than that seen for other actin complexes, while the gap volume index is higher (Table 10.1).

To evaluate the extent of consistency with the interface region identified by the footprinting experiment, we calculated a parameter defined as the footprinting interface consistency score (FICS) (see Section 10.5.8). The footprinting experiments defined a set of residues protected as a function of complex formation (24 actin residues and 1 GS1 residue, Table 10.2). The probe residues that were recognized as interface residues in the model numbered 6, including the actin residues Tyr143, Tyr166, Tyr169, Leu346, and Leu349, and the GS1 residue Phe80. Thus, the FICS score for actin for this model is 0.21 and for GS1 is 1.0.

Although the use of footprinting to provide the choice of driving force for binding did generate some models that were of interest, this was not sufficient to provide a definitive choice of model. Thus, to further improve the docking we set up "attract" constraints, using selected footprinting probe residues from Table 10.2 that are positioned in the cleft between subdomains 1 and 3 (Tyr143, Tyr166, Tyr169, Leu349, Phe352, and Met355). The rank of the structure closest to the crystal structure improved to 2 from 6, cluster size increased to 69 from 33, and the RMSD of the GS1 decreased to 6.1 from 7.4 Å (Fig. 10.5b). The N-terminus is found interacting with C-terminus regions of actin in this model consistent with crystal structure and with footprinting data. The interface segments identified for actin in the model are 23–25, 133, 143–149, 166–169, 345–355, and 373–375, which overlap with those identified by footprinting data (119–147, 157–178, 292–312, 316–326, 337–359, and 374–375) and with those in the crystal structure (23–25, 143–148, 167–169, 292–296, 334, 341–355, and 373). The interface segments identified for GS1 in the model are 1–8, 24–27, 63, 71–94, and 125, which span the peptide 72–84 detected by footprinting data and overlap with all of the interface segments of the crystal structure (1–8, 25–26, 63–65, 71–86, 94–96, and 124–125). One out of the two predicted H-bonds at the interface (actin-GS1 numbers), Thr350–Phe25, is conserved in the crystal structure. Three additional actin probe residues (Phe352, Met355, and Phe375) became buried in the "attract" model as compared to the "no constraint" model due to the appearance of the GS1 N-terminus interaction with the actin C-terminus as discussed above. For this model, the FICS score for actin improved to 0.33 from 0.21.

Analysis of the FICS scores for the structural models suggests the importance of solution dynamics in interpreting the footprinting data. Typically, we detect more probe residues as being protected upon complex formation by footprinting than are seen to be structurally buried in the interface of the model. Probe residues from within and around the interface are likely to experience oxidation events due to their dynamic fluctuations, but are observed as inaccessible in the free and bound forms of the static crystal structure. For example, the actin probe residues Met119, Met123, and Phe124 (from the peptide 119–147), which are found protected upon the complex formation, are completely buried in the free (SASA = 8.6, 0, and 0 Å^2, respectively) and bound (SASA = 8.8, 0, and 0 Å^2, respectively) forms, and are also located in the proximity of the interface region, precisely at the backside of the cleft, which is not entirely overlapping with the GS1 binding site as GS1 binds at the front half of the cleft. Complex formation likely suppresses this dynamic behavior, such that oxidation is reduced although the precise residue in question does not form a contact. Another probe residue from the same peptide, Ty143 (SASA = 63.9 Å^2), is recognized as within the interface region in the model (SASA = 10 Å^2) and is located within one of the hinge helices. Thus, our hypothesized interface may include closely lying accessible residues, and their protection arises due to a binding-induced restriction of their conformational flexibility, without affecting the overall backbone conformation. For example, the actin probe residues Met305–Pro307 (from peptide 292–312), Pro322, and Met325 (from peptide 316–326), which are found protected upon actin/GS1 complex formation, lie just outside of, but in close proximity to, the interface (precisely at the front side of subdomain 3). Their protection can be attributed to differences in the side chain dynamics, which is not reflected in the crystal structure or in the rigid body docking strategy. Therefore, we expect FICS values not to approach 1.0 and use them only for comparing different models.

To further develop and test the method, we set up "block" constraints (without attract constraint) in the docking step in a separate run. From the footprinting probe residues that were found unchanged upon complex formation, several residues that are sufficiently exposed and somewhat remote from the hypothesized binding interface were selected for "block" constraints (these included 21, 40, 67, 69, 73, 79, 87, 88, 91, 101, 102, 190, 200, 201, and 202). The model closest to the crystal structure is now top scored with an RMSD of 6.7 Å (Fig. 10.5c). The RMSD of the interface is 2.9 Å, and the value of the FICS is 0.29 for actin and 1.0 for GS1. As compared to the "attract" model, the footprinting residue Tyr166 of actin is not found at the interface of the "block" model due to the location of the residue being near the boundary region of the interface. A small orientational difference of the "block" model from the "attract" model causes this residue to fall out of the interface. The interface segments are very similar to those for the attract model (actin: 23–25, 143–148, 167–169, 341–356, and 372–375; GS1: 1–8, 24–27, 63, 71–86, 94, and 125). The two H-bonds predicted at the interface are Ser145–Gln71 and Thr351–Phe25, which exhibit a high degree of conservation with two H-bonds in the crystal structure, Ser144–Gln71 and Thr350–Phe25. The cluster size is increased to 65 from 33 of the "no constraint model." Upon incorporating both "attract" and "block" constraints together, the model obtained has an increased cluster size. Slight variations in cluster size (84–93) and rank (1–3) are observed run to run due to overpopulation of other interfaces when both constraints are used together and the filtering is retaining hits at a relatively proportional rate (Comeau, 2007). When the attract and/or block constraints are used, a larger number of docked conformations near "attract" residues and nonblocked residues are retained in the initial docking stage, which results in a high ranking for that particular conformation. These data reveal that docking alone cannot distinguish between models. When models generated using attract, block, or both are analyzed, a single candidate in overall agreement with the crystal structure is provided.

Finally, we examined the effect of clustering radius on the simulations. The clustering radius is dependent on the size of the ligand (Comeau et al., 2004a, 2004b; Comeau, 2007). However, there is no clear statistical data on the precise correlation between residue length of the ligand and clustering radius (Comeau, 2007). We have used the default clustering radius of 9 Å for docking actin with DNaseI whose residue length is 260 and with profilin whose residue length is 139, and obtained reliable results (RMSDs with crystal structures are 3.2 and 4.3 Å, respectively). The residue length of GS1 is 125, which is lower than those of DNaseI and profilin. Therefore, we decided to observe the effect of reducing the clustering radius for docking actin/GS1. In addition, as discussed earlier, footprinting revealed the involvement of the hinge helices of actin (from the cleft between subdomains 1 and 3) and the long actin binding helix of GS1 in the binding. The geometry of the groove shaped by these hinge helices of actin provides a very narrow space for the positioning of actin binding helix of GS1. Thus, from the perspective of the available information on the binding geometry also, there is indication for the need of a lower clustering radius to sample only the structurally close neighbors for clustering. Therefore, we decreased the clustering radius from the default value of 9 to 8, 7, and 6 Å. In all cases, the closest model to crystal structure is found top scored with RMSD of GS1 falling below 5 Å. The RMSD is 3.4 Å with 8 Å radius, 4.6 Å with 7 Å radius, and 2.5 Å with 6 Å radius. Thus, in carrying out the docking, the clustering radius must be tailored to the modeling problems in question. When a small binding partner is used, a smaller clustering radius is required.

We also looked at the effect of using the incorrect energy filter (electrostatic free energy filter as opposed to desolvation free energy filter) with "no," "attract," and "block" constraints for docking actin/GS1. In these docking simulations, no models with GS1 at the cleft between subdomains 1 and 3 showed up in the top 10 models without using any constraints. With attract or block constraints, 8th-ranked and 10th-ranked models obtained have GS1 located somewhat close to the subdomains 1 and 3 cleft. However, their configurations are very different from the native conformation; their RMSDs are 17.0 and 25.5 Å, respectively, using attract constraints and 46.5 and 41.6 Å using block constraints. This indicates that although we target specific regions to have more ligand clustering in the docking step through attract and block constraints, use of an improper energy filter causes rejection or poor rankings for these conformations owing to a high energy score calculated for the specified energy type. Therefore, the correct energy filter is required for the correct prediction to be top scoring, which is feasible only if there is experimental information about the binding surfaces. Overall, the results indicate that when the interface can be defined using attract and block constraints and the energy filtering is correctly defined, the model from the docking calculation that is most consistent with the experimental data, which has received a high scoring and often being placed as the topmost model, is the one that is actually most consistent with the crystallographic structure.

10.3.5 Docking Complex of Unknown Crystal Structure Using Footprinting Constraints

Since footprinting data revealed an allosteric conformational change for actin as a result of cofilin binding that could be attributed to the closure of its nucleotide cleft, we constructed at first a model of G-actin in the putative "bound form" (closed nucleotide cleft conformation) by homology modeling using SWISS-MODEL (Schwede et al., 2003). The details are provided elsewhere (Kamal et al., 2007). Since footprinting data did not detect observable conformational changes for cofilin as a result of binding to G-actin, we used the available crystal structure of cofilin (1COF) (Fedorov et al., 1997) for docking to the modeled G-actin structure. The interface regions of actin and cofilin as revealed by the footprinting indicate oppositely charged electrostatic surface patterns (Fig. 10.2b, f, and g). Therefore, we chose electrostatic free energy filtering of the docked structures for actin/cofilin, as compared to desolvation; this is consistent with the known pH-dependent binding of the proteins. We first docked actin and cofilin without incorporating any experimental constraints. The first-ranked structure (Fig. 10.6a) (cluster size 43) had neither the cofilin side of the interfacial residues nor the actin side of the interfacial residues as identified by footprinting. The second-ranked structure (cluster size 30) is consistent with the experimental data of the interface residues (Fig. 10.6b, see discussion below). The FICS value obtained is 0.10 for actin and 0.2 for cofilin. Actin probe residues His87 and Tyr91 and cofilin probe residue Met99 from the 26 protected residues shown in bold in Table 10.3 (21 actin residues and 5 cofilin residues) are recognized as interface residues in the model. A single H-bond (actin–cofilin numbers),

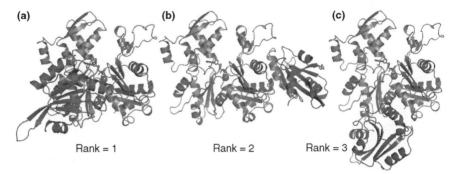

FIGURE 10.6 Three top-ranked G-actin/cofilin models from the computational modeling strategy without incorporating experimental constraints. (a) First-ranked model, (b) second-ranked model, and (c) third-ranked model. Red-colored region in actin marks the interface segments identified by footprinting (Kamal et al., 2007), while that in cofilin marks the interface segments identified by footprinting (Guan et al., 2002), mutagenesis (Lappalainen et al., 1997), and NMR (Pope et al., 2004). (See the color version of this figure in Color Plates section.)

Glu99–Lys105, is predicted to be at the interface. The third-ranked structure (Fig. 10.6b) has cofilin at the previously assumed binding location, for example, cleft between subdomains 1 and 3 (cluster size 23).

Next we incorporated "attract" constraints in the docking step using seven proposed interfacial residues of G-actin from the footprinting data, which are solvent exposed and/or located at the surface, including Trp79 (SASA $= 53 \, \text{Å}^2$), His87 ($89 \, \text{Å}^2$), His88 ($33 \, \text{Å}^2$), Tyr91 ($72 \, \text{Å}^2$), Tyr362 ($13 \, \text{Å}^2$), Pro367 ($46 \, \text{Å}^2$), and His371 ($28 \, \text{Å}^2$). A similar conformation to the "no constraint" model seen in Fig. 10.6b, but with a slightly different orientation (RMSD of the cofilin of the "attract" model with that of the "no constraint" model is 3.8 Å), is top ranked (cluster size 57) (Fig. 10.7) and the conventional model (Fig. 10.6c) ranks seventh (cluster size 20). The similarity between the top-ranked "attract" model (Fig. 10.7a) and the second-ranked

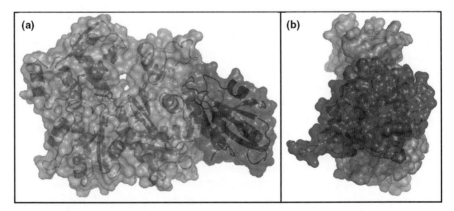

FIGURE 10.7 Three-dimensional model of G-actin/cofilin complex. (a) Top-scored G-actin/cofilin model from the docking constrained with the footprinting data. (b) Side view (rotated 90° to the left) of the model. (See the color version of this figure in Color Plates section.)

"no constraint" model (Fig. 10.6b) suggests that this particular binding mode has such high propensity for shape complementarity as would be found in a native interface structure. The FICS value of actin improved to 0.19 from 0.10 for this model, while that of cofilin remained the same at 0.20 (actin residues Trp79, His87, His88, and Tyr91; cofilin residue Met99). The actin segments identified at the interface of the model are 4–5, 48–54, 79–100, 118–128, and 359–363 and the cofilin segments identified are 59–60, 87–110, 117–126, and 134–138. These interface segments overlap with those identified by the footprinting (actin: 1–18, 51–61, 69–84, 85–95, 96–113, 118–125, and 360–372; cofilin: 4–20, 10–17, 83–96, 91–105, and 106–117). Although the involvement of cofilin N-terminus in the G-actin binding is predicted by footprinting (Leu13) (Kamal et al., 2007), mutation (Met1–Gly5) (Lappalainen et al., 1997), and NMR (Ala7) (Pope et al., 2004), the model does not reflect this fact. This is due to the missing N-terminus segment of 1–6, due to which the nearby residues such as Val7 and Leu13 of the N-terminus do not see a net change in SASA value from free to the modeled complex. However, the location and directionality of this segment clearly indicate its involvement in the binding (Figs. 10.6b and 10.7). In addition to N-terminus, a second segment of cofilin involved with the G-actin binding spans the long α3 helix (Ser89–Leu112), which is considered the actin binding motif of cofilin/ADF family of proteins, common with other homologous proteins such as GS1. Footprinting identifies Pro94, Met99, Leu108, and Leu112 from this region. Mutation (Lappalainen et al., 1997) and NMR (Pope et al., 2004) recognize residues Val95, Arg96, Lys98, Met99, Tyr101, Ser103, Leu108, and Arg109. The residues recognized by the model are Thr87, Ser89, Asp91, Thr92, Ala93, Val95, Arg96, Lys98, Met99, Ala102, Ser103, Lys105, Asp106, Arg109, and Arg110. This is the center of cofilin's actin binding surface and interacts with maximum number of actin segments (Gly48–Ser52, Trp79–Asn92, Glu125–Asn128, and Lys359–Asp363). Some of the charged cofilin interface residues identified by mutation (Lappalainen et al., 1997) and NMR (Pope et al., 2004), which could not be detected by the footprinting (possibly due to the decay of the modified charged residues), namely, Gln120, Asp123, and Glu126, from β6–α4 loop, are recognized as interface residues in the model. However, footprinting does detect a hydrophobic residue Phe124 from this loop as not belonging to the interface. Consistently, this residue is not recognized as interface residue in the model due to its pointing away from the surface, deeply buried within the protein matrix. Additional interface residues recognized from this region in the model are Thr117, Asp118, Val119, Thr122, and Ser125. These residues are followed by several residues from the α4 helix (C-terminus), namely, Glu134, Arg135, and Arg138, which are also found to be part of the interface in the model. These regions (Thr117–Glu126 and Glu134–Arg138) are found interacting with the Glu93–Glu100 segment of actin. An additional region of interaction was suggested by the model as part of G-actin binding site that is not identified by footprinting, mutation, or NMR; residues Glu59 and Asn60 form the tip of the β4–α2 loop, which interacts with the lower side of the actin's D-loop, especially with the Gln49 residue. The number of H-bonds predicted at the interface is nine and are Gln49–Asn60, His87–Ser89, Asn92–Asp91, Arg95–Asp123, Arg95–Ser125, Glu100–Arg135, Glu100–Arg138, Asn128–Lys105, and Lys359–Asp106. This substantial number of H-bonds at the interface that are comparable to those seen in the crystal structures of actin/DNaseI, actin/profiling, and actin/GS1, suggests a high degree of accuracy of this model of actin/cofilin to be the native structure.

With the incorporation of "block" constraint for regions within subdomain 3 (peptides 284–290 and 323–327) and subdomain 4 (peptides 198–203, 224–234, and 241–248) that are unlikely to be at the binding surface as shown by the footprinting data, similar structure that ranked first with "attract" constraint (RMSD between the two structures is only 1.4 Å) ranks again first with a cluster size of 48. The FICS value of actin is 0.14 and that of cofilin is 0.20. The lower actin FICS value for the "block" structure compared to "attract" structure (0.19) is because of the exclusion of Trp79 from the interface, which is located at the rear end of the cleft helix and belongs to the boundary region of the interface. The interface regions of the "block" model are very similar to the "attract" model (actin: 4, 48–52, 80–100, 125–128, and 359–363; cofilin: 59–60, 89–110, 118–126, and 135–138). The number of H-bonds predicted at the interface is 11. All of the H-bonds predicted for the "attract" model, except Gln49–Asn60, are conserved in this model. In addition, it has extra two H-bonds: His87 has an additional bond with Thr122 apart from Ser89 and Glu100 has additional bond with the second nitrogen atom of Arg138. Upon incorporating both "attract" and "block" constraints together, the same overall structure (RMSD compared to the attract structure is 1.0 Å and compared to that of the block structure is 1.7 Å) ranks first with an increased cluster size 68. SASA calculations of actin and cofilin that engage in the binding interface of the putative model (attract and block models) derived are in the range reported for other actin/ABP structures (Table 10.1). Other interface parameters are also within the range reported (Table 10.1).

Next we tried the effect of clustering radius on the docking results using the attract constraints. Decreasing the clustering radius to 8 Å did not change the ranking or the RMSD to any appreciable extent relative to the initial model (1.4 Å with the model of 9 Å clustering radius), although the cluster size varied. Upon decreasing the clustering radius still further to 7 Å, the RMSD of the model increased to 4.6 Å. Decreasing the clustering radius to 6 Å not only increased the RMSD further (5.4 Å), but had also changed the orientation of the bound cofilin such that the residues at the C-terminus of actin do not interact with cofilin, which is not consistent with footprinting data and would reduce the FICS score. Cofilin's size is similar to that of profilin (residue length is 139); in the latter case we obtained reliable results with 9 Å clustering radius (4.3 Å RMSD) (Comeau et al., 2004a, 2004b). In addition, the putative model of actin/cofilin shows a broad convex/concave interface, and such surfaces provide better results with a clustering radius of 8–9 Å in general (Comeau, 2007). In such cases, a narrower clustering radius results in a lower sampling of structural neighbors for clustering, which may result in a model having appreciable deviation from the native structure.

10.4 METHOD SUMMARY

Due to technical limitations of crystallography, NMR, and cryo-EM, bridging the gap between the protein sequence and structure requires the application of computational methods. However, a main drawback of the computational methods lies in distinguishing correct models from false positives. Incorporation of experimental data with computational modeling methods reduces the computational search time and facilitates correct predictions. There are many computational methods reported in the literature that incorporate experimental data such as NMR, mutation, cryo-EM and so on in the structure determination of individual proteins as well as protein binary complexes. In this chapter, we have described a method to generate three-dimensional model of protein binary complexes using structural mass spectrometry data. Using radiolytic footprinting approaches, we monitor the surface accessibility of the reactive residues in proteins; this accessibility can change due to allosteric effects of ligand binding or due to solvent exclusion at macromolecular interfaces.

Without the constraints provided by footprinting data, we were able to generate accurate models of actin/DNaseI and actin/profilin using docking alone. For actin/profilin, the top-scoring model is consistent with the crystal structure, whereas for actin/DNaseI, the second-ranked structure matched with the crystal structure. In the latter case, the top-scoring model deviates from the crystal structure by nearly 30 Å RMSD. In addition, these models could only be provided if the correct energy filtering (desolvation or electrostatics) was chosen. When incorrect energy filtering was used, the models generated were more than 10 Å RMSD from the respective crystal structures. Overall, the probability of the correct structure being top scored by ClusPro is only 27%, based on a study involving 48 pairs of interacting proteins. Therefore, picking the correct model from the multiple models generated by docking algorithms is very difficult without additional experimental information.

We used radiolytic footprinting data as experimental constraints to derive a model for actin/GS1 complex consistent with the known crystal structure and of actin/cofilin complex of unknown crystal or NMR structure. First, the binding (interface) region of the larger protein (actin in this case) identified by footprinting is examined and surface residues that are sufficiently solvent exposed in that binding partner are chosen as "attract" residues to feed into the docking step. Surface residues from the regions outside the binding are chosen as "block" residues to feed into the docking step. Attract and block constraints are set up in separate runs or in combination. Attract constraint provides positive weights to the chosen residues in computing the van der Waals energy during the docking step; block constraints prevent docking in these regions by providing a high positive value of the energy parameter in the docking step. Second, the binding regions of individual proteins identified by the footprinting are examined for the nature of the electrostatic surface pattern; based on these data, either electrostatic or desovlation free energy filter option is chosen in the docking program. This step retains the top 2000 energy-minimized conformations for the complex out of the 20,000 docked conformations that have the best shape complementarity scores. For actin/GS1 docking, there is a low probability of finding the correct structure as top scoring when constraints in the docking step are not used or when the incorrect energy filter is used. However, when the correct energy filter is used, and attract, block, or a combination of both constraints are used, the models that ranked second, first, and first, respectively, were those that showed an interface structure similar to that of the crystal structure.

We also derived a model for G-actin/cofilin complex of unknown crystal or NMR structure based on footprinting data. A conformation consistent with footprinting data was top scored with "attract," "block," or "attract + block" constraints and with the correct energy filter. Observation of a substantial number of H-bonds (9–12 in number) at the interface is consistent with the reported charge-dependent binding of cofilin to G-actin (Pavlov et al., 2006). Our model also satisfies the findings of mutational and NMR data on the involvement of cofilin's charged residues Gln120, Asp123, and Glu126 (from β6–α4 loop) in G-actin binding.

Overall, the method outlined in Scheme 10.1 provides a rational and successful approach to generating and selecting the most accurate structural models when docking binary complexes. The current capabilities of the ClusPro server allow receptor molecule heavy atom (non-H-atoms) size of 9700 and ligand molecule heavy atom size of 3700. Also, attract and block constraints are allowed only for the receptor molecule residues. However, improvements in this and other docking algorithms are likely in the near future (Tovchigrechko and Vakser, 2006).

10.5 EXPERIMENTAL METHODS

10.5.1 Radiolysis

Radiolysis experiments, in which protein samples are exposed to a white synchrotron X-ray beam for intervals from 0 to 200 ms, are performed at the X-28C beamline of the National Synchrotron Light Source, Brookhaven National Laboratory (Upton, New York), as previously described (Guan et al., 2002, 2003). On these timescales, oxidative modifications dominate the chemistry compared to cross-linking events and cleavage (Davies and Dean, 1997) and effects on the global protein structure are minimal.

10.5.2 Mass Spectrometry

Radiolyzed samples are subjected to enzymatic proteolysis (Guan et al., 2002, 2003) and the resulting peptide mixtures are separated and analyzed using a coupled, high-pressure liquid chromatography–electrospray ion-source mass spectrometer (HPLC–ESI MS) equipped with quadrupole ion trap (Waters 2690 microflow HPLC–Finnigan LCQ classic ESI MS, Dionex nanoflow HPLC–Finnigan LCQ DecaXP nanospray MS). The oxidized probe sites within each peptide are confirmed by tandem mass spectrometry (MS/MS) (Guan and Chance, 2005; Takamoto and Chance, 2006). First-order rate constants of modification are derived from dose–response curves (Guan and Chance, 2005; Takamoto and Chance, 2006). The detailed procedures have been previously described (Guan and Chance, 2005; Takamoto and Chance, 2006).

10.5.3 Solvent-Accessible Surface Area Calculation

The solvent-accessible surface areas of amino acids were calculated using the computerprogram GETAREA 1.1 (http://www.pauli.utmb.edu/cgi-bin/get_a_form.tcl)

from the crystal structures with a probe radius of 1.4 Å with additional atomic parameter database and residue-type library entries. SASA is defined as the surface mapped out by the center of the probe as if it were rolled around the van der Waals surface of the protein.

10.5.4 Homology Modeling

Actin in the cofilin-bound form, precisely in the closed nucleotide cleft conformation, was obtained by homology modeling using SWISS-MODEL protein modeling server (http://swissmodel.expasy.org/) (Schwede et al., 2003), inputting the rabbit actin sequence as "target" and the 3D structure of bovine actin (2BTF chain A, 2.55 Å resolution) as "template."

10.5.5 Protein–Protein Docking

Docking of ligand protein to receptor protein, energy filtering, clustering, and ranking were done using the ClusPro Web server (http://nrc.bu.edu/cluster). Docking with DOT (Daughter of Turnip) program (Katchalski-Katzir et al., 1992; Ten Eyck et al., 1995), provided in the ClusPro, is executed using a surface complementarity fit of the rotating molecule (ligand protein, which is ABP in the present study) on the fixed molecule (receptor protein, which is actin in the present study) (Katchalski-Katzir et al., 1992), which is based on van der Waals contact energy minimization. This employs a 128 Å × 128 Å × 128 Å grid with a spacing of 1 Å. The receptor protein is placed on the center of the 128 Å^3 cube. Each heavy atom of the protein is surrounded by an inner repulsive layer of 1.5 Å and an outer attractive layer that extends 4 Å beyond the repulsive core. Technically, each heavy atom of the receptor protein is given an energy score of $+1000$ within the repulsive layer and of -1.0 within the attractive layer. The ligand atoms are viewed as point charges with a value of $+1.0$ and cannot be modified. The goal, then, is to minimize the "energy" score by maximizing the number of attractive layer–ligand interactions, while minimizing the number of repulsive layer–ligand interactions. Using a predefined list of 13,000 rotations of ligand protein within the 128 Å^3 around the receptor protein, 2.7×10^{10} structures are evaluated, retaining the 20,000 structures with the best shape complementarity scores. These structures are then energy filtered (electrostatics or desolvation) to retain top 2000 structures, which are then clustered (9 Å clustering radius) and ranked according to their cluster sizes. For details of the desolvation and electrostatics energy function, refer Comeau et al. (2004a, 2004b). Clustering is accomplished based on the work of Shortle et al. (1998), where the native conformation is found to be that with the highest number of structural neighbors (e.g., highest cluster sizes).

10.5.6 Docking with Experimental Constraints

Constraints, namely, "attract" and "block" can be set up within the docking protocol (Comeau et al., 2004a, 2004b) individually in separate runs or in combination.

"Attract" constraint provides positive weighting to docking at the experimentally derived interfacial residues of receptor protein, and "block" constraint blocks the docking outside this interfacial area. Technically, when attracting to certain regions, the repulsive layer is decreased by 1 Å, leaving only a 0.5-Å repulsive core on the atom and changing the energy score of the attractive layer from -1.0 to -2.0. When blocking certain regions of the protein, the attractive layer of all selected atoms is removed by changing the value from -1.0 to 0.0. The details of these constraints are provided in the documentation of the ClusPro (Comeau et al., 2004a, 2004b). Finally, the data from models are compared with biological and biochemical data to understand the functional implications of the model.

10.5.7 Electrostatic Potential Surface Mapping and Calculation of Interface Parameters

Electrostatic potential surfaces of proteins were mapped using DeepView. The root mean square deviation of the actin binding protein and of the interface of the models with the crystal structures or with other structural models are calculated using DeepView. Actin molecules are aligned at first using the option "fit molecules from selection," followed by calculating the RMSD of the $C\alpha$ atoms of the ligand proteins and of the interface residues, respectively. Interface SASA, percentage of polar atoms, hydrogen bonds/salt bridges, planarity, and gap volume index are derived using protein–protein interaction server V 1.5. (http://www.biochem.ucl.ac.uk/bsm/PP/server/) (Jones and Thornton, 1996). Interface residues are all those residues that showed a SASA decrease of 1 Å^2 or more upon complexation (Jones and Thornton, 1996). Planarity of the interface is defined as the RMSD of the atoms forming the interface plane, which is the best fit plane through the three-dimensional coordinates of the atoms in the interface using principal component analysis (Jones and Thornton, 1996). Gap volume index is defined as the ratio of the volume of the gaps between the two interacting proteins at the interface, as calculated using a program SURFNET, to the interface SASA. The gap volume index is a measure of the complementarity of the interacting surfaces.

10.5.8 Footprinting Interface Consistency Score

Footprinting interface consistency score is the fraction of the total number of footprinting probe residues that are hypothesized to form the binding surface based on an independent analysis of the footprinting data compared (denominator) with the number that are actually found in the interface of the docked model (numerator). As such, this score can vary between 0 (no common residues) and 1 (all residues match).

ACKNOWLEDGMENT

This work was supported by U.S. National Institutes of Health (NIH) grant NIBIB-P41-01979. I would like to thank Dr. Mark Chance for all the useful discussions.

REFERENCES

Agbulut, O., Huet, A., Niederländer, N., Puceat, M., Menasché, P., and Coirault, C., 2007. Green fluorescent protein impairs actin–myosin interactions by binding to the actin-binding site of myosin. *J Biol Chem* 282, 10465–10471.

Baker, D. and Sali, A., 2001. Protein structure prediction and structural genomics. *Science* 294, 93–96.

Blanchoin, L. and Pollard, T. D., 1998. Interaction of actin monomers with Acanthamoeba actophorin (ADF/cofilin) and profilin. *J Biol Chem* 273, 25106–25111.

Bonanno, J. B., Almo, S. C., Bresnick, A., Chance, M. R., Fiser, A., Swaminathan, S., Jiang, J., Studier, F. W., Shapiro, L., Lima, C. D., Gaasterland, T. M., Sali, A., Bain, K., Feil, I., Gao, X., Lorimer, D., Ramos, A., Sauder, J. M., Wasserman, S. R., Emtage, S., D'Amico, K. L., and Burley, S. K., 2005. New York-Structural GenomiX Research Consortium (NYSGXRC): a large scale center for the protein structure initiative. *J Struct Funct Genomics* 6, 225–232.

Chance, M. R., Fiser, A., Sali, A., Pieper, U., Eswar, N., Xu, G., Fajardo, J. E., Radhakannan, T., and Marinkovic, N., 2004. High-throughput computational and experimental techniques in structural genomics. *Genome Res* 14, 2145–2154.

Chandonia, J. M. and Brenner, S. E., 2005. Implications of structural genomics target selection strategies: Pfam5000, whole genome, and random approaches. *Proteins* 58, 166–179.

Chiu, W., Baker, M. L., and Almo, S. C., 2006. Structural biology of cellular machines. *Trends Cell Biol* 16, 144–150.

Comeau, S. R., 2007. Development of high-throughput tools for the automated generation of protein–protein complexes. In: *Bioinformatics Program*, Boston, MA: Boston University. 162 pp.

Comeau, S. R., Gatchell, D. W., Vajda, S., and Camacho, C. J., 2004a. ClusPro: a fully automated algorithm for protein–protein docking. *Nucleic Acids Res* 32, W96–W99.

Comeau, S. R., Gatchell, D. W., Vajda, S., and Camacho, C. J., 2004b. ClusPro: an automated docking and discrimination method for the prediction of protein complexes. *Bioinformatics* 20, 45–50.

Comeau, S. R., Vajda, S., and Camacho, C. J., 2005. Performance of the first protein docking server ClusPro in CAPRI rounds 3–5. *Proteins* 60, 239–244.

Davies, M. J. and Dean, R. T., 1997. *Radical-Mediated Protein Oxidation: From Chemistry to Medicine*. Oxford: Oxford University Press.

Dominguez, R., 2004. Actin-binding proteins: a unifying hypothesis. *Trends Biochem Sci* 29, 572–578.

Ewing, R. M., Chu, P., Elisma, F., Li, H., Taylor, P., Climie, S., McBroom-Cerajewski, L., Robinson, M. D., O'Connor, L., Li, M., Taylor, R., Dharsee, M., Ho, Y., Heilbut, A., Moore, L., Zhang, S., Ornatsky, O., Bukhman, Y. V., Ethier, M., Sheng, Y., Vasilescu, J., Abu-Farha, M., Lambert, J. P., Duewel, H. S., Stewart, I. I., Kuehl, B., Hogue, K., Colwill, K., Gladwish, K., Muskat, B., Kinach, R., Adams, S. L., Moran, M. F., Morin, G. B., Topaloglou, T., and Figeys, D., 2007. Large-scale mapping of human protein–protein interactions by mass spectrometry. *Mol Syst Biol* 3, 89.

Fedorov, A. A., Lappalainen, P., Fedorov, E. V., Drubin, D. G., and Almo, S. C., 1997. Structure determination of yeast cofilin. *Nat Struct Biol* 4, 366–369.

Gavin, A. C., Aloy, P., Grandi, P., Krause, R., Boesche, M., Marzioch, M., Rau, C., Jensen, L. J., Bastuck, S., Dumpelfeld, B., Edelmann, A., Heurtier, M. A., Hoffman, V., Hoefert, C.,

Klein, K., Hudak, M., Michon, A. M., Schelder, M., Schirle, M., Remor, M., Rudi, T., Hooper, S., Bauer, A., Bouwmeester, T., Casari, G., Drewes, G., Neubauer, G., Rick, J. M., Kuster, B., Bork, P., Russell, R. B., and Superti-Furga, G., 2006. Proteome survey reveals modularity of the yeast cell machinery. *Nature* 440, 631–636.

Goldsmith, S. C., Guan, J. Q., Almo, S., and Chance, M., 2001. Synchrotron protein footprinting: a technique to investigate protein–protein interactions. *J Biomol Struct Dyn* 19, 405–418.

Gruschus, J. M., Greene, L. E., Eisenberg, E., and Ferretti, J. A., 2004. Experimentally biased model structure of the Hsc70/auxilin complex: substrate transfer and interdomain structural change. *Protein Sci* 13, 2029–2044.

Guan, J. Q., Almo, S. C., and Chance, M. R., 2004. Synchrotron radiolysis and mass spectrometry: a new approach to research on the actin cytoskeleton. *Acct Chem Res* 37, 221–229.

Guan, J. Q., Almo, S. C., Reisler, E., and Chance, M. R., 2003. Structural reorganization of proteins revealed by radiolysis and mass spectrometry: G-actin solution structure is divalent cation dependent. *Biochemistry* 42, 11992–12000.

Guan, J. Q. and Chance, M. R., 2005. Structural proteomics of macromolecular assemblies using oxidative footprinting and mass spectrometry. *Trends Biochem Sci* 30, 583–592.

Guan, J. Q., Takamoto, K., Almo, S. C., Reisler, E., and Chance, M. R., 2005. Structure and dynamics of the actin filament. *Biochemistry* 44, 3166–3175.

Guan, J. Q., Vorobiev, S., Almo, S. C., and Chance, M. R., 2002. Mapping the G-actin binding surface of cofilin using synchrotron protein footprinting. *Biochemistry* 41, 5765–5775.

Halperin, I., Ma, B., Wolfson, H., and Nussinov, R., 2002. Principles of docking: an overview of search algorithms and a guide to scoring functions. *Proteins* 47, 409–443.

Janin, J., Henrick, K., Moult, J., Eyck, L. T., Sternberg, M. J., Vajda, S., Vakser, I., and Wodak, S. J., 2003. Critical assessment of predicted interactions. *Proteins* 52, 2–9.

Jones, S. and Thornton, J. M., 1996. Principles of protein–protein interactions. *Proc Natl Acad Sci USA* 93, 13–20.

Kamal, J. K. A., Benchaar, S. A., Takamoto, K., Reisler, E., and Chance, M. R., 2007. Three-dimensional structure of cofilin bound to monomeric actin derived by structural mass spectrometry data. *Proc Natl Acad Sci USA* 104, 7910–7915.

Katchalski-Katzir, E., Shariv, I., Eisenstein, M., Friesem, A. A., Aflalo, C., and Vakser, I. A., 1992. Molecular surface recognition: determination of geometric fit between proteins and their ligands by correlation techniques. *Proc Natl Acad Sci USA* 89, 2195–2199.

Kontoyianni, M., McClellan, L. M., and Sokol, G. S., 2004. Evaluation of docking performance: comparative data on docking algorithms. *J Med Chem* 47, 558–565.

Lappalainen, P., Fedorov, E. V., Fedorov, A. A., Almo, S. C., and Drubin, D. G., 1997. Essential functions and actin-binding surfaces of yeast cofilin revealed by systematic mutagenesis. *EMBO J* 16, 5520–5530.

Mannherz, H. G., Ballweber, E., Galla, M., Villard, S., Granier, C., Steegborn, C., Schmidtmann, A., Jaquet, K., Pope, B., and Weeds, A. G., 2007. Mapping the ADF/cofilin binding site on monomeric actin by competitive cross-linking and peptide array: evidence for a second binding site on monomeric actin. *J Mol Biol* 366, 745–755.

McLaughlin, P. J., Gooch, J. T., Mannherz, H. G., and Weeds, A. G., 1993. Structure of gelsolin segment 1–actin complex and the mechanism of filament severing. *Nature* 364, 685–692.

Mendez, R., Leplae, R., De Maria, L., and Wodak, S. J., 2003. Assessment of blind predictions of protein–protein interactions: current status of docking methods. *Proteins* 52, 51–67.

Pavlov, D., Muhlrad, A., Cooper, J., Wear, M., Reisler, E., 2006. Severing of F-actin by yeast cofilin is pH-independent. *Cell Motil Cytoskeleton* 63(9), 533–542.

Perola, E., Walters, W. P., and Charifson, P. S., 2004. A detailed comparison of current docking and scoring methods on systems of pharmaceutical relevance. *Proteins* 56, 235–249.

Pons, T., Gonzalez, B., Ceciliani, F., and Galizzi, A., 2006. FlgM anti-sigma factors: identification of novel members of the family, evolutionary analysis, homology modeling, and analysis of sequence-structure-function relationships. *J Mol Model* 12, 973–983.

Pope, B. J., Zierler-Gould, K. M., Kuhne, R., Weeds, A. G., and Ball, L. J., 2004. Solution structure of human cofilin: actin binding, pH sensitivity, and relationship to actin-depolymerizing factor. *J Biol Chem* 279, 4840–4848.

Reboul, J., Vaglio, P., Rual, J. F., Lamesch, P., Martinez, M., Armstrong, C. M., Li, S., Jacotot, L., Bertin, N., Janky, R., Moore, T., Hudson, J. R., Jr., Hartley, J. L., Brasch, M. A., Vandenhaute, J., Boulton, S., Endress, G. A., Jenna, S., Chevet, E., Papasitropoulos, V., Tolias, P. P., Ptacek, J., Snyder, M., Huang, R., Chance, M. R., Lee, H., Doucette-Stamm, L., Hill, D. E., and Vidal, M., 2003. C. elegans ORFeome version 1.1: experimental verification of the genome annotation and resource for proteome-scale protein expression. *Nat Genet* 34, 35–41.

Sali, A., Glaeser, R., Earnest, T., and Baumeister, W., 2003. From words to literature in structural proteomics. *Nature* 422, 216–225.

Schwede, T., Kopp, J., Guex, N., and Peitsch, M. C., 2003. SWISS-MODEL: an automated protein homology-modeling server. *Nucleic Acids Res* 31, 3381–3385.

Shortle, D., Simons, K. T., and Baker, D., 1998. Clustering of low energy conformations near the native structures of small proteins. *Proc Natl Acad Sci USA* 95, 11158–11162.

Takamoto, K. and Chance, M. R., 2006. Radiolytic protein footprinting with mass spectrometry to probe the structure of macromolecular complexes. *Annu Rev Biophys Biomol Struct* 35, 251–276.

Ten Eyck, L. F., Mandell, J., Roberts, V. A., and Pique, M. E., 1995. Surveying molecular interactions with DOT. In: Hayes, A. and Simmons, M., editors. *Proceedings of the 1995 ACM/IEEE Supercomputing Conference*, New York: ACM Press.

Tovchigrechko, A., and Vakser, I. A., 2006. GRAMM-X public web server for protein-protein docking. *Nucleic Acids Res* 34, W310–314.

Tyers, M., and Mann, M., 2003. From genomics to proteomics. *Nature* 422, 193–197.

Vitkup, D., Melamud, E., Moult, J., and Sander, C., 2001. Completeness in structure genomics. *Nat Struct Biol* 8, 559–566.

von Mering, C., Krause, R., Snel, B., Cornell, M., Oliver, S. G., Fields, S., and Bork, P., 2002. Comparative assessment of large-scale data sets of protein–protein interactions. *Nature* 417, 399–403.

Wriggers, W., Tang, J. X., Azuma, T., Marks, P. W., and Janmey, P. A., 1998. Cofilin and gelsolin segment-1: molecular dynamics simulation and biochemical analysis predict a similar actin binding mode. *J Mol Biol* 282, 921–932.

CHAPTER 11

Studies of Intact Proteins and Protein Complexes: ESI MS Approaches

IGOR A. KALTASHOV, RINAT R. ABZALIMOV, and AGYA K. FRIMPONG

Department of Chemistry, University of Massachusetts at Amherst, Amherst, MA, USA

STEPHEN J. EYLES

Department of Polymer Science and Engineering, University of Massachusetts at Amherst, MA, USA

11.1 INTRODUCTION

The exceptional success of proteomics and bioinformatics in recent years has resulted in unimaginable growth of information on the composition of complex networks of proteins interacting at the cellular level and beyond. Coupled with the avalanche of new high-resolution protein structures, these advances raise hopes that rational manipulation of these networks could lead to the development of novel therapeutic strategies. However, possessing an inventory of protein structures and simple knowledge of their interaction partners is often insufficient to understand how the components of sophisticated biological machinery work together. It is becoming increasingly clear that not only the most stable protein structures but also the multiplicity of all thermodynamically available conformations, as well as the transitions among these states, are critical determinants of protein–protein interactions. Contrary to earlier beliefs, conformational heterogeneity and large-scale dynamics are important and vital characteristics of such interactions (Dunker et al., 2002; Haynes et al., 2006; Hilser and Thompson, 2007; Xie et al., 2007).

Traditionally, a stable, well defined structure was considered an absolute requirement for biological function (e.g., interaction with other biopolymers or small ligands). However, limitations of the "binding as rigid docking" paradigm are now becoming increasingly apparent. *Folding via adaptation* to a binding partner (Griffith and Kaltashov, 2003) is one of the examples of *folding–binding coupling* that was initially predicted by the Nussinov group, who used the *folding funnel* concept to rationalize protein binding mechanisms (Tsai et al., 999; Kumar et al., 2000), and was

Mass Spectrometry Analysis for Protein–Protein Interactions and Dynamics, Edited by Mark Chance
Copyright © 2008 John Wiley & Sons, Inc.

recently considered by Wolynes and coworkers (Shoemaker et al., 2000; Papoian and Wolynes, 2003). Indeed, the results of our own research emphasize the importance of structural plasticity both in modulating functional properties (Gumerov et al., 2003; Xiao and Kaltashov, 2005) and assembly of multisubunit proteins (Griffith and Kaltashov, 2003, 2007). Furthermore, a growing number of proteins are found to be either partially or fully unstructured under native conditions, both *in vitro* and *in vivo* (Stirling et al., 2003). Traditionally, such lack of structure has had negative connotations, since nonnative proteins were viewed as species lacking any functional activity and prone to aggregation. It is now becoming apparent that *intrinsic disorder* (i.e., anomalous flexibility exhibited under native conditions) is quite ubiquitous in nature and is vital for the function of many proteins (Dunker et al., 2001, 2002).

It is, however, important to realize that delicate balance between order and chaos must be maintained to ensure the proper functioning and indeed the very survival of a cell. Protein misfolding is linked to a variety of pathological conditions, including Alzheimer's disease and prion disorders (Thompson and Barrow, 2002). Another area where the issues of protein misfolding and aggregation have profound practical importance is the biopharmaceutical industry (Weng and DeLisi, 2002). As the share of biologics (mostly protein pharmaceuticals) among the new drugs has reached 25% and is steadily growing, it is becoming clear that the efficacy of protein drugs is directly linked to their stability during manufacturing, storage, and administration. Since conformation and dynamics are key to protein stability, characterization of higher order structure and conformational heterogeneity becomes one of the major analytical tasks in biotechnology.

The central role of macromolecular higher order structure and interactions in fields as diverse as biophysics, structural biology, and biotechnology places a premium on the ability to characterize them. However, experimental investigation of architecture and conformational heterogeneity of proteins, and their associations with each other, remains a challenging task. X-ray crystallography, by definition, is biased toward the static structures of the most stable state(s). Although recently there have been dramatic technological improvements in this field, particularly in cryo-crystallography and time-resolved crystallography, allowing observation of small-scale dynamic events in many cases (Petsko and Ringe, 2000; Moffat, 2001; Rajagopal et al., 2004), large-scale dynamics and protein states with high degrees of disorder evade analysis by these techniques. Large-scale conformational changes that result in significant changes in protein shape can be detected using solution scattering methods (Wall et al., 2000; Doniach, 2001). However, despite dramatic recent progress in using this technology to provide crude characterization of protein shape in monodisperse systems (Koch et al., 2003), characterization of protein higher order structure remains difficult. Furthermore, the utility of solution scattering methods is often limited by the requirement to carry out measurements in very concentrated protein solutions.

Spectroscopic methods, such as fluorescence, circular dichroism (CD) (Sreerama and Woody, 2004) and FTIR (Barth and Zscherp, 2002), are often used to detect significant conformational changes. However, the structural information provided in such measurements is limited to either cumulative changes of the secondary structure (far-UV CD and FTIR), interactions among aromatic residues (intrinsic fluorescence, near-UV CD), or local environments of extrinsic fluorescent probes. Likewise, time-resolved spectroscopic techniques (such as fluorescence spectroscopy) characterize protein dynamic events with unsurpassed temporal resolution and with incredible sensitivity (Neuweiler and Sauer, 2004), but provide only rather limited structural information (e.g., changes in the local environment of aromatic residues).

Nuclear magnetic resonance (NMR) remains the only spectroscopic technique capable of providing a wealth of both structural and dynamic information on proteins (Ishima and Torchia, 2000; Clore and Schwieters, 2002; Kempf and Loria, 2003; Schwalbe, 2003). However, despite its spectacular success in recent years, the utility of high-resolution NMR remains limited in many cases by its low tolerance to paramagnetic ligands and cofactors and a practical upper molecular weight limitation of ca. 30 kDa. Furthermore, NMR analysis usually requires high protein concentration in solution and produces structural data averaged across the entire protein ensemble, unable to make a clear distinction between various protein states that may coexist in solution at equilibrium.

Mass spectrometry (MS) has emerged relatively recently as an attractive alternative in the studies of protein architecture and dynamics, capable of providing information on protein conformation at various levels (Kaltashov and Eyles, 2002a, 2005). It has become a potent experimental tool in biophysics that, in many cases, provides valuable information on various aspects of protein behavior in solution. A particularly unforgiving limitation inherent to most experimental techniques used to probe macromolecular structure and dynamics is the extreme difficulty in characterizing behavior of individual polymers in multicomponent systems, which arises due to inevitable signal interference from different species. One of the most unique features of MS, which makes it a particularly attractive option in the studies of proteins, is its ability to deal with inhomogeneous samples containing mixtures of biopolymers and small ligands. Significant progress in MS instrumentation in recent years has resulted in enhanced ability of modern MS to afford distinction among analytes with similar masses, thereby allowing the signal interference problem to be avoided even in the least favorable circumstances (He et al., 2004).

Most of the biophysical methods to characterize protein conformation and dynamics presented in this book rely on using certain chemical reactions to place labels on a protein surface followed by quenching the reaction and localization of labels by means of MS. However, several MS-based experimental techniques offer an opportunity to study protein behavior in solution directly with no sample workup prior to MS analysis. This can be done either by carrying out the labeling reaction online during the sample injection to electrospray ionization mass spectrometry (ESI MS) (e.g., by using the sample injection syringe as an H/D exchange reactor) or by relying on certain intrinsic properties of electrospray-generated ions (such as charge state distributions) as reporters of macromolecular properties in solution. Elimination of the sample-handling steps prior to MS analysis not only shortens the analysis time, but often allows one to avoid artifacts associated with the quenching step, as well as sample storage and/or cleanup prior to MS measurements. Although the techniques based on direct ESI MS measurements have their own limitations (most notably those related to solvent compatibility with the ESI process), in many cases they provide unique information on various aspects of protein structure and behavior in solution.

11.2 TERTIARY STRUCTURE INTEGRITY AND CONFORMATIONAL HETEROGENEITY (CHARGE STATE DISTRIBUTIONS)

Multiple charging of macromolecular analytes is a unique feature of ESI MS, which makes it distinct from other ionization methods used in mass spectrometry to transfer polar and thermally labile species from condensed phase to vacuum. Although FAB and matrix-assisted laser desorption/ionization (MALDI) mass spectra of some biopolymers may also contain contributions from ionic species carrying more than one elementary charge, they usually constitute only a small fraction of the overall ionic signal and the charge density of such species (expressed as a number of charges per 1 kDa of macromolecular mass) is very low compared to ESI-generated ions. Protonation is the most common form of multiple (positive) charging of polypeptides in ESI MS, although formation of alkali metal and ammonium adducts is also often observed. Generation of multiple protonated polypeptide ions by ESI is a convoluted process, whose outcome does not depend on the protonation state of amino acid side chains in solution (Kelly et al., 1992). The major determinant of the extent of protonation is the physical size of the protein molecule in solution (or, more precisely, its solvent-exposed surface area), although other factors may also influence the number of charges carried by biopolymer ions into the gas phase. Since the surface of a polypeptide chain in solution is determined by its higher order structure, the extent of biopolymer charging observed in ESI mass spectra contains information on its conformation in solution. Therefore, analysis of protein ion charge state distributions in ESI mass spectra often provides a simple yet very effective means of evaluating the integrity of higher order structure of proteins and their complexes and also assessing their conformational heterogeneity.

The potential of ESI MS as a means of detecting large-scale conformational transitions in solution was realized soon after the introduction of this ionization technique, when dramatic changes of charge state distributions of protein ions were linked to protein denaturation (Chowdhury et al., 1990; Loo et al., 1991). Natively folded proteins give rise to ions carrying a relatively small number of charges, as their compact shapes in solution do not allow a significant number of protons to be accommodated on the surface upon transition from solution to the gas phase. As a result, protein ion peaks in ESI mass spectra acquired under near-native conditions (i.e., in aqueous solutions kept at neutral pH in the absence of denaturants) typically appear in the high m/z region of ESI mass spectra. Charge state distributions in such cases are almost always narrow. In contrast, nonnative (partially or fully unfolded) protein conformers give rise to ions carrying a significantly larger number of charges, as many more protons can be accommodated on the surface of a protein once it loses its compactness (compare panels (a) and (c) in Fig. 11.1).

If both native and denatured states of the protein coexist in solution under equilibrium, the protein ion charge state distributions in the ESI spectra are bimodal (Fig. 11.1b). Therefore, dramatic changes of protein ion charge state distributions often serve as gauges of large-scale conformational changes (Konermann and Douglas, 1997). This feature is widely used to study protein dynamics in processes ranging from folding (Konermann and Douglas, 1998; Grandori, 2002; Borysik et al., 2004) to ligand

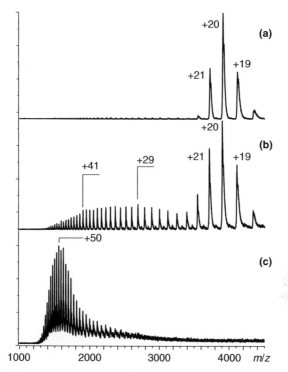

FIGURE 11.1   ESI mass spectra of an 80 kDa protein transferrin acquired under near-native (10 mM ammonium acetate, pH 7.0, panel a), mildly denaturing (10 mM ammonium acetate, pH adjusted to 5.0, panel b), and strongly denaturing (water/methanol/acetic acid, 47:50:3 v:v:v, panel c) conditions. Emergence of nonnative (partially unfolded) states is evident in (b) as the charge state distribution becomes bimodal. Further unfolding of the protein (population of significantly less compact states) is manifested in (c) by a dramatic increase of the abundance of highly charged protein ions. Adapted, with permission from Elsevier Ltd., from Abzalimov et al., 2006.

binding (Yu et al., 1993; Gumerov and Kaltashov, 2001; Low et al., 2002; van den Bremer et al., 2002) to protein assembly (Griffith and Kaltashov, 2003, 2007; Simmons et al., 2004) and their interaction with other biopolymers (Kamadurai et al., 2003).

As long as a fraction of protein molecules in solution retain their native fold, ionic signal corresponding to these molecules is usually characterized by a nearly constant average charge. On the contrary, the portion of ionic signal representing nonnative (less compact) states is much more heterogeneous and evolves as the solution conditions change (e.g., compare panels (b) and (c) in Fig. 11.1). The reason for such behavior is that most proteins have multiple nonnative conformations, not just a single random coil state. Very often differences in the solvent-exposed surface areas among such states are not significant enough to give rise to distinct ionic signals in ESI mass spectra. As a result, charge state distributions corresponding to various protein conformational isomers may be unresolved or poorly resolved (i.e., two or more different conformers may give rise to ions carrying the same number of charges). Therefore, monitoring changes in the relative abundance of individual protein ions typically does not allow characterization of the transitions among various nonnative states in a quantitative fashion.

The problem of overlapping ionic signals can be circumvented using chemometric tools (Dobo and Kaltashov, 2001; Mohimen et al., 2003). In brief, an array of ESI MS data is acquired over a range of near-native and denaturing conditions to ensure adequate sampling of various protein states and significant variation of their fractional concentrations. The total number of protein conformers sampled in the course of the experiment is determined by subjecting the entire set of collected mass spectra to singular value decomposition (SVD) (Malinowski, 2002). SVD of the data array yields a number of independent components, which (apart from the noise) are responsible for the observed variations of protein ion charge state distributions. This

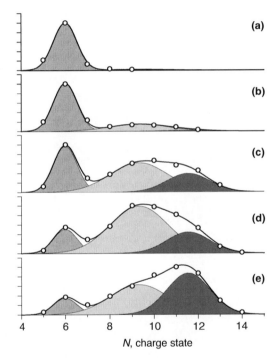

FIGURE 11.2 Charge state distributions of ubiquitin (Ub) in ESI mass spectra acquired in 10 mM ammonium acetate, pH 7.0 (a); 10 mM ammonium acetate, pH 7.0, and 60% methanol by volume (b); 10 mM ammonium acetate, pH adjusted to 2.0 with acetic acid (c); 10 mM ammonium acetate, pH adjusted to 2.0 with acetic acid, and 60% methanol by volume (d); H_2O/methanol (40:60, v:v), pH adjusted to 2.0 with acetic acid (e). The charge state distributions were deconvoluted by using a chemometric approach described in (Mohimen et al., 2003). The three conformers contributing to the overall ionic signal are assigned as the native state (blue), the A-state (yellow), and the random coil (red). Reprinted from: Kaltashov, I. A., Eyles, S. J., Mohimen, A., Hoerner, J. K., Abzalimov, R. R., and Griffith W. P. Analysis of partially folded proteins by electrospray ionization mass spectrometry. In: Methods in Protein Structure and Stability Analysis: NMR and EPR Spectroscopies, Mass-Spectrometry and Protein Imaging. Uversky, V. ed. Hauppauge, New York: Nova Science Publishers, Inc., 2008, pp. 175–196.

number is equal to the number of protein conformers whose geometries are different enough to allow at least some distinction to be made as far as their individual contributions to the overall charge state distributions. A set of basis functions is then constructed, each representing a "pure signal" of a certain conformer and the entire data array is fitted using this set, yielding ionic profiles of individual protein states over the range of solution conditions (Mohimen et al., 2003). The contribution of each conformer to the total ionic signal is determined by using a supervised minimization routine. This procedure allows any set of asymmetric charge state distributions to be represented as sums of a limited number of symmetric basis functions, each corresponding to a particular protein conformer (Fig. 11.2).

Lack of a sufficient number of basic residues has been cited in the past as a factor limiting the extent of multiple charging and, therefore, preventing detection of protein unfolding in solution (Felitsyn et al., 2002). In a recent study we investigated the possibility of false negatives in the unfolding of proteins with low pI. A model protein used in these studies was pepsin, a gastric protein whose basic side chains (four) are greatly outnumbered by acidic ones (forty one) (Abzalimov et al., 2006). Despite the small number of basic sites, the highest charge of pepsin ions generated under denaturing conditions (neutral and basic pH) exceeds the number of available basic residues by nearly a factor of 10. Furthermore, the extent of multiple charging of pepsin polycations in ESI MS under these conditions is not lower than that of polyanions (Fig. 11.3). This surprising finding clearly demonstrates that the extent of

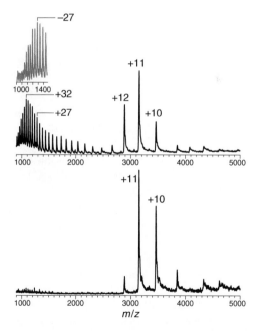

FIGURE 11.3   ESI mass spectra (positive ions) of pepsin in 10 mM ammonium acetate, 50% methanol by volume, pH adjusted to 10 (top), and 10 mM ammonium acetate, pH 6 (bottom). Inset shows the low m/z region of negative ion mass spectrum acquired at pH 10. Adapted with permission from Elsevier Ltd., from Abzalimov et al., 2006.

multiple charging is determined by the protein geometry in solution, not the number of available basic sites, and rules out the possibility of false negatives when charge state distributions are relied upon as indicators of unfolding in solution (Abzalimov et al., 2006).

11.3 QUATERNARY STRUCTURE INTEGRITY AND COMPOSITION OF NON-COVALENT COMPLEXES

Another unique feature of ESI MS that makes it extremely attractive as a tool to study protein interactions in solution directly is its ability to preserve noncovalent complexes formed by proteins interacting with each other, small ligands, or indeed other biopolymers (Loo, 1997, 2000; Benesch and Robinson, 2006). The existing body of knowledge suggests that many types of intra- and intermolecular interactions in proteins that are responsible for integrity of noncovalent associations in solution (with the notable exception of hydrophobic interactions) are preserved *in vacuo* (Hoaglund-Hyzer et al., 1999; Jarrold, 2000). Therefore, noncovalent macromolecular complexes often survive the transition from solution to the gas phase when mild desolvation conditions are employed in the ESI MS interface. In many cases, this allows the composition of macromolecular assemblies to be determined reliably and with minimal sample consumption.

Above and beyond the ability to characterize quaternary structure at the level of binding stoichiometry, ESI MS is capable of providing a wealth of complementary information related to protein interactions in solution. For example, we have recently demonstrated that the correlation between the extent of multiple charging of protein ions in ESI MS and the degree of protein compactness in solution can be used to provide estimates of solvent-exposed surface area (SASA) of natively folded proteins and their complexes (Kaltashov and Mohimen, 2005). Although evaluation of SASA on the basis of measurements of average charges of protein ions in ESI MS cannot rival other established techniques as far as measurement precision is concerned, it may be extremely useful for characterizing protein assemblies in solution that are not amenable to analysis using traditional biophysical tools due to their transient nature or heterogeneous character. One such example is shown in Fig. 11.4, in which the surface area of a nonglobular protein assembly (an octameric form of a sickle cell hemoglobin, HbS) is estimated by measuring the average charge of protein ions in ESI MS and using the charge–surface correlation (Kaltashov and Mohimen, 2005) as a "calibration curve." The octameric form of HbS is often viewed as a precursor to HbS polymerization in red blood cells, a process leading to erythrocyte deformation (Manning et al., 1998). Therefore, it appears that the charge–surface correlation can be used to provide reasonable estimates of solvent-shielded surface at protein–protein interfaces within macromolecular assemblies in solution.

Furthermore, charge state distributions of protein assemblies, subassemblies, and individual subunits in ESI mass spectra can be used to shed light on protein interaction processes by characterizing conformational heterogeneity of each player in the assembly process. In this case, mass measurement provides a means to make a

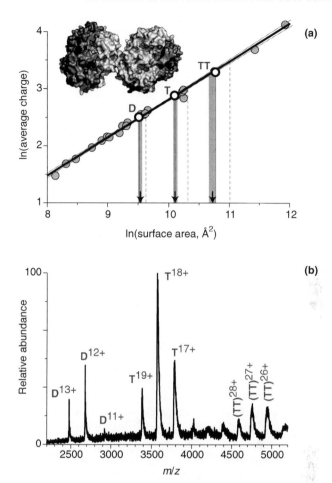

FIGURE 11.4 The use of charge–surface correlation for natively folded proteins (a) as a calibration curve for estimation of surface area of the octameric species of human sickle cell hemoglobin (HbS). The ESI mass spectrum of HbS (40 μM solution at neutral pH) is shown in panel (b). The evaluation of surface areas of dimers (D), tetramers (T), and octamers (TT) was based on the experimentally measured average number of charges accommodated by the respective protein ions (open circles in panel a). Shaded circles on the graph in panel (a) represent the set of protein ions (ranging from a 5 kDa insulin to 0.5 MDa ferritin) used to establish the charge–surface correlation and construct the calibration curve. None of the sickle cell hemoglobin species was used to construct the calibration curve. Vertical lines projected from the x-axis represent the numerical values of surfaces of the protein species based upon crystal structures of oligomers (solid) and simple summation of the surfaces of the constituent monomeric species (dashed). Arrows indicate the numerical value of surfaces estimated based on the ESI MS data ($1.40 \cdot 10^4$ Å^2 for D, $2.48 \cdot 10^4$ Å^2 for T, and $4.56 \cdot 10^4$ Å^2 for TT), with shaded areas representing confidence intervals for such estimates. Adapted with permission from Kaltashov and Mohimen 2005. Copyright 2008 American Chemical Society.

distinction among different species that become populated in the course of protein–protein interaction, while the distinction among different conformations of the same species is based upon the appearance of its ionic charge state distributions. This makes ESI MS unrivaled in its ability to characterize large-scale protein conformational dynamics in highly heterogeneous mixtures, a feature that we have used successfully to highlight an important role of structural plasticity in the assembly processes of normal and α-thalassemic hemoglobin tetramers, Hb A (Griffith and Kaltashov, 2003) and Hb H (Griffith and Kaltashov, 2007).

11.4 FUNCTIONAL COMPETENCE

While various proteomic approaches have been very successful in revealing the identity of interacting partners at the critical nodes of complex interactomes, understanding the mechanisms of such interactions at the molecular level requires detailed knowledge of higher order structure and conformational dynamics of all players involved in these processes. The ability of ESI MS to characterize biopolymers and their assemblies at various levels makes it an indispensable tool in such studies. The utility of direct ESI MS measurements extends well beyond proteins, allowing functionally important interactions to be probed in processes ranging from enzyme catalysis to gene expression. Figure 11.5 shows ESI mass spectra of NikR, a regulatory metalloprotein that controls expression of proteins facilitating nickel uptake in several bacteria (Chivers and Sauer, 2002). Since this metal is a cofactor of the *H. pylori* urease, an enzyme essential for chronic colonization of the hostile acidic environment in the host stomach (van Vliet et al., 2004), NikR is an attractive therapeutic target in the treatment of ulcers and related conditions. Both apo and holo forms of this protein are folded and form stable tetrameric structures, as revealed by ESI MS (Fig. 11.5). However, only the nickel-saturated form of NikR has the ability to bind strongly to the operator DNA region, as evidenced by the appearance of the abundant ionic species at m/z above 4,500, whose mass corresponds to the protein–DNA complex. Importantly, ESI MS data provide clear evidence that the apoprotein also has the ability to associate weakly with the operator. Metal binding dramatically enhances the DNA binding ability of NikR (presumably by stabilizing the binding-competent conformer) and results in a dramatic shift of equilibrium in the association reaction

$$NikR + dsDNA \rightleftharpoons NikR \cdot dsDNA.$$

This example highlights the range of opportunities offered by direct ESI MS measurements as a means of monitoring biological interactions at an incredible level of detail.

Direct ESI MS measurements are readily applicable to the analysis of protein higher order structure and interactions in highly heterogeneous systems, making them an indispensable tool in the studies of enzymatic reactions and their modulation by both intrinsic (e.g., higher order structure) and extrinsic factors (e.g., inhibitors).

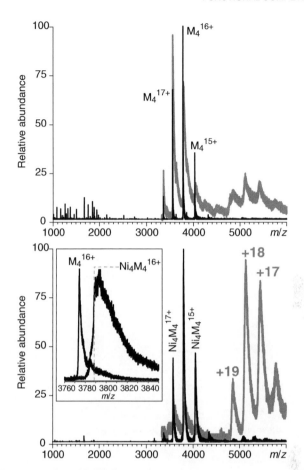

FIGURE 11.5   Modulation of NikR interaction with the operator DNA sequence by nickel. Although both apo (top) and holo (bottom) form of NikR form stable tetrameric structures, binding of Ni^{2+} to this protein is required to form a stable protein–DNA complex (mass spectra of protein/DNA mixtures are shown in gray on each panel). This event *in vivo* arrests the expression of proteins responsible for nickel uptake by the bacterium. Modeling of this process *in vitro* is illustrated with ESI mass spectra of protein–DNA mixtures obtained by diluting premixed protein/DNA (1:1 molar ratio) stock solutions to a final concentration of 10 µM in 100 µM ammonium acetate. As expected, only nickel-saturated protein binds strongly to the DNA operator sequence. The inset shows peak shapes of tetrameric protein ions for apo and holo forms of NikR. Reproduced with permission from Springer Science and Business Media from Kaltashov et al., 2006.

This is illustrated in Fig. 11.6 with the results of our recent study of pepsin inactivation, where loss of proteolytic activity can be clearly linked to either inhibitor binding to or a conformational transition within the enzyme (Frimpong et al., 2007). According to the results of the analysis of pepsin ion charge state distributions in ESI MS, native pepsin is still present in solution at pH as high as 6.5, although accumulation of another compact (inactive) state is evident at much lower pH

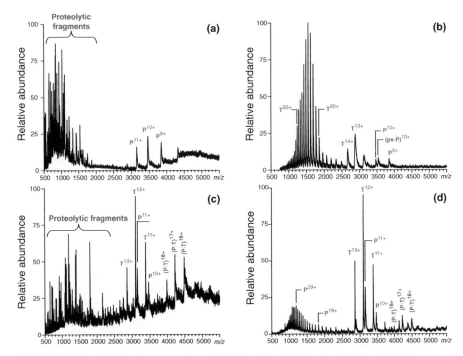

FIGURE 11.6   ESI mass spectra of pepsin (P) incubated with the N-lobe of human serum transferrin (T) for 5 min at room temperature in solutions whose pH was kept at 5.0 (a), 6.4 (c), and 9.5 (d). Panel (b) shows a mass spectrum of P/T mixture incubated at pH 1.6 in the presence of excess pepstatin (pS). The P/T molar ratio was kept at 4:1 in (a) and (b) to increase the abundance of pepsin ions (intensity of pepsin ions was very weak at 1:1 enzyme/substrate molar ratio due to the protein ion signal suppression by ions at low m/z). The absence of proteolytic fragments in panels b and d indicates inactivation of pepsin. Loss of the enzymatic activity in neutral and basic solutions is due to a transition from the native state to a compact inactive conformation (d). Pepsin inactivation under near-physiological conditions (b) is due to the inhibitor binding (see also Fig. 11.9).

(Frimpong et al., 2007). Incubation of a model substrate protein (N-lobe of human serum transferrin, T) with pepsin at pH <6 results in quick and efficient proteolysis (Fig. 11.6a), unless pepsin inhibitor pepstatin (ps) is present in solution (Fig. 11.6b). ESI MS measurements also provide clear evidence that pepsin retains its enzymatic activity at pH as high as 6.5, although a fraction of the substrate molecules remain intact under these conditions (Fig. 11.6c). Complex formation between the substrate and the inactive state of pepsin is also evident in the mass spectrum under these conditions. No proteolytic activity is observed at pH above 6.5 (Fig. 11.6d).

Probing pepsin activity as a function of pH unequivocally confirms the results of the chemometric analysis of conformational transitions occurring within this protein as the pH is raised from highly acidic to neutral. Inactivation of pepsin in mildly acidic solutions is partial due to the coexistence of two states in equilibrium, native (active) and compact (inactive). The complete demise of the native state of pepsin above neutral pH results in complete inactivation of the enzyme, even though a significant fraction of the protein molecules still populates a compact (and presumably significantly structured) conformation (Frimpong et al., 2007).

11.5 FLEXIBILITY MAPS AND BINDING INTERFACES

Monitoring mass and charge state distributions provides a very efficient and relatively simple means to observe the evolution of protein higher order structure in solution at low resolution. However, direct ESI MS measurements can also be employed to characterize protein conformation, dynamics, and interaction at a higher level of structural detail. Combination of hydrogen/deuterium exchange (HDX) with mass spectrometric (MS) detection has become a powerful biophysical tool capable of probing protein structure and dynamics in solution under a variety of conditions (Kaltashov and Eyles, 2002; Wales and Engen, 2006) and on different timescales (Konermann and Simmons, 2003). MS offers several important advantages compared to other means of monitoring the progress of HDX reactions, such as NMR. First, the superior sensitivity of ESI and MALDI MS allows many HDX MS experiments to be carried out using only minute quantities of proteins. In many cases, this enables studies of protein behavior at, or even below, their endogenous concentrations without encountering artifacts such as protein aggregation in solution. Second, ESI MS greatly outperforms high-field NMR in its ability to handle larger proteins and their complexes and, unlike high-field NMR, it is tolerant to paramagnetic ligands. Third, under certain conditions MS detection allows protein characterization to be carried out in a conformer-specific fashion, providing a means of making clear distinction between various equilibrium intermediates of the same protein (Simmons and Konermann, 2002; Zhang et al., 2002; Hoerner et al., 2005; Pan et al., 2005; Xiao et al., 2005).

Site-specific information in HDX MS experiments is usually obtained by quenching the exchange reaction followed by protein digestion, rapid separation and MS analysis of deuterium content of proteolytic fragments (Engen and Smith, 2001; Wales and Engen, 2006). This methodology (described in detail in Chapters 2 and 7 of this book) remains the most popular choice in HDX MS studies, as it can be implemented using a variety of (often inexpensive) platforms (Mandell et al., 1998; Woods and Hamuro, 2001; Wu et al., 2006) and in some favorable cases allows the deuteration patterns to be mapped down to the single amide level (Del Mar et al., 2005). However, it is possible in many cases to bypass the proteolytic step by inducing fragmentation of protein ions in the gas phase (Eyles et al., 2000; Kaltashov and Eyles, 2002b). Such direct HDX MS measurements (sometimes dubbed "top-down" HDX MS) eliminate the need for several steps prior to MS analysis (e.g., proteolysis and separation of peptic fragments), while still providing a wealth of information on flexibility and solvent exposure of various protein segments. This alternative methodology is gaining popularity and has been applied to characterize dynamics and structural features of proteins in their native (Hoerner et al., 2004) and nonnative (Hoerner et al., 2005; Kaltashov, 2005) states, as well as to elucidate the role of protein dynamics in ligand binding (Xiao and Kaltashov, 2005).

One unique advantage of HDX MS/MS experiments that has not been exploited as yet is its ability to generate fragment ions from a particular conformer with a specific level of deuterium incorporation. Under certain conditions, HDX MS allows a distinction to be made among various protein conformations based on the differences in backbone amide protection levels (Xiao et al., 2005). An example is shown in Fig. 11.7, where the isotopic profile of a deuterated protein (a stable mutant of Cellular Retinoic Acid Binding Protein I, wt^*-CRABP I) is recorded following a 10-min exposure to the ^1H−based solvent at pH 3.0. The bimodal appearance of the isotopic distribution of the molecular ion (top trace in Fig. 11.7a) clearly indicates the presence of at least two conformers with different degrees of backbone protection. Collisional activation of the entire protein ion population generates a set of fragment ions with very convoluted isotopic distributions (top trace in Fig. 11.7b). This makes it difficult to relate the protection patterns of individual fragments to the structures of individual protein conformers (the very same problem would arise should the structural studies be carried out using proteolytic digestion of protein under slow exchange conditions, as the peptic fragments derived from protein molecules populating different conformations would be inevitably mixed with each other). However, mass selection of precursor ions with a specific level of deuterium content

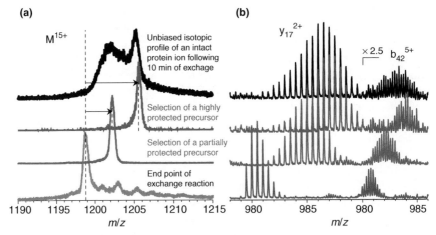

FIGURE 11.7 Application of direct HDX MS/MS measurements for characterization of local dynamics in wt^*-CRABP I in a conformer-specific fashion. The black trace in panel a shows the isotopic distribution of the deuterated protein following a 10-min exposure to $^1H_2O/CH_3CO_2N^1H_4$ at pH 3.5 (the gray trace at the bottom of the panel corresponds to the end point of this exchange reaction). CAD of the entire ionic population generates fragment ions whose ^2H levels reflect protection of the corresponding segments in both conformers, whose distinct contributions cannot be resolved (black trace in panel b). However, precursor ion mass selection prior to fragmentation allows the ^2H content of these protein segments to be probed separately in a highly protected (blue traces) and partially protected conformer (red traces). (See the color version of this figure in Color Plate section.)

allows the top–down HDX MS measurements to be carried out in a conformation-specific fashion, taking full advantage of the HDX MS ability to observe distinct conformers. For example, selective activation of protein ions representing a highly protected conformation can be achieved by mass-selecting a narrow population of molecular ions (e.g., m/z 1205–1206 for a $+15$ charge state, the blue trace in Fig. 11.7a) prior to the activation event. Likewise, mass selection and consecutive activation of a narrow population of protein ions with significantly lower deuterium content (e.g., m/z 1201–1203, the red trace in Fig. 11.7a) generates a set of fragment ions that can be used to probe the protection pattern within nonnative protein states.

Indeed, the isotopic distributions of fragment ions generated from these two distinct precursor ion populations are very telling as far as the structure of the protein states visualized in HDX MS experiments is concerned. For example, deuterium content of the b_{42} fragment derived from a highly protected precursor (the blue trace in Fig. 11.7b) corresponds to a significant level of protection in the N-terminal region of the protein consisting of two structural elements (β-strand 1 and α-helix I) of the so-called portal region of CRABP I (Xiao and Kaltashov, 2005). At the same time, deuterium content of the b_{42} fragment derived from precursor ions representing nonnative protein states (the red trace in Fig. 11.7b) matches that of a fully unstructured polypeptide chain at pH 3 and room temperature.

This behavior is in sharp contrast with the protection patterns displayed by the C-terminal fragments. For example, significant levels of deuterium retention by y_{17} fragments regardless of their origin (red and blue traces in Fig. 11.7b) indicates noticeable protection of this segment in both native and nonnative states of CRABP I. Although the space limitations do not allow us to provide a detailed analysis of HDX MS/MS data, it is clear that this methodology holds great promise for characterization of structure and dynamics of individual populations of conformers of proteins coexisting in solution, a capability unique to this experimental strategy.

11.6   GAS PHASE ION CHEMISTRY AND ITS INFLUENCE ON THE MEASUREMENT OF PROTEIN PROPERTIES IN SOLUTION

While ESI MS has been extremely successful in revealing the most intimate details of protein structure and interactions in solution, it is important to remember that MS measurements are carried out in the absence of solvent, and such a dramatic change of the macromolecular environment surely affects their properties in often unexpected ways (Wolynes, 1995). Therefore, prior to applying any MS-based method to probe macromolecular properties in solution, the following questions must be addressed: Are there any processes occurring in the gas phase prior to MS detection that may influence the measured ionic parameters? If so, what effect will such interference have on the measurements of macromolecular properties in solution? Finally, what can be done to minimize these effects?

Of particular concern are phenomena related to dissociation of noncovalent complexes in the gas phase, which are often cited as reasons for somewhat skeptical assessment of direct ESI MS measurements of protein binding in solution. Indeed, incorrect identification of gas phase fragmentation products in ESI MS as species originating from solution would compromise conclusions of any study aimed at the characterization of protein quaternary structure, as well as mechanisms of protein association in solution. However, numerous studies have demonstrated that the pathways of dissociation of protein noncovalent complexes in the gas phase are very different from those in solution (Sobott et al., 2003; Abzalimov et al., 2006; Benesch and Robinson, 2006). Gas phase dissociation of a noncovalent protein complex ion generated by ESI under native conditions proceeds via the so-called asymmetric charge partitioning, where a monomer ion carrying a disproportionately large fraction of the total charge is expelled from the assembly.

While the exact mechanism of asymmetric charge partitioning remains a subject of continuous inquiry (Felitsyn et al., 2001; Jurchen et al., 2004; Abzalimov et al., 2006; Sinelnikov et al., 2007), it allows those fragment ions generated solely from gas phase processes to be readily distinguished from ions representing various stages of noncovalent complex assembly or disassembly in solution. This is illustrated in Fig. 11.8, where gas phase dissociation of tetrameric ions of hemoglobin $(\alpha^*\beta^*)^{17+}$ results in the formation of highly charged monomers α^{n+} and β^{k+} and complementary trimer ions. The most abundant signal generated by trimeric fragments corresponds to an exotic species consisting of three globin chains and four (!) heme groups (see the inset in Fig. 11.8). This fragmentation behavior is absolutely inconsistent with hemoglobin tetramer assembly/disassembly pathways in solution, which proceed via the formation of dimers (Griffith and Kaltashov, 2003). Therefore,

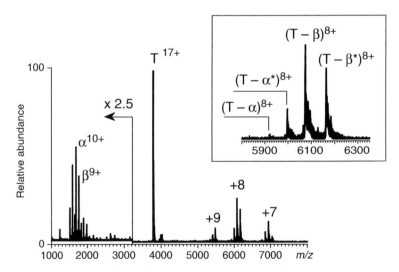

FIGURE 11.8 CAD mass spectrum of a tetrameric species of sickle cell human hemoglobin, HbS (charge state $+17$) acquired with a hybrid quadrupole TOF mass spectrometer. Ionic species labeled in the spectra are tetramers (T) and their fragments corresponding to losses of α- or β-globins (asterisks denote a heme group). Inset shows a zoomed view of the trimeric fragments region in the spectrum (charge state $+8$). Adapted, with permission from Elsevier Ltd., from Abzalimov et al. (2006).

a distinction between gas phase fragmentation and noncovalent complex dissociation in solution is usually straightforward.

Seemingly counterintuitively, gas phase dissociation of protein–small ligand complexes presents a more significant challenge vis-à-vis assigning the origin of subcomplex ions compared to polypeptide assemblies (Clark and Konermann, 2004; Sun et al., 2007). If a departing ligand does not possess sufficient physical dimensions to accommodate multiple charges, its dissociation from the host protein would not result in asymmetric charge partitioning. Therefore, the resulting gas phase fragment ions might not be easily distinguishable from those generated in solution. However, as long as at least one charge can be placed on a ligand, its dissociation from the protein in the gas phase will affect the charge state distribution of the latter, providing a means to identify gas phase dissociation products. This is illustrated in Fig. 11.9, which shows the ESI mass spectrum of a pepsin–pepstatin mixture acquired under conditions highly favorable for the formation of the enzyme–inhibitor complex. Although the peaks arising from the apoprotein are prominent in this spectrum, a close examination of the charge state distribution of pepsin ions reveals a shift to a lower number of charges compared to the mass spectrum of pepsin acquired in the absence of pepstatin. This apparent charge reduction clearly indicates that the ligand-free pepsin ions in the ESI mass spectrum of the pepsin–pepstatin mixture are

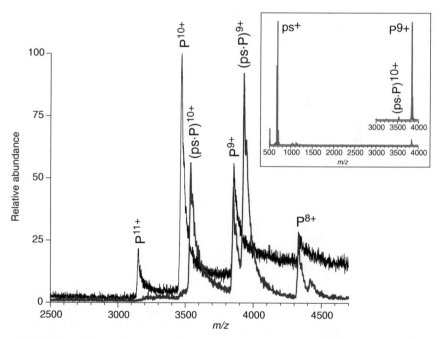

FIGURE 11.9 ESI mass spectrum of pepsin (P) acquired at pH 2.3 (black trace). The brown trace shows a mass spectrum of pepsin incubated with excess of pepstatin (ps). Inset: CAD mass spectrum of $(ps \cdot P)^{10+}$ ion suggests that enzyme–inhibitor complex dissociation in the gas phase proceeds via charge separation. (See the color version of this figure in Color Plate section.)

produced upon dissociation of the enzyme–inhibitor complex in the gas phase, which proceeds via energetically favorable charge separation.

Although asymmetric charge partitioning is beneficial when direct ESI MS is used to characterize protein quaternary structure, under certain circumstances it may give rise to false positive signals of protein unfolding. For example, formation of metastable protein aggregates either in the bulk solution or in the electrosprayed droplets, their consequent transfer to the gas phase and asymmetric dissociation may give rise to a population of highly charged ions (Abzalimov et al., 2006). The presence of such ions in ESI mass spectra usually indicates the loss of native structure in solution (*vide supra*). Therefore, studies of large-scale conformational dynamics in solution by monitoring protein ion charge state distributions should be carried out at low protein concentrations to minimize the occurrence of false positive signals of protein unfolding. Charge transfer reactions involving protein ions in the gas phase may also lead to alterations of charge state distributions not related to conformational changes in solution (Gumerov et al., 2002).

Gas phase processes may also influence the accuracy and reliability of HDX MS measurements, especially in those cases when polypeptide ion fragmentation in the gas phase is used to probe local patterns of deuterium distribution along the backbone (Johnson et al., 1995; Demmers et al., 2002). The results of our past studies suggest that the occurrence and the extent of the so-called hydrogen scrambling are influenced by a variety of experimental parameters, most important of which is the ion activation rate (Hoerner et al., 2004). More recently, we studied the effect of various experimental factors on the extent of hydrogen scrambling in HDX MS/MS experiments. Hydrogen scrambling does not occur when the charge state of the precursor protein ions selected for fragmentation is high, as fragment ions derived from both N- and C-terminal parts of the protein are equally unaffected by scrambling. However, spatial distribution of deuterium atoms obtained by fragmenting low charge density protein ions is consistent with a very high degree of scrambling prior to the dissociation events. The extent of hydrogen scrambling is also high when multistage fragmentation is used to probe deuterium incorporation locally. Taken together, the experimental results provide a coherent picture of intramolecular processes occurring prior to the dissociation event and provide guidance for the design of experiments whose outcome is unaffected by hydrogen scrambling (*manuscript in preparation*).

11.7   CHALLENGES AND FUTURE OUTLOOK

As both the scope and number of applications of direct ESI MS to probing macromolecular behavior in solution are rapidly expanding, it becomes ever more important to address the challenges that still limit the utility of this powerful technique in some areas of protein research. For example, we already mentioned that a very important advantage offered by direct ESI MS is its ability to deal with complex multicomponent systems. Indeed, ion peaks corresponding to different protein components of the mixture will generally have different m/z values and, therefore, will not interfere with each other. However, in some unfavorable situations, signal interference does present a problem even for ESI MS measurements, as high degrees of structural heterogeneity (e.g., those frequently encountered among highly glycosylated proteins) lead to significant broadening of ion peaks and, in extreme cases, overlap of peaks corresponding to different species and charge states.

A similar problem is expected to be encountered when direct ESI MS measurements are applied to characterize conformational properties and binding preferences of synthetically modified proteins (e.g., PEGylation). Protein conjugation with synthetic polymer chains is becoming increasingly popular as a way to enhance solubility, bioavailability, and circulation time of biopharmaceuticals (Harris and Chess, 2003). However, even "monodisperse" and relatively low molecular weight polymer chains exhibit a significant degree of heterogeneity compared to similar-sized biopolymers whose production is genetically controlled (Fig. 11.10). This degree of heterogeneity will be further amplified due to multiple conjugations of a single protein molecule with several polymer chains. Clearly, a meaningful interpretation of the results of direct ESI MS measurements in these cases will require utilization of sophisticated mathematical tools (e.g., maximum entropy-based methods to resolve individual ionic species and charge states in the mass spectra prior to applying chemometric tools).

Another aspect of direct ESI MS measurements that must be mentioned here is the requirement that the solvent must be compatible with the ESI process. This limits the repertoire of electrolyte systems that can be used to buffer/adjust the ionic strength of protein solutions to several volatile systems (with ammonium acetate and ammonium bicarbonate being the most popular ones). Although this limitation was considered as very unforgiving in the past, significant improvements in ESI interface design,

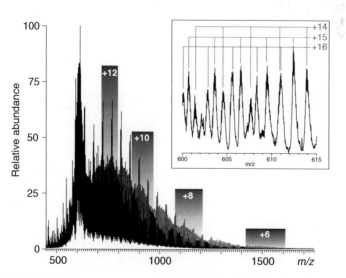

FIGURE 11.10 ESI mass spectrum of PEG-8000 in 10 mM ammonium acetate.

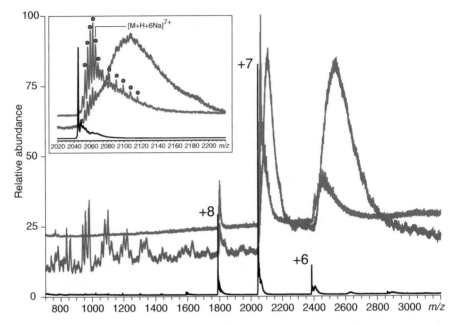

FIGURE 11.11 ESI MS of 10 μM solution of chicken egg white lysozyme acquired in 100 mM ammonium bicarbonate in the absence of nonvolatile electrolytes (black trace) and in the presence of 50 mM NaCl (blue and red traces). Inset shows zoomed views of lysozyme ions carrying seven charges. Peaks labeled with red circles correspond to $[M + (7-n) H + nNa]^{+7}$ ions. Peaks marked with blue circles correspond to clusters $[M + (7-n) H + nNa + kNaCl]^{+7}$. The blue trace represents a mass spectrum acquired under mild conditions in the ESI MS interface; significant desolvation of protein ions was achieved by increasing their collisional energy in the interface region (red trace) (See the color version of this figure in Color Plate section).

and the emergence of nano-ESI technique in particular, provide reasonable grounds for optimism. Higher tolerance of nano-ESI (compared to ESI) for salt and nonvolatile buffers was noted some time ago, at least for small peptides (Juraschek et al., 1999). The presence of concentrated NaCl in protein solutions inevitably leads to spectral quality deterioration in nano-ESI MS, giving rise to abundant $Na^+(NaCl)_n$ clusters in the low m/z region and protein–NaCl adducts at higher m/z (Fig. 11.11). Their presence in the mass spectra makes the accurate mass assignment a very challenging task, and is also likely to compromise the ability of direct ESI MS approaches to analyze heterogeneous mixtures by making distinction among various components on the basis of differences in their mass. However, even relatively mild collisional activation of ions in the ESI interface results in facile dissociation of the cluster species, yielding easily interpretable MS data (see the red trace in Fig. 11.11). Further optimization of experimental conditions is likely to expand the scope of direct MS measurements to include conditions that more closely mimic the *in-vivo* environment.

It appears certain that further improvements in experimental methodology and expansion of the scope of direct ESI MS characterization of biopolymer behavior in solution will provide an urgently needed means to evaluate higher order structure, probe conformational dynamics, and monitor interactions for a variety of macromolecules, including species of both biotic and abiotic origin. Undoubtedly, this will greatly benefit multiple fields not only in traditional biochemical disciplines, but also in the emerging fields at the interface of biology, medicine, chemistry, and physics, such as the design of biopharmaceuticals and bioinspired materials.

ACKNOWLEDGMENTS

Preparation of this chapter was supported by grants from the National Institutes of Health (R01 GM061666) and the National Science Foundation (CHE-0406302). Prof. Roman A. Zubarev (Biomedicinska Centrum, Uppsala University) is gratefully acknowledged for hosting one of the authors (I.K.) during his Sabbatical Leave from UMass-Amherst and help with LTQ MS and LTQ FT MS work.

REFERENCES

Abzalimov, R. R., Frimpong, A. K., and Kaltashov, I. A., 2006. Gas-phase processes and measurements of macromolecular properties in solution: on the possibility of false positive and false negative signals of protein unfolding. *Int J Mass Spectrom* 253, 207–216.

Barth, A. and Zscherp, C., 2002. What vibrations tell us about proteins. *Q Rev Biophys* 35, 369–430.

Benesch, J. L. and Robinson, C. V., 2006. Mass spectrometry of macromolecular assemblies: preservation and dissociation. *Curr Opin Struct Biol* 16, 245–251.

Borysik, A. J., Radford, S. E., and Ashcroft, A. E., 2004. Co-populated conformational ensembles of β2-microglobulin uncovered quantitatively by electrospray ionization mass spectrometry. *J Biol Chem* 279, 27069–27077.

Chivers, P. T. and Sauer, R. T., 2002. NikR repressor: high-affinity nickel binding to the C-terminal domain regulates binding to operator DNA. *Chem Biol* 9, 1141–1148.

Chowdhury, S. K., Katta, V., and Chait, B. T., 1990. Probing conformational changes in proteins by mass spectrometry. *J Am Chem Soc* 112, 9012–9013.

Clark, S. M. and Konermann, L., 2004. Determination of ligand–protein dissociation constants by electrospray mass spectrometry-based diffusion measurements. *Anal Chem* 76, 7077–7083.

Clore, G. M. and Schwieters, C. D., 2002. Theoretical and computational advances in biomolecular NMR spectroscopy. *Curr Opin Struct Biol* 12, 146–153.

Del Mar, C., Greenbaum, E. A., Mayne, L., Englander, S. W., and Woods, V. L., Jr., 2005. Structure and properties of α-synuclein and other amyloids determined at the amino acid level. *Proc Natl Acad Sci USA* 102, 15477–15482.

Demmers, J. A., Rijkers, D. T., Haverkamp, J., Killian, J. A., and Heck, A. J., 2002. Factors affecting gas-phase deuterium scrambling in peptide ions and their implications for protein structure determination. *J Am Chem Soc* 124, 1191–11198.

Dobo, A. and Kaltashov, I. A., 2001. Detection of multiple protein conformational ensembles in solution via deconvolution of charge state distributions in ESI MS. *Anal Chem* 73, 4763–4773.

Doniach, S., 2001. Changes in biomolecular conformation seen by small angle X-ray scattering. *Chem Rev* 101, 1763–1778.

Dunker, A. K., Lawson, J. D., Brown, C. J., Williams, R. M., Romero, P., Oh, J. S., Oldfield, C. J., Campen, A. M., Ratliff, C. M., Hipps, K. W., Ausio, J., Nissen, M. S., Reeves, R., Kang, C., Kissinger, C. R., Bailey, R. W., Griswold, M. D., Chiu, W., Garner, E. C., and Obradovic, Z., 2001. Intrinsically disordered protein. *J Mol Graph Model* 19, 26–59.

Dunker, A. K., Brown, C. J., Lawson, J. D., Iakoucheva, L. M., and Obradovic, Z., 2002. Intrinsic disorder and protein function. *Biochemistry* 41, 6573–6582.

Engen, J. R. and Smith, D. L., 2001. Investigating protein structure and dynamics by hydrogen exchange MS. *Anal Chem* 73, 256A–265A.

Eyles, S. J., Speir, P., Kruppa, G., Gierasch, L. M., and Kaltashov, I. A., 2000. Protein conformational stability probed by Fourier transform ion cyclotron resonance mass spectrometry. *J Am Chem Soc* 122, 495–500.

Felitsyn, N., Kitova, E. N., and Klassen, J. S., 2001. Thermal decomposition of a gaseous multiprotein complex studied by blackbody infrared radiative dissociation: investigating the origin of the asymmetric dissociation behaviour. *Anal Chem* 73, 4647–4661.

Felitsyn, N., Peschke, M., and Kebarle, P., 2002. Origin and number of charges observed on multiply-protonated native proteins produced by ESI. *Int J Mass Spectrom* 219, 39–62.

Frimpong, A. K., Abzalimov, R. R., Eyles, S. J., and Kaltashov, I. A., 2007. Gas-phase interference-free analysis of protein ion charge-state distributions: detection of small-scale conformational transitions accompanying pepsin inactivation. *Anal Chem* 79, 4154–4161.

Grandori, R., 2002. Detecting equilibrium cytochrome *c* folding intermediates by electrospray ionisation mass spectrometry: two partially folded forms populate the molten-globule state. *Protein Sci* 11, 453–458.

Griffith, W. P. and Kaltashov, I. A., 2003. Highly asymmetric interactions between globin chains during hemoglobin assembly revealed by electrospray ionization mass spectrometry. *Biochemistry* 42, 10024–10033.

Griffith, W. P. and Kaltashov, I. A., 2007. Protein conformational heterogeneity as a binding catalyst: ESI-MS study of hemoglobin H formation. *Biochemistry* 46, 2020–2026.

Gumerov, D. R. and Kaltashov, I. A., 2001. Dynamics of iron release from transferrin N-lobe studied by electrospray ionization mass spectrometry. *Anal Chem* 73, 2565–2570.

Gumerov, D. R., Dobo, A., and Kaltashov, I. A., 2002. Protein-ion charge-state distributions in electrospray ionization mass spectrometry: distinguishing conformational contributions from masking effects. *Eur J Mass Spectrom* 8, 123–129.

Gumerov, D. R., Mason, A. B., and Kaltashov, I. A., 2003. Interlobe communication in human serum transferrin: metal binding and conformational dynamics investigated by electrospray ionization mass spectrometry. *Biochemistry* 42, 5421–5428.

Harris, J. M. and Chess, R. B., 2003. Effect of PEGylation on pharmaceuticals. *Nat Rev Drug Discov* 2, 214–221.

Haynes, C., Oldfield, C. J., Ji, F., Klitgord, N., Cusick, M. E., Radivojac, P., Uversky, V. N., Vidal, M., and Iakoucheva, L. M., 2006. Intrinsic disorder is a common feature of hub proteins from four eukaryotic interactomes. *PLoS Comp Biol* 2, e100.

He, F., Emmett, M. R., Hakansson, K., Hendrickson, C. L., and Marshall, A. G., 2004. Theoretical and experimental prospects for protein identification based solely on accurate mass measurement. *J Proteome Res* 3, 61–67.

Hilser, V. J. and Thompson, E. B., 2007. Intrinsic disorder as a mechanism to optimize allosteric coupling in proteins. *Proc Natl Acad Sci USA* 104, 8311–8315.

Hoaglund-Hyzer, C. S., Counterman, A. E., and Clemmer, D. E., 1999. Anhydrous protein ions. *Chem Rev* 99, 3037–3080.

Hoerner, J. K., Xiao, H., Dobo, A., and Kaltashov, I. A., 2004. Is there hydrogen scrambling in the gas phase? Energetic and structural determinants of proton mobility within protein ions. *J Am Chem Soc* 126, 7709–7717.

Hoerner, J. K., Xiao, H., and Kaltashov, I. A., 2005. Structural and dynamic characteristics of a partially folded state of ubiquitin revealed by hydrogen exchange mass spectrometry. *Biochemistry* 44, 11286–11294.

Ishima, R. and Torchia, D. A., 2000. Protein dynamics from NMR. *Nat Struct Biol* 7, 740–743.

Jarrold, M. F., 2000. Peptides and proteins in the vapor phase. *Annu Rev Phys Chem* 51, 179–207.

Johnson, R. S., Krylov, D., and Walsh, K. A., 1995. Proton mobility within electrosprayed peptide ions. *J Mass Spectrom* 30, 386–387.

Juraschek, R., Dulcks, T., and Karas, M., 1999. Nanoelectrospray: more than just a minimized-flow electrospray ionization source. *J Am Soc Mass Spectrom* 10, 300–308.

Jurchen, J. C., Garcia, D. E., and Williams, E. R., 2004. Further studies on the origins of asymmetric charge partitioning in protein homodimers. *J Am Soc Mass Spectrom* 15, 1408–1415.

Kaltashov, I. A., 2005. Probing protein dynamics and function under native and mildly denaturing conditions with hydrogen exchange and mass spectrometry. *Int J Mass Spectrom* 240, 249–259.

Kaltashov, I. A. and Eyles, S. J., 2002a. Studies of biomolecular conformations and conformational dynamics by mass spectrometry. *Mass Spectrom Rev* 21, 37–71.

Kaltashov, I. A. and Eyles, S. J., 2002b. Crossing the phase boundary to study protein dynamics and function: combination of amide hydrogen exchange in solution and ion fragmentation in the gas phase. *J Mass Spectrom* 37, 557–565.

Kaltashov, I. A. and Eyles, S. J., 2005. Mass Spectrometry in Molecular Biophysics: Conformation and Dynamics of Biomolecules. Hoboken, N. J.: John Wiley and Sons, Inc.

Kaltashov, I. A. and Mohimen, A., 2005. Estimates of protein surface areas in solution by electrospray ionization mass spectrometry. *Anal Chem* 77, 5370–5379.

Kaltashov, I. A., Zhang, M., Eyles, S. J., and Abzalimov, R. R., 2006. Investigation of structure, dynamics and function of metalloproteins with electrospray ionization mass spectrometry. *Anal Bioanal Chem* 386, 472–481.

Kaltashov, I. A., Eyles, S. J., Mohimen, A., Hoerner, J. K., Abzalimov, R. R., and Griffith, W. P., 2008. Analysis of partially folded proteins by electrospray ionization mass spectrometry. In: Uversky, V., editor. *Methods in Protein Structure and Stability Analysis: NMR and EPR Spectroscopies, Mass-Spectrometry and Protein Imaging.* Hauppauge, NY: Nova Science Publishers, Inc. pp. 175–196.

Kamadurai, H. B., Subramaniam, S., Jones, R. B., Green-Church, K. B., and Foster, M. P., 2003. Protein folding coupled to DNA binding in the catalytic domain of bacteriophage lambda integrase detected by mass spectrometry. *Protein Sci* 12, 620–626.

Kelly, M. A., Vestling, M. M., Fenselau, C. C., and Smith, P. B., 1992. Electrospray analysis of proteins: a comparison of positive-ion and negative-ion mass spectra at high and low pH. *Org Mass Spectrom* 27, 1143–1147.

Kempf, J. G. and Loria, J. P., 2003. Protein dynamics from solution NMR: theory and applications. *Cell Biochem Biophys* 37, 187–211.

Koch, M. H., Vachette, P., and Svergun, D. I., 2003. Small-angle scattering: a view on the properties, structures and structural changes of biological macromolecules in solution. *Q Rev Biophys* 36, 147–227.

Konermann, L. and Douglas, D. J., 1997. Acid-induced unfolding of cytochrome *c* at different methanol concentrations: electrospray ionization mass spectrometry specifically monitors changes in the tertiary structure. *Biochemistry* 36, 12296–12302.

Konermann, L. and Douglas, D. J., 1998. Equilibrium unfolding of proteins monitored by electrospray ionization mass spectrometry: distinguishing two-state from multi-state transitions. *Rapid Commun Mass Spectrom* 12, 435–442.

Konermann, L. and Simmons, D. A., 2003. Protein-folding kinetics and mechanisms studied by pulse-labeling and mass spectrometry. *Mass Spectrom Rev* 22, 1–26.

Kumar, S., Ma, B., Tsai, C. J., Sinha, N., and Nussinov, R., 2000. Folding and binding cascades: dynamic landscapes and population shifts. *Protein Sci* 9, 10–19.

Loo, J. A., 1997. Studying noncovalent protein complexes by electrospray ionization mass spectrometry. *Mass Spectrom Rev* 16, 1–23.

Loo, J. A., 2000. Electrospray ionization mass spectrometry: a technology for studying noncovalent macromolecular complexes. *Int J Mass Spectrom* 200, 175–186.

Loo, J. A., Loo, R. R., Udseth, H. R., Edmonds, C. G., and Smith, R. D., 1991. Solvent-induced conformational changes of polypeptides probed by electrospray-ionization mass spectrometry. *Rapid Commun Mass Spectrom* 5, 101–105.

Low, L. Y., Hernandez, H., Robinson, C. V., O'Brien, R., Grossmann, J. G., Ladbury, J. E., and Luisi, B., 2002. Metal-dependent folding and stability of nuclear hormone receptor DNA-binding domains. *J Mol Biol* 319, 87–106.

Malinowski, E. R., 2002. *Factor Analysis in Chemistry,* 3rd ed. New York: John Wiley and Sons, Inc.

Mandell, J. G., Falick, A. M., and Komives, E. A., 1998. Measurement of amide hydrogen exchange by MALDI-TOF mass spectrometry. *Anal Chem* 70, 3987–3995.

Manning, J. M., Dumoulin, A., Li, X., and Manning, L. R., 1998. Normal and abnormal protein subunit interactions in hemoglobins. *J Biol Chem* 273, 19359–19362.

Moffat, K., 2001. Time-resolved biochemical crystallography: a mechanistic perspective. *Chem Rev* 101, 1569–1581.

Mohimen, A., Dobo, A., Hoerner, J. K., and Kaltashov, I. A., 2003. A chemometric approach to detection and characterization of multiple protein conformers in solution using electrospray ionization mass spectrometry. *Anal Chem* 75, 4139–4147.

Neuweiler, H. and Sauer, M., 2004. Using photoinduced charge transfer reactions to study conformational dynamics of biopolymers at the single-molecule level. *Curr Pharm Biotechnol* 5, 285–298.

Pan, J., Wilson, D. J., and Konermann, L., 2005. Pulsed hydrogen exchange and electrospray charge-state distribution as complementary probes of protein structure in kinetic experiments: implications for ubiquitin folding. *Biochemistry* 44, 8627–8633.

Papoian, G. A. and Wolynes, P. G., 2003. The physics and bioinformatics of binding and folding: an energy landscape perspective. *Biopolymers* 68, 333–349.

Petsko, G. A. and Ringe, D., 2000. Observation of unstable species in enzyme-catalyzed transformations using protein crystallography. *Curr Opin Chem Biol* 4, 89–94.

Rajagopal, S., Kostov, K. S., and Moffat, K., 2004. Analytical trapping: extraction of time-independent structures from time-dependent crystallographic data. *J Struct Biol* 147, 211–222.

Schwalbe, H., 2003. Kurt Wüthrich, the ETH Zurich, and the development of NMR spectroscopy for the investigation of structure, dynamics, and folding of proteins. *Chembiochem* 4, 135–142.

Shoemaker, B. A., Portman, J. J., and Wolynes, P. G., 2000. Speeding molecular recognition by using the folding funnel: the fly-casting mechanism. *Proc Natl Acad Sci USA* 97, 8868–8873.

Simmons, D. A. and Konermann, L., 2002. Characterization of transient protein folding intermediates during myoglobin reconstitution by time-resolved electrospray mass spectrometry with on-line isotopic pulse labeling. *Biochemistry* 41, 1906–1914.

Simmons, D. A., Wilson, D. J., Lajoie, G. A., Doherty-Kirby, A., and Konermann, L., 2004. Subunit disassembly and unfolding kinetics of hemoglobin studied by time-resolved electrospray mass spectrometry. *Biochemistry* 43, 14792–14801.

Sinelnikov, I., Kitova, E. N., and Klassen, J. S., 2007. Influence of Coulombic repulsion on the dissociation pathways and energetics of multiprotein complexes in the gas phase. *J Am Soc Mass Spectrom* 18, 617–631.

Sobott, F., McCammon, M. G., and Robinson, C. V., 2003. Gas-phase dissociation pathways of a tetrameric protein complex. *Int J Mass Spectrom* 230, 193–200.

Sreerama, N. and Woody, R. W., 2004. Computation and analysis of protein circular dichroism spectra. *Methods Enzymol* 383, 318–351.

Stirling, P. C., Lundin, V. F., and Leroux, M. R., 2003. Getting a grip on non-native proteins. *EMBO Rep* 4, 565–570.

Sun, J., Kitova, E. N., and Klassen, J. S., 2007. Method for stabilizing protein–ligand complexes in nanoelectrospray ionization mass spectrometry. *Anal Chem* 79, 416–425.

Thompson, A. J. and Barrow, C. J., 2002. Protein conformational misfolding and amyloid formation: characteristics of a new class of disorders that include Alzheimer's and Prion diseases. *Curr Med Chem* 9, 1751–1762.

Tsai, C.-J., Kumar, S., Ma, B., and Nussinov, R., 999. Folding funnels, binding funnels, and protein function. *Protein Sci* 8, 1181–1190.

van den Bremer, E. T., Jiskoot, W., James, R., Moore, G. R., Kleanthous, C., Heck, A. J., and Maier, C. S., 2002. Probing metal ion binding and conformational properties of the colicin E9 endonuclease by electrospray ionization time-of-flight mass spectrometry. *Protein Sci* 11, 1738–1752.

van Vliet, A. H. M., Ernst, F. D., and Kusters, J. G., 2004. NikR-mediated regulation of Helicobacter pylori acid adaptation. *Trends Microbiol* 12, 489–494.

Wales, T. E. and Engen, J. R., 2006. Hydrogen exchange mass spectrometry for the analysis of protein dynamics. *Mass Spectrom Rev* 25, 158–170.

Wall, M. E., Gallagher, S. C., and Trewhella, J., 2000. Large-scale shape changes in proteins and macromolecular complexes. *Annu Rev Phys Chem* 51, 355–380.

Weng, Z. and DeLisi, C., 2002. Protein therapeutics: promises and challenges for the 21st century. *Trends Biotechnol* 20, 29–35.

Wolynes, P. G., 1995. Biomolecular folding in vacuo? *Proc Natl Acad Sci USA* 92, 2426–2427.

Woods, V. L., Jr. and Hamuro, Y., 2001. High resolution, high-throughput amide deuterium exchange-mass spectrometry (DXMS) determination of protein binding site structure and dynamics: utility in pharmaceutical design. *J Cell Biochem* 37S 89–98.

Wu, Y., Engen, J. R., and Hobbins, W. B., 2006. Ultra performance liquid chromatography (UPLC) further improves hydrogen/deuterium exchange mass spectrometry. *J Am Soc Mass Spectrom* 17, 163–167.

Xiao, H. and Kaltashov, I. A., 2005. Transient structural disorder as a facilitator of protein–ligand binding: native H/D exchange-mass spectrometry study of cellular retinoic acid binding protein I. *J Am Soc Mass Spectrom* 16, 869–879.

Xiao, H., Hoerner, J. K., Eyles, S. J., Dobo, A., Voigtman, E., Mel'cuk, A. I., and Kaltashov, I. A., 2005. Mapping protein energy landscapes with amide hydrogen exchange and mass spectrometry: I. A generalized model for a two-state protein and comparison with experiment. *Protein Sci* 14, 543–557.

Xie, H., Vucetic, S., Iakoucheva, L. M., Oldfield, C. J., Dunker, A. K., Uversky, V. N., and Obradovic, Z., 2007. Functional anthology of intrinsic disorder. 1. Biological processes and functions of proteins with long disordered regions. *J Proteome Res* 6, 1882–1898.

Yu, X., Wojciechowski, M., and Fenselau, C., 1993. Assessment of metals in reconstituted metallothioneins by electrospray mass spectrometry. *Anal Chem* 65, 1355–1359.

Zhang, Y. H., Yan, X., Maier, C. S., Schimerlik, M. I., and Deinzer, M. L., 2002. Conformational analysis of intermediates involved in the *in vitro* folding pathways of recombinant human macrophage colony stimulating factor beta by sulfhydryl group trapping and hydrogen/deuterium pulsed labeling. *Biochemistry* 41, 15495–15504.

■■■■■■ CHAPTER 12

Two Approaches to Mass Spectrometric Protein Footprinting: PLIMSTEX and FPOP

MICHAEL L. GROSS

Department of Chemistry, Washington University in St. Louis, St. Louis, MO, USA

MEI M. ZHU

Millennium Pharmaceuticals, Inc., Cambridge, MA, USA

DAVID M. HAMBLY

Amgen, Inc., Seattle, WA, USA

12.1 INTRODUCTION: PROTEIN – LIGAND INTERACTIONS BY MASS SPECTROMETRY, TITRATION, AND HYDROGEN/DEUTERIUM AMIDE EXCHANGE AND FAST PHOTOCHEMICAL OXIDATION OF PROTEINS

Determination of protein – protein and protein – ligand interactions provides insight into protein properties *in vitro* and *in vivo*. In drug discovery, for example, target proteins important in a given disease are identified, and libraries of small-molecule drug candidates are screened to identify those that bind. Strong binders are drug candidates. Mass spectrometric methods continue to be developed to screen libraries of thousands of compounds often by measuring the extent of ligand-binding to a protein.

The purpose of the methods described here is to provide a more refined view and elicit details of binding (e.g., binding constants, binding sites, regions of the protein that change upon binding) that can be achieved by direct mass spectrometry. The methods are not suitable for testing large libraries or for rapid screening. Rather, the intention is to probe more deeply than can be done with direct mass spectrometric methods. Furthermore, we view these methods as a complement to analytical proteomics, and we expect they can be invoked once a protein has been identified to

Mass Spectrometry Analysis for Protein – Protein Interactions and Dynamics, Edited by Mark Chance
Copyright © 2008 John Wiley & Sons, Inc.

provide more information long before solution NMR and X-ray crystallography can be employed. Although success can be achieved in the determination of protein – ligand binding affinities by spectroscopic, calorimetric, and other methods, the large amounts of specifically labeled ligand or protein required often impose limitations. Moreover, one may turn to additional spectroscopic or reaction probes, denaturants, or measurements of equilibrium concentrations following a separation, which may perturb the equilibrium. It is important to add to this arsenal of methods particularly if new methods are convenient, simple, rapid, informative, and sensitive.

Mass spectrometry is now playing a role in the characterization of protein structure, differential expression, dynamics, and functions (Loo, 1997; Yates, 1998; Engen and Smith, 2001; Kaltashov and Eyles, 2002). Several mass spectrometry-based methods are now available for the investigation of protein – ligand binding (Johnson et al., 2002; Powell et al., 2002; Breuker, 2004; Clark and Konermann, 2004; Muckenschnabel et al., 2004; Schermann et al., 2005; Slon-Usakiewicz et al., 2005). Recently, we developed a method (Zhu et al., 2003) to quantify protein – ligand interactions in solution by mass spectrometry, titration, and H/D exchange (PLIMSTEX). This strategy can determine the conformational change, binding stoichiometry, and affinity in a variety of protein – ligand interactions including those involving small molecules, metal ions, peptides, and proteins (Zhu et al., 2003, 2005). To extract binding constants from the titration curves, appropriate curve fitting and modeling is needed, and we recently described this modeling (Zhu, 2004). A complementary approach to PLIMSTEX is kinetics of H/D exchange, and one can couple kinetics with equilibrium titrations (PLIMSTEX) to investigate a variety of effects on binding. Examples are the study of media and ionic strength (Zhu et al., 2003), species specificity, mutations on protein – ligand binding, and systematic changes in ligands (Zhu, 2004).

PLIMSTEX probes the backbone of a protein by following H/D exchange of the amide linkages. Another opportunity for chemical marking or footprinting a protein is to modify side chains instead of amide linkages in the protein backbone. The use of acetylation of lysines with analysis by mass spectrometry is one of the early approaches (Steiner et al., 1991). This idea can be made more powerful and effective by using a reagent that is more general than that achieved by acetylation. Furthermore, if developed properly, the approach can be highly complementary to H/D exchange. If appropriate reagents are used, the change in solvent accessibility or conformation will alter the chemical reactivity of the target protein, enabling the location and affinity of the protein – ligand interaction to be determined. Hydroxyl radical is a reagent that has the appropriate properties. It is reactive with a number of amino acid residues, principally hydrophobic ones. It can be generated in solution by the Fenton reagent. The most fruitful development, however, is by Chance and Brenowitz (Guan et al., 2005), who pioneered the use of hydroxyl radicals for modifying various amino acid residues to determine sites of protein interactions and to follow RNA folding (Takamoto et al., 2004) and protein conformational changes (Liu et al., 2003). In their approach, described in Chapter 4 of this book, synchrotron radiation is used to ionize water molecules of a protein solution to produce the radical cation of water. The cation sheds a proton to form the OH radical ($H_2O^{+\bullet} \rightarrow {^\bullet}OH + H^+$). The radical reacts rapidly with solvent-accessible functional groups that contain sulfur or are aromatic. Less reactive are the aliphatic side chains found in leucine, isoleucine, and valine. Our goal is to incorporate into this procedure a means of generating radicals that is more accessible than a synchrotron and is faster, generating and quenching the hydroxyl radicals in less than a microsecond while allowing significant oxidation of the protein's, solvent-exposed residues (Hambly and Gross, 2005). We call our approach FPOP (fast photochemical oxidation of proteins), whereby we employ a pulsed laser to generate OH radicals by photolyzing HOOH (hydrogen peroxide) on the nanosecond timescale and control their lifetime and concentration in solution with scavengers.

For this approach to be successful, one needs a rapid, reliable, and sensitive method for locating the amino acids that are "footprinted" by this or any other related method. The emergence of modern mass spectrometry interfaced to high-performance chromatography is that method. Old ideas of chemical modification can be made modern by the utilization of mass-spectrometry-based proteomic methods of analysis.

In the following sections, we describe the basis for PLIMSTEX and FPOP, their development, and some applications.

12.2 PROTEIN–LIGAND INTERACTIONS BY MASS SPECTROMETRY, TITRATION, AND H/D AMIDE EXCHANGE (PLIMSTEX)

12.2.1 General Protocol for PLIMSTEX

The PLIMSTEX experiment starts with equilibrating a protein with different concentrations of ligand in aqueous buffer solutions, adding D_2O containing buffer and salts in the same concentration as in the starting solution, and allowing the system to reach a steady state for a time determined by previous experiments to map out the kinetics of exchange (e.g., approximately 1 h). At the appropriate time, the exchange is quenched by decreasing the pH to $\sim$2.5 and cooling to near 0 °C. The solution is then loaded on a small reverse-phase column held at ice bath temperature and desalted by washing with ice-cold, aqueous formic acid (pH $\sim$2.5), a procedure that also exchanges the nonamide sites back to hydrogen in the immobilized protein. Rapid elution with high organic composition solvent delivers the protein to an electrospray ionization (ESI) source. Although the initial studies used ESI (ion trap or quadrupole/time-of-flight analyzers) in the positive-ion mode, we are modifying the method whereby MALDI MS can be used for mass analysis.

12.2.2 Titration Curves

Quenching and desalting cause the ligand(s) to dissociate, and the molecular mass of the liberated protein is measured, allowing the number of deuteriums taken up by solvent-accessible amides to be determined by taking the difference between the centroided molecular mass of the exchanged protein and the corresponding mass of

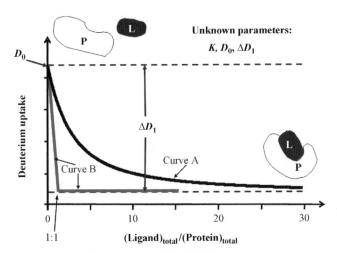

FIGURE 12.1   Schematic illustration of a PLIMSTEX curve for one-to-one protein:ligand binding. (P is protein, L is ligand, K is the binding constant for binding one ligand, D_0 is the deuterium level for apoprotein, and ΔD_1 is the difference between the average deuterium level of one-ligand-bound protein complex and that of the apoprotein.)

the unexchanged or native protein. The extent of exchange is then determined at various ligand concentrations, and a plot of the deuterium uptake versus the total ligand concentration gives the titration or PLIMSTEX curve (Fig. 12.1).

Like all titration experiments for determining affinity, PLIMSTEX curves are sensitive to the total protein concentration. When the protein is titrated at high concentration ($\sim$100 times the $1/K_a$ or K_d), a "sharp break" curve is obtained, and the ratio of [ligand]$_{total}$ to [protein]$_{total}$ at the break clearly indicates binding stoichiometry (Fig. 12.1, curve B). To quantify affinity, PLIMSTEX requires change to occur in the extent of H/D exchange during a titration at a protein concentration comparable to the dissociation constant K_d (Fig. 12.1, curve A). The change may be due to conformational or stability differences between the apo- and ligand-bound proteins, and undoubtedly the difference arises from an increase in hydrogen bonding for those amides somehow involved in the ligand binding.

Modeling of the titration data, using a 1:n protein:ligand sequential binding model, where n is the number of binding sites for the same ligand, yields a best fit to the data and also affords the affinity constant and the number of amides (ΔD_i) for which hydrogen bonding is increased. In the model, we assume that (1) the ligand binds the protein in a stepwise fashion and that (2) the H/D exchange of each amide is independent. To establish the best fit, we use a nonlinear least squares (NLLS) regression (Bates and Watts, 1988) to fit the titration data as a function (Eq. 12.1) of the total ligand concentration ([Lig$_T$]). The best fit gives the overall binding constants (β_i, which is the product of the stepwise macroscopic binding constants K_i, where $i = 1$ to n), and the species deuterium shifts (D_0 and ΔD_i, $i = 1$ to n), where D_0 is the shift in the molecular weight of the apoprotein caused by H/D exchange. We use the statistical power of the entire data set to determine by the model the value of D_0 (the deuterium uptake of the apoprotein). The term ΔD_i is weighted by its binding fraction [Prot$-$Lig$_i$]/[Prot$_T$], which is a function of [Lig$_T$] and β_i ($i = 1-n$), the latter of which is the product of all the stepwise equilibrium binding constants ($\beta_i = K_1 K_2 \ldots K_i$). Usually ΔD_i is positive, indicating that binding of the ligand leads to more hydrogen bonding (less D uptake as compared to the apo form) although occasionally the ΔD_i is negative, suggesting the complex has less hydrogen bonding relative to its apo form. When ΔD_i is approximately zero, little conformational change apparently occurs upon binding, although changes in one part of the protein may be compensated by changes in another. If no net change occurs, PLIMSTEX may not be appropriate for determining the corresponding equilibrium constant (β_i).

$$\Delta D(\beta_1, \ldots, \beta_n, D_0, \Delta D_1, \ldots, \Delta D_n, [\mathrm{Lig_T}]) = D_0 - \sum_{i=1}^{n} \Delta D_i \frac{[\mathrm{Prot-Lig}_i]}{[\mathrm{Prot_T}]} \quad (12.1)$$

12.3 APPLICATIONS OF PLIMSTEX

12.3.1 Determination of Association Constant (K$_a$), Stoichiometry (n), and Protection (ΔD_i)

In the development of the PLIMSTEX approach to determine binding constants (K_a), we chose some well-known protein$-$ligand binding systems as tests. One system was the binding of Mg^{2+} to GDP-bound human ras protein, another was the binding of Ca^{2+} to apo-calmodulin (CaM), a third was the binding of a fatty acid carboxylate to intestinal fatty acid binding protein (IFABP), and the fourth was the binding of peptides to Ca^{2+}-saturated calmodulin (holo-CaM)]. We also extended PLIMSTEX to protein$-$protein interactions involving self-associations of various insulins (Chitta et al., 2004, 2006). These are widely studied systems, and their individual K values range from 10^4 to $10^8 \, \mathrm{M}^{-1}$.

The binding constants determined by PLIMSTEX for the test cases are within a factor of 6 of those reported previously using conventional methods (Table 12.1). All tests gave positive ΔD_i values, indicating an increase in H-bonding (more protection) with ligand binding (Table 12.1). The ΔD values in the case of insulin represent changes in the solvent accessibility of the oligomer compared to that of monomer.

12.3.2 Ras$-$GDP Interacting with Mg^{2+}: A 1:1 Protein:Metal Ion Interaction

To illustrate more completely the approach, let us consider human p21$^{\mathrm{Ha\text{-}ras}}$, a 21-kDa protein, which is important in controlling cellular growth and acts as a molecular switch in signal transduction pathways by cycling between its biologically active ras$-$GTP and inactive ras$-$GDP forms (Wittinghofer and Pai, 1991). Mg^{2+} is

TABLE 12.1 List of Affinity Constants Determined by Using PLIMSTEX and Compared with Literature Values

Protein (C_{total}) + Ligand (1 to n)	ΔD_i^a	PLIMSTEX[a] $K_a(M^{-1})$	$\dfrac{K_a(\text{Literature})^b}{K_a(\text{PLIMSTEX})}$
Human ras − GDP (1.5 μM) + Mg^{2+} (1 to 1)	25.6 ± 0.6^c	K_1: $(4.1 \pm 0.2) \times 10^4$	1.7^d
Porcine apo-CaM (15 μM) + Ca^{2+} (1 to 4)	12.6 ± 0.3^e	K_3: $(7 \pm 2) \times 10^4$	K_3: 0.6^f
		K_4: $(1.1 \pm 0.4) \times 10^5$	K_4: 2.8^f
		K_3K_4: $(9 \pm 1) \times 10^9 \, M^{-2}$	K_3K_4: 1.4^f
Rat IFABP (0.3 μM) + oleate (1 to 1)	13.8 ± 0.7^c	K_1: $(2.6 \pm 0.6) \times 10^6$	1.2^g
Porcine holo-CaM (0.15 μM) + melittin (1 to 1)	29.3 ± 0.8^c	K_1: $(5.4 \pm 0.9) \times 10^7$	6.1^h or 0.2^i
r-Human insulin + r-human insulin (mono- to di- to hexamer)	14 ± 2^j	K_{12}: $(7 \pm 1.2) \times 10^5$	K_{12}: 0.2^k
	23 ± 3^l	K_{26}: $(2 \pm 0.7) \times 10^9$	K_{26}: 0.2^k

[a]Each protein − ligand titration was done in duplicate. Values were determined by fitting the average data at similar conditions. A subsampling method was used to evaluate the second-order statistics of the parameters.
[b]K_i (Literature) determined under comparable experimental conditions (e.g., similar pH, ionic strength, if available) were selected.
[c]ΔD_1.
[d]From Zhang and Matthews (1998).
[e]ΔD_4.
[f]From Linse et al. (1991).
[g]From Kurian et al. (1996).
[h]From Comte et al. (1983) for CaM from bovine brain.
[i]From Yao and Squier (1996) for CaM from wheat germ.
[j]ΔD_{12}.
[k]From Pocker and Biswas (1981).
[l]ΔD_{26}.

an essential cofactor for the ras family of small GTPases (Sprang and Coleman, 1998; Zhang et al., 2000). We chose the C-terminal truncated p21^{Ha-ras} (residues 1 − 166), which retains crucial kinetic and structural properties (de Vos et al., 1988; John et al., 1989; Pai et al., 1989). Our goal was to test the ability of PLIMSTEX to investigate the binding of Mg^{2+} to ras − GDP to form a ternary complex; that is, the ras − GDP binary complex was treated as the "apo" protein and Mg^{2+} was treated as the ligand.

The first step in applying PLIMSTEX to a protein/ligand interaction is to choose an appropriate H/D exchange time for the titration. We did this by determining the forward H/D exchange kinetics of ras − GDP (1.5 μM) as a function of [Mg^{2+}]. As the ratio of [Mg^{2+}]:[ras − GDP]$_{total}$ increases, more binary ras − GDP complex is transformed to the ras − GDP − Mg^{2+} ternary complex. A fit of the kinetic data shows that the number of fast-exchanging amide hydrogens decreases monotonically from 79 to 65, whereas the number of slow-exchanging hydrogens increases from 48 to 76. Fast-exchanging hydrogens are likely to be weakly hydrogen bonded, whereas slow ones are protected by high-order structure and strong hydrogen bonding. The shift from fast to slow amide exchangers upon Mg^{2+} binding is consistent with a global folding of ras into a more compact and stable, less solvent-accessible form. After 3 h of exchange, the deuterium uptake levels off for three concentration ratios of metal ion to protein, indicating a steady-state H/D exchange. Therefore, we chose 3 h as the exchange time for the PLIMSTEX titration. These data can be obtained rapidly; as a result, they may be useful in guiding *ab initio* folding calculations for an unknown protein.

As Mg^{2+} is added to 1.5 µM ras—GDP and a complex forms, the deuterium uptake decreases. A fit of the PLIMSTEX curve occurs to afford K_a, ΔD_1, and D_0 for ras—GDP as $(4.1 \pm 0.2) \times 10^4 M^{-1}$, 25.6 ± 0.6, and 105.7 ± 0.5, respectively (root mean square of the error between the predicted and the experimental deuterium shifts is 0.7 Da). Detailed descriptions of the H/D exchange kinetics and the PLIMSTEX curve can be found in Zhu et al. (2007). A relatively weak 1/1 interaction between Mg^{2+} and ras—GDP occurs to cause ~26 backbone amide protons of the binary complex to become protected upon binding with Mg^{2+}. The value of the affinity constant determined in this way agrees fairly well with the literature value $(6.9 \times 10^4 M^{-1})$, which was determined by using circular dichroism (CD) in an equilibrium unfolding stability study (Zhang and Matthews, 1998). PLIMSTEX also yields a global measure of the number of amide hydrogens that become more hydrogen bonded upon metal ion binding. This number, $\Delta D_1 = 26$, is in agreement with the kinetic results that show that the number of the slow-exchanging amide protons of ras in its ternary complex ras—GDP—Mg^{2+} (76 protons) minus those for the binary complex ras—GDP (48 protons) is 28.

12.3.3 The Interactions of Apo-Calmodulin with Ca^{2+}: A 1:4 Protein: Metal Ion Interaction

We wished to apply PLIMSTEX to a more complicated system than one involving 1:1 binding. Therefore, we examined its applicability to the more challenging 1/4 protein—metal ion binding of calmodulin and Ca^{2+}. CaM binds to Ca^{2+} and undergoes conformational changes that enable it to bind to and activate other target proteins as a part of cell metabolism (Klee, 1988; Weinstein and Mehler, 1994). The challenge is to learn whether we can determine the conformational changes, binding stoichiometry, and binding constants for Ca^{2+} interactions with calmodulin as a function of electrolyte identity and ionic strength (Zhu et al., 2003).

A titration of rat CaM with Ca^{2+} shows that the complex is more stable (more hydrogen bonded) than the apoprotein (Fig. 12.2). Although the interactions with the first and second calcium ions cause little perturbation to CaM's conformation, the formation of CaM—$4Ca^{2+}$ species largely determines the shape of the titration curve, indicating that this last binding step causes the largest conformational change in the stepwise Ca^{2+} binding. As a consequence, PLIMSTEX cannot determine K_1 and K_2 from the titration. We took these constants from fluorescence studies done

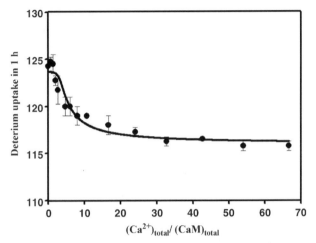

FIGURE 12.2 Ca titration for 15 µM of rat calmodulin in 50 mM HEPES, and 100 mM KCl (pH 7.4, $T = 21.5°C$, 90% D_2O). Error bars are based on the deviation from two sets of LCQ MS data. The solid curve is the best fit for the average data using the four-parameter model.

under comparable pH and ionic strength (Linse et al., 1991; VanScyoc et al., 2002). Modeling the titration curve gave β_3 and β_4, from which we could calculate K_3 and K_4. We found that $K_3 = 3.4 \times 10^5 \, M^{-1}$ and $K_4 = 3 \times 10^4 \, M^{-1}$, which agree with the literature K_3 and K_4 within a factor of 6. Owing to the cooperative binding of the last two Ca^{2+} to the mammalian CaM, variations were observed in the reported individual K_3 and K_4 values. Nevertheless, the product of the first two or the last two macroscopic binding constants was similar in many studies (Linse et al., 1991). PLIMSTEX gives an average affinity for binding the last two Ca^{2+} ions ($\sqrt{K_3 K_4}$) that agrees with the literature value (VanScyoc et al., 2002) within a factor of 3. No nonspecific binding of more than four Ca^{2+} ions is seen in the titration. If nonspecific binding occurs, it causes no significant conformational changes in the protein. The results show that PLIMSTEX can determine the intermediate binding species and related binding constants, which are often difficult to obtain by most other mass-spectrometry-based methods. Its outcome is not confused by any nonspecific binding, which is seen in direct ESI measurements.

12.3.4 Applications in Biologically Relevant Media

Cellular solutions consist of nonvolatile buffers and salts at relatively high ionic strength. These conditions are inimical to investigating protein — ligand complexes by direct ESI or MALDI under physiologically relevant conditions. Direct methods almost always use solutions of low ionic strength containing volatile electrolytes (e.g., ammonium acetate) that are not buffers and a significant fraction of organic solvent content (e.g., methanol, acetonitrile). These conditions are not physiologically relevant, and affinity constants determined in a direct way (i.e., by measuring the abundance of the complex and the starting protein by ESI) may not be the same as

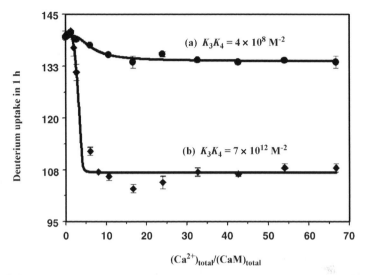

FIGURE 12.3 Ca titration of 15-μM porcine CaM in two different media (99% D_2O) (Zhu et al., 2003):(a) 50-mM HEPES/100 mM KCl, apparent pH 7.4; (b) 2-mM NH_4OAc, apparent pH 7.0. (Error bars were based on two sets of LCQ MS titration data. Solid curves are from the four-parameter model and are the best fit for the average data.)

those in cellular solutions. On the other hand, the PLIMSTEX approach to affinity measurements will often lead to high sensitivity because the measurements are carried out at low pH (needed to quench the exchange) where the protein has been liberated from the complex and sprays to give strong signals. Furthermore, buffer components, salts, metal cations, and other ligands do not interfere because they are removed by online chromatography prior to MS analysis. Removal of metal cations, for example, reduces cationization in the ESI process, improving the mass resolving power and sensitivity. As a result, protein/ligand binding can be explored not only at low ionic strength (e.g., in ammonium acetate at low millimolar concentration) but also under conditions with strong buffers and relatively high ionic strength (e.g., 150 mM).

PLIMSTEX is well suited to study affinity as a function of various parameters (e.g., pH, ionic strength, and temperature) and to carry out experiments that would be difficult or impossible to conduct using direct MS methods. An example is our investigation of CaM interacting with Ca^{2+} in low ionic-strength media (2 mM NH_4OAc) and progressively in higher ionic strength media (50 mM HEPES with 100 mM KCl). The titration curves shift to higher $[Ca^{2+}]_{total}$ as the ionic strength increases. The values for K_3 and K_4 for Ca^{2+} decrease by approximately four orders of magnitude with increases in ionic strength (Fig. 12.3), consistent with results from other methods (Haiech et al., 1981; Linse et al., 1991).

12.3.5 The Interaction of Holo-CaM and Peptides

Peptide and protein binding studies of CaM often use melittin, a small, 26-residue, hydrophobic peptide from bee venom, as a model. This peptide binds to Ca^{2+}-

saturated CaM (holo-CaM) in a 1:1 ratio, but the K_d values are uncertain, varying between 3 (Comte et al., 1983) and 110 nM (Yao and Squier, 1996).

PLIMSTEX determinations of affinity, like that of most methods, must be done at a concentration that is comparable to the K_d value (see Fig. 12.1, curve A). When the concentration of the protein is too high, "sharp break" curves are obtained (see Fig. 12.1, curve B); these are suitable for determining stoichiometry or even purity of the protein by titrating with a ligand of known concentration, but they are not appropriate for obtaining a binding constant. The binding stoichiometry for holo-CaM:milettin is 1:1, as determined by titrating a relatively high concentration of holo-CaM (15 μM) with melittin (Fig. 12.4). The result is in accord with the known change of CaM's structure from an open dumbbell shape for apo (Scaloni et al., 1998) to a closed globular shape with both domains interacting with the peptide as it binds to melittin. We found that the binding of mastoparan, which is a 14-amino acid residue peptide from the wasp and is approximately half the size of melittin, causes even more CaM protection than that of melittin. The number of amide hydrogens that are protected is greater than the number of residues in mastoparan, ruling out a direct block of the surface amides and indicating significant conformational change with the binding. These results indicate that PLIMSTEX can be used to follow conformational changes associated with (Zhu et al., 2004) protein – peptide binding.

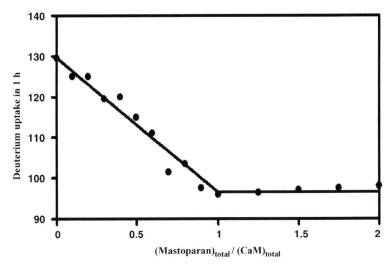

FIGURE 12.4 Sharp break PLIMSTEX curves at high protein concentration (Zhu et al., 2004). [Mastoparan (a 16-amino acid peptide) titration of 15-μM Ca^{2+}-saturated porcine calmodulin (CaM – 4Ca) in 50 mM HEPES, 100 mM KCl, 0.49 mM Ca^{2+}, 99% D_2O, apparent pH 7.4. Data points are based on the average of two runs for each titration system, and the breaking point clearly indicates one-to-one protein – ligand binding stochiometry.]

12.4 SELF-ASSOCIATION OF INSULIN: A PROTEIN/PROTEIN INTERACTION

Protein/protein interactions are essential in cell regulation (Scott and Pawson, 2000) and are key to understanding those human diseases that arise from unwanted protein — protein interactions (Staub et al., 1996). Insulin, a protein with 51 residues in two chains (Adams et al., 1969), is a good model system for testing whether a PLIMSTEX-like approach can determine protein self-association. The oligomer state of insulin is important for proper treatment of type I diabetes. The large size of the two Zn^{2+}-containing hexamers of insulin prevents its efficient absorption into the blood stream (Brange et al., 1997), whereas aggregation is prevented by using insulin analogs that are stable in monomeric form (Brange and Volund, 1999).

To obtain the needed data, solutions of various concentrations of insulin are prepared and amide H/D exchange is initiated, followed by quenching the exchange, and the new molecular mass of the exchanged protein is determined. Upon quenching, the oligomers dissociate into monomers, but the increase in the mass of the monomer gives a weighted average of the increase in the mass of various oligomers. These data can be used to obtain a species-specific deuterium number for each oligomer and to calculate the association constants for the oligomerization. To fit the insulin self-association data, we modified the modeling, recognizing that both ligand and protein are the same (Chitta et al., 2004, 2006). To illustrate, we see that the deuterium uptake for r-human insulin decreases gradually with increasing concentration of insulin (see the points in Fig. 12.5). As the protein concentration increases, oligomerization increases, and more amide hydrogens become slower for exchange (more protection occurs).

Fitting the experimental data with the SIMSTEX model gives, when the best fit is achieved, the solid curve in Fig. 12.5. The K_a's calculated for a model in which

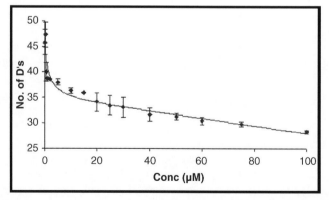

FIGURE 12.5 Plot of the uptake of deuterium as a function of solution concentration of r-human insulin. The points are the experimental data, and the curves are the theoretical fit from which equilibrium constants for self-association can be obtained (see text). This figure reproduced, with permission from Elsevier Ltd., from Chitta et al., 2006.

monomers interact to give dimers and dimers to give hexamers agree with literature values (from sedimentation equilibrium and CD) within a factor of 5 (Pocker and Biswas, 1981).

We took a similar approach to investigate insulin mutants that may be considered as replacements for recombinant insulin in human therapy. For lispro insulin (in which positions P28 and K29 in human insulin are reversed), and for several other insulin mutants, PLIMSTEX clearly can distinguish the self-association properties and binding constants of lispro and r-human insulins (Chitta et al., 2006). Lispro insulin is less associated in these experiments, and this is consistent with its pharmaceutical property to be faster acting than human insulin.

12.5 FEATURES OF PLIMSTEX

PLIMSTEX can determine the association constant K_i, stoichiometry, and protection (ΔD_i) in protein — ligand interaction. It is a general mass-spectrometry-based method that is applicable to a wide range of protein — ligand binding cases including binding of metal ions, small organic molecules, peptides, and proteins. When PLIMSTEX data are considered along with kinetic measurements of H/D exchange, we find insights into protein structure and its changes with ligand interactions.

Owing to the high sensitivity of mass spectrometers and the chromatographic desalting and concentrating procedure in the protocol, we can apply PLIMSTEX to a wide range of protein concentrations by simply adjusting the amount of solution injected into the mass spectrometer. Small quantities (high picomole) and low concentration (nanomolar) of proteins are sufficient to obtain each point in a titration or kinetic run.

ESI and MALDI (Yates, 1998; Ferguson and Smith, 2003) ionization methods can introduce noncovalent complexes to the gas phase (Loo, 1997, 2000; Hillenkamp, 1998). If the assumption that the gas-phase ion abundances for the complex and the apo protein, at known excess concentration of ligand, are directly related to their equilibrium concentrations in solution is correct, then one can use them to determine the relative and absolute binding affinities. This approach has been extensively applied; one area of considerable success is that of carbohydrate/protein interactions as explored by Klassen and coworkers (Wang et al., 2005). Unfortunately, ESI is discriminatory, and peak intensities, especially when measuring a system at equilibrium, may not be reliable (Liang et al., 2003; Sojo et al., 2003). Furthermore, it is difficult to predict a priori whether the approach is valid. Electrostatic forces in complexes are strengthened in the solventless environment of the mass spectrometer, making electrostatically bound protein — ligand complexes more stable in the gas phase than in solution. Binding that is largely governed by hydrophobic interactions in solution, however, weakens in the vacuum of a mass spectrometer, and complexes bound by hydrophobic forces break apart to an unpredictable extent, leading to incorrect affinities (Robinson et al., 1996; Wu et al., 1997; Loo, 2000). It may be possible to correct for fragmentation of a noncovalent complex in the gas phase by using response factors that relate the mass spectrometer signal to the concentration of the complex in solution and ultimately give the correct stability of the complex, but this complicates the procedure.

An additional problem for all direct methods is that they cannot use high ionic strength and nonvolatile buffers, which are needed to simulate physiological conditions, because ESI does not work under these conditions. One usually wants the affinity values in water, but ESI is best done with solutions that have some organic solvent content (e.g., acetonitrile, methanol), and the affinity is likely to be different in these media.

These problems are not an issue for PLIMSTEX because it uses a single mass spectrometric parameter, m/z, to determine affinity. As a ligand is added to the protein under conditions of H/D exchange, the m/z value of the protein changes, and this is followed by ESI (or MALDI) mass spectrometry. Signal intensities for the protein and the complex are not required. Thus, PLIMSTEX takes advantage of the increasing ability of mass spectrometers to measure accurately m/z. SUPREX, another method for measuring the free energies of binding from H/D exchange rates during unfolding (for some examples, see Powell et al. (2002) and Dhavan et al. (2003), also takes a single parameter from the mass spectrum (i.e., m/z) and avoids the complications of relying on ESI signal intensities, but this approach does require a denaturant.

Many conventional methods, particularly NMR and fluorescence, require that the protein be specially labeled, so that signals that are a measure of concentration can be seen. PLIMSTEX, like other MS-based methods, uses changes in hydrogen/ deuterium exchange of amide hydrogens that are present in all protein systems; therefore, no special labeling is necessary.

Some biophysical approaches require physical separations of the free ligand or protein from the protein−ligand binding system; for example, affinity chromato- graphy, size exclusion chromatography, and ultra filtration. Other methods (e.g., circular dichroism, various spectroscopic approaches) (van Mierlo and Steensma, 2000; McLaughlin and Jackson, 2002; Greenfield, 2004) as well as SUPREX (Powell and Fitzgerald, 2003; Powell et al., 2002)) require denaturants, and they may perturb the original binding equilibrium. PLIMSTEX obviously requires no separation of the equilibrium constituents and does not use a denaturant.

More and more H/D exchange studies now incorporate digestion of the protein under quench conditions and measurement of the extent of H/D exchange of constituent peptides. Usually this is done in a time-dependent manner to afford kinetics of exchange at the peptide level (Zhang and Smith, 1993; Smith and Dharmasiri, 1998; Engen et al., 1999; Wang et al., 1999; Woods and Hamuro, 2001; Hamuro et al., 2002; Wales and Engen, 2006). The paradigm we are establishing is that once the binding affinity and protection of an intact protein are determined by PLIMSTEX, the experiment can be repeated with pepsin digestion to give insight into the peptide level of those regions of the protein that are involved in the binding. A key future direction for expanding PLIMSTEX is to provide a higher resolved view than the global view of the whole protein that is currently obtained, allowing PLIMSTEX to give a view of the protein that currently emerges from NMR and X-ray methods.

12.6 FAST PHOTOCHEMICAL OXIDATION OF PROTEINS: AN EXAMPLE OF FAST PROTEIN FOOTPRINTING

Another approach, radical footprinting, uses reactions of a reactive species (e.g., •OH) with amino acid side chains. It is complementary to PLIMSTEX because it exploits side chain reactivity instead of exchange of backbone amide hydrogens. It introduces a nonreversible modification (a radical probe reacts to form a stable covalent bond) instead of making a reversible change to the protein. As such, it can be readily expanded to locate the amino acid residue that has reacted, taking advantage of all proteases including a number that work at pH 7. Thus, the locations of cleavage sites are predictable, which is a limitation with pepsin, the enzyme of choice for PLIMSTEX and related H/D exchange methods. Additionally, fast photochemical oxidation of proteins has the ability to locate specific side chains that are buried when complexed with a ligand because the chemical probe, such as H/D exchange, reacts rapidly with a solvent-exposed residue but reacts much more slowly with a side chain that has little or no solvent exposure. A chemical reaction method such as hydroxyl-radical footprinting probes specific functions (e.g., hydrophobic and sulfur-containing amino acid side chains), whereas H/D exchange probes the exchange rate of every backbone amide hydrogen in the protein.

12.6.1 Hydroxyl Radicals as a Probe

The hydroxyl radical is a small, highly reactive probe that is formed in water and reacts most rapidly with hydrophobic residues (Hawkins and Davies, 2001). This radical has the potential to be an ideal probe for protein:protein interactions because the aromatic amino acids, tyrosine, tryptophan, and phenylalanine, are most likely to be found at a protein interface (Bogan and Thorn, 1998; Dhavan et al., 2003).

The hydroxyl radical reacts at nearly a diffusion limited rate ($k \sim 1 \times 10^{10} - 5 \times 10^9 \, M^{-1} s^{-1}$) with aromatic amino acids, methionine, and cysteine (Davies, 1997). Most other side chains are $10 - 100$ times less reactive, making the hydroxyl radical relatively specific for residues that are typically located at protein:protein interfaces yet sufficiently reactive to give a "snapshot" view of the protein. This method has been successfully used by others to study protein:DNA and protein:metal interactions (Dhavan et al., 2003; Kiselar et al., 2003; Shcherbakova et al., 2004).

Hydroxyl radicals can be generated chemically by using the Fenton reagent (Fenton and Jackson, 1899); Fe^{2+} reduces H_2O_2 to hydroxide and hydroxyl radical, but the reaction is slow when compared to other methods for making radicals. Hydroxyl radicals can also be generated by radiation; for example, synchrotron radiation cleaves water into a proton, electron, and hydroxyl radical (Guan et al., 2004), and UV light homolytically cleaves H_2O_2 into two hydroxyl radicals (Urey et al., 1929; Schiffman et al., 1993; Sharp et al., 2003). Given that previous publications describing these methods report the need for tens of milliseconds to minutes, we reasoned that a fast, pulsed UV laser should enable fast photolysis of hydrogen peroxide into hydroxyl radicals. The laser pulse cleaves hydrogen peroxide into two hydroxyl radicals, which either react with the protein side chains or recombine to re-form H_2O_2 (Fig. 12.6). Given that these two pathways are rapid, the radical concentration drops to below 1 μM within approximately 100 μs, as we determined by kinetic calculations using the known rate constant for hydroxyl radical recombination (Buxton et al., 1988). The protein oxidation profile achieved in this time frame shows that considerable oxidation (note the peaks separated by ~16 Da in Fig. 12.8b) occurs on this timescale.

By adding excess chemical quencher to the system and then carrying out the irradiation, the radicals would react with the quencher according to first-order kinetics. If 20 mM phenylalanine was added to the system, kinetic predictions show that the radicals would be essentially consumed within 70 ns of the laser pulse, whereas the use of 20 mM glutamine results in reaction of all radicals within 1 μs of the laser pulse (Fig. 12.6). Given that protein secondary structure packing does not unfold faster than 10 μs, for even the fastest systems studied thus far (Gilmanshin et al., 1997; Gulotta et al., 2001; Vu et al., 2004; Hambly et al., 2005), a 1-μs reaction timescale eliminates concerns about protein unfolding as a result of oxidation.

The other means of eliminating concerns about protein unfolding is to conduct the radical footprinting under conditions that are nearly "single hit." This is the approach of Chance and coworkers (Xu and Chance, 2007) who used synchrotron radiation to produce OH radicals from the ionization of water. "Single-hit" conditions would produce a smaller extent of modifications, which would be more difficult to detect than those produced by more extensive oxidation. The synchrotron method approaches "single-hit" conditions and still produces sufficient modifications that can be detected by LC/MS and LC/MS/MS. Furthermore, the synchrotron approach

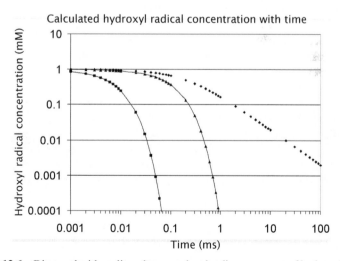

FIGURE 12.6 Diamond with no line: the second-order disappearance of hydroxyl radical as a result of self-reaction to form H_2O_2. Square: the first-order disappearance of hydroxyl radical as a result of reaction with excess phenylalanine (20 mM). Triangle: the first-order disappearance of hydroxyl radical as a result of reaction with excess glutamine (20 mM). This figure reproduced, with permission from Elsevier Ltd., from Hambly and Gross, 2005.

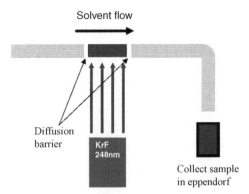

FIGURE 12.7   Diagram of the experimental setup showing the laser pulse irradiating the sample solution in a UV transparent window. The laser beam and flow rate are coordinated to ensure that there is a diffusion buffer between irradiated sample and not yet irradiated sample.

is also relatively fast, requiring in some cases a minimum of 10 ms to complete the reaction (Maleknia et al., 2004).

12.6.2   Fast Hydroxyl-Radical Footprinting

To ensure that there is sufficient protein to be oxidized and then analyzed, we employed a flow system coupled with a pulsed laser to produce OH radicals from hydrogen peroxide. The protein is mixed with 15 mM H_2O_2, and the solution passed through fused silica tubing, irradiated, at a point where the polyimide coating is removed to afford a UV transparent window, and collected (Fig. 12.7). To ensure that no fraction of irradiated protein is oxidized a second time, the laser frequency can be matched to the solvent flow rate and tubing diameter. Furthermore, by designing the flow so that ~20% of the protein solution is not irradiated, one can be assured that all oxidations are from the first pulse of light and there is always unreacted protein to serve as a reference point. To observe the oxidations (Fig. 12.7) on the protein, we load approximately 5 pmol (80 ng) of protein onto a small trap column and desalt the protein with water. The mixture of oxidized and unoxidized protein is then eluted into a mass spectrometer (e.g., QTOF or other spectrometer suitable for protein molecular mass measurements), and the extent of modification (oxidation) is determined (Fig. 12.8). Clearly, this footprinting is not done under "single-hit" conditions as there is considerable oxidation of the protein. The expectation, however, is that the modification is done sufficiently rapidly so that a "snapshot" of the protein, uncomplicated by oxidation-induced protein unfolding, is obtained.

12.6.3   Locating the Sites of Radical Reaction

To determine the sites of reaction for the radicals, it is necessary to probe separately the solvent-exposed side chains of the protein alone and of the protein:ligand complex. Side chains that are modified in the protein alone, and not in the complex, indicate areas of decreased solvent accessibility owing to ligand binding. After the protein is modified, standard proteomic analytical methods can be applied to pinpoint the oxidized amino acids, as is also done when synchrotron radiation is used to produce the radicals. We loaded the digest onto a C_{18} nanocolumn (using an Eksigent Nano 1D-LC) and analyzed the peptides in a data-dependent manner with an ion trap/FT mass spectrometer combination (LTQ-FT). The FT mass spectrometer of the hybrid affords maximum dynamic range and sensitivity, whereas the ion trap enables both unmodified parent peptides and modified peptides to be sequenced in the same LC/MS/MS experiment. Good dynamic range is important because an oxidized peptide may represent less than 1% of the parent, unmodified protein. Using a data-dependent analysis, we observed that nearly all peptides for which we saw molecular ion signals by FTMS can be sequenced in the MS/MS part of the experiment. As many oxidations produce low-abundance products, using a trapping mass spectrometer allows the ions to be collected prior to fragmenting them.

To locate the oxidized side chains, the product-ion spectra of oxidized precursor ions are compared to those of the corresponding unoxidized peptide. Most oxidations can be located by searching for peptides whose m/z values are $+16$, $+32$, $+48$, -22, or $+5$ Da compared to those of the precursor ion. These mass shifts correspond to adding one, two, or three oxygen atoms to any residue, except for histidine, which undergoes other side reactions, producing mainly $+5$ and -22 end products. The mass accuracy of the FTMS adds certainty to the analysis by ensuring that only peptides whose molecular weights are within 5 ppm of the expected modified peptide are considered. Using the product-ion spectra of both unmodified and modified peptides allows, in nearly all cases, the oxidized side chain to be determined.

12.6.4 Application of FPOP to Apomyoglobin

We tested the fast radical footprinting reaction by applying it to apomyoglobin, a protein that is well characterized in the holo form (Evans and Brayer, 1990) and is often used as a model in the protein-folding field (Dyson and Wright, 2004; Nishimura et al., 2005; Nishimura et al., 2006). Fast radical footprinting in the presence of 20 mM phenylalanine, a concentration 2000 times greater than that of the protein, should be complete in $\sim$70 ns (Hambly et al., 2005). Indeed, virtually no protein oxidation occurs with this high concentration of scavenger (Fig. 12.8a). In the presence of 20 mM glutamine, the reaction duration now increases to $\sim$1 μs (Fig. 12.8b) whereas in the absence of scavenger, the duration is greater than 100 μs (Fig. 12.8c), and significant oxidation occurs. It should also be noted that the reaction will be sensitive to protein concentration as well as scavenger concentration. Figure 12.9 demonstrates that the relative abundance of oxidized apomyoglobin decreases as the apomyoglobin concentration is raised. All other parameters are maintained the same. Following analysis of the intact protein is the digestion of the mixture of oxidized proteins and analysis of the resulting peptides by LC/MS/MS to determine sites of oxidation.

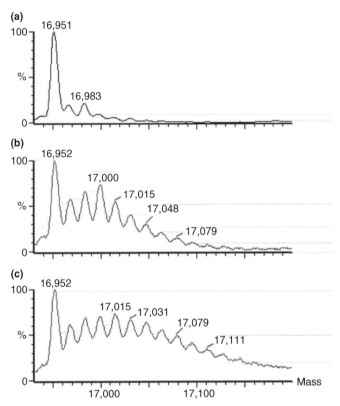

FIGURE 12.8 (a) Laser irradiation of 10-μM apomyoglobin in 10 mM NaH$_2$PO$_4$, pH 7.8, and 20 mM phenylalanine as a scavenger. (b) Oxidation of 10-μM apomyoglobin in 10 mM NaH$_2$PO$_4$, pH 7.8, 15 mM H$_2$O$_2$, and 20 mM glutamine as a scavenger limiting the reaction to 1 μs. (c) Oxidation of 10 μM apomyoglobin in 10 mM NaH$_2$PO$_4$, pH 7.8, 15 mM H$_2$O$_2$ with no scavenger resulting in up to 100 μs reaction duration.

For example, let us consider the myoglobin C-terminal tryptic peptide (K) YKELGFQG at m/z 941.4744, which can be identified with a mass error of 2 ppm, well within the 5 ppm error specification of the FT mass spectrometer (Fig. 12.10a). A peptide representing oxidation of this peptide appears at m/z 957.4697 (mass error of 2 ppm). The selected ion chromatogram of these two peptides shows that the modified peptide is ~200 times less abundant than the oxidized peptide, underscoring the need for the good dynamic range that is afforded by the FT mass spectrometer coupled to an ion trap. Although the mass accuracy strongly indicates that the minor peptide is an oxidized form of the "native" peptide, giving an m/z of 941, one cannot be certain of the oxidation site because the peptide contains tyrosine, phenylalanine, and leucine, all of which can be oxidized. A selected region of the product-ion spectrum of the m/z 941.47 ion (Fig. 12.10b) shows that both the b$_6$ and b$_7$ ions can be observed along with other fragments arising from water losses. In the product-ion spectrum of

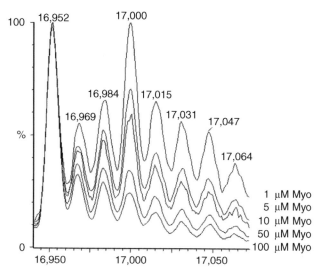

FIGURE 12.9   Deconvoluted mass spectra of oxidized myoglobin as the protein concentration is varied from 1 μM apomyoglobin (myo) to 100-μM apomyoglobin.

the m/z 957.47 ion (Fig. 12.10c), both b_6 and b_7 ions are shifted up in mass by 16 Da, whereas the y_7 ion does not show any mass shift. The presence of the unmodified y_7 ion indicates that the product ion $[KELGFQG + H]^+$ is not modified, whereas the presence of the $+16$ modification of the b_6 daughter ion confirms that the tyrosine is the modified residue.

To test whether the hypothesis that only solvent-accessible side chains are oxidized, we calculated side chain solvent accessibility using the X-ray structure and a 1.1 Å probe in the program GetArea 1.1, which is available on the Web (Fraczkiewicz and Braun, 1998). The data (Table 12.2) indicate that all methionine and tryptophan residues are labeled (oxidized) even if they are almost completely buried. Both tyrosines are modified, whereas the phenylalanine residues are selectively modified based on solvent exposure. Phe138, however, appears to be an exception, made more interesting considering that the adjacent leucine 137 is oxidized.

To understand this phenomenon, we reconsidered the crystal structure of holomyoglobin (pdb = 1WLA) and found that much of the ligand binding pocket of apomyoglobin is hydrophobic, containing predominantly leucine and isoleucine residues. None of these get modified in the OH radical reactions. The phenylalanine is located at the very recesses of this pocket and has 32 Å^2 of solvent-accessible surface area when the porphyrin is removed, provided the pocket remains open. Consistent with the large hydrophobic surface area of the pocket, the data indicate that the pocket closes, excluding water and increasing the entropy of the system when compared to that of the water-filled hydrophobic pocket. To rule out that the modified Phe in the peptide cannot be observed by LC/MS, we examined apomyoglobin by generating

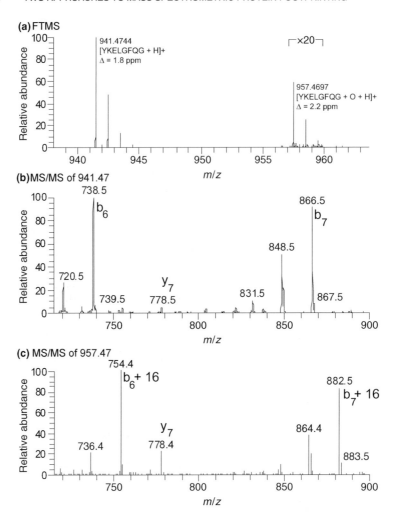

FIGURE 12.10 (a) A mass spectrum taken by FTMS showing the $[M + H]^+$ of the MASCOT-identified peptide YKELGFQG and a likely oxidation, as determined by accurate mass. (b) Partial product-ion spectrum of the m/z 941.47 ion, confirming the peptide sequence. (c) Partial product-ion spectrum of the m/z 957.47 ion, confirming an oxidation of tyrosine as indicated by an unmodified y_7 ion and complementary oxidized (mass shifted) b_6 and b_7 ions.

hydroxyl radicals continuously for 30 s (unpublished data). This experiment shows that the Phe138 can be modified and detected, whereas Leu137 is not oxidized. The evidence supports an explanation that the lack of oxidation on Phe138 occurs because the ligand binding site closes if the heme is removed. The oxidation of Phe138 in the 30-s experiment occurs because the protein is sampling the open conformation over the much longer time experiment, allowing OH radicals to enter and modify the Phe side chain. The results also demonstrate that the 1-μs oxidation method ensures that no protein conformational changes occur on this short timescale with apomyoglobin.

TABLE 12.2 **The Reactivities with OH Radicals of Various Residues in Apomyoglobin and Their Side Chain Solvent Exposure, as Ordered by Residue Reactivity (Trp > Tyr > Met > Phe > His)**

Type	A.A.#	Solvent Exposure ($\mathring{A}^2$)	Oxidized	Type	A.A.#	Solvent Exposure ($\mathring{A}^2$)	Oxidized
Trp	7	17	Y	Phe	151	34	Y
Trp	14	14	Y	His	24	11	Y
Tyr	103	20	Y	His	36	37	Y
Tyr	146	12	Y	His	48	90	Y
Met	55	11	Y	His	64	36	Y
Met	131	5	Y	His	81	104	Y
Phe	33	2		His	82	8	
Phe	43	62	Y	His	93	37	Y
Phe	46	4		His	97	50	nd
Phe	106	33	Y	His	113	61	Y
Phe	123	7		His	116	60	Y
Phe	138	(32)		His	119	27	Y

Solvent exposure values in parantheses are for residues located in the core of the hydrophobic pocket and may not be solvent exposed. Peptides that did not bind the column are annotated with nd.

Additional oxidations not listed in Table 12.2 include leucine 86 with 4 $\mathring{A}^2$ and leucine 89 with 51 $\mathring{A}^2$ of solvent-accessible surface area. The oxidation on leucine 86 is puzzling because numerous other leucines of significantly more solvent exposure are not modified. In NMR studies of apomyoglobin, however, Eliezer noted that the F-helix, comprising residues $83-100$, gives no observable NMR signal (Eliezer and Wright, 1996). The lack of an NMR signal indicates that the nuclei are adopting multiple conformations on the submillisecond timescale, dispersing the signal. Therefore, the F-helix, comprising leucine residues 86 and 89 is conformationally dynamic, and fast radical footprinting demonstrates that some of the helix is significantly solvent exposed, at least for a time adequate for it to be modified by radical reactions.

The fast radical footprinting methodology has demonstrated that in situations where X-ray crystallography and NMR are hampered by conformational flexibility, the power of mass spectrometry can be leveraged to obtain additional information about protein structure and function. In a sequel paper (Hambly and Gross, 2007), we described a demonstration that holomyoglobin is conformationally constrained at the F-helix and adopts a more rigid conformation when the porphyrin is ligated to the protein.

12.7 FEATURES OF FPOP

FPOP offers the advantage of any method that employs a reagent making an irreversible change to probe protein interactions; that is, the sites of reaction can be readily determined by digestion and LC/MS/MS analysis (a standard analytical proteomic approach). As with PLIMSTEX, the hydroxyl radical method measures a change in the m/z and, therefore, is not susceptible to variations in ESI efficiency. With the ability to conduct any post-reaction processing, the fast radical footprinting method can be used under any solution conditions including high salt, with denaturants, and at low concentrations (concentrations as low as 100 nM of protein are possible). Concomitant with these advantages, current technology permits minute sample amounts to be used. In our experiments, we loaded 50 pmol (800 ng) of apomyoglobin for footprinting. The sequencing stage requires only a few picomoles; thus, this experiment can be carried out on less than 80 ng of protein, a quantity less than the amount that can be observed on an SDS-PAGE gel using Coomassie staining. Given that analytical proteomic methods are being utilized, this method can be carried out in solutions containing a mixture of proteins, although multiple proteins in solution might seriously complicate data analysis. Additionally, no specifically labeled proteins are required; this method, like PLIMSTEX, enables significant information to be gleaned from wild-type proteins, and no mutants are needed. FPOP, when properly carried out with suitable chemical scavengers in solution, should produce oxidized proteins that are modified before the protein complex changes conformation as a result of the changes caused by modifying the protein or its substrate.

We can imagine that FPOP, like PLIMSTEX, can be used to determine K_d. Furthermore, we see the prospects of developing dose-dependent radical footprinting using different scavengers and concentrations to vary and then quantify the loss of unmodified peptide signal and the increase in the amount of modified peptide. Prospects exist that the reaction timescale can be established by using scavengers that react with OH radicals with known rate constants. At the present time, our calculations are worst case scenarios, so the reaction with 20-mM glutamine may be complete significantly before 1 µs.

12.8   FUTURE

Both PLIMSTEX and FPOP require more validations before they will be accepted by the biophysical community. Although they were originally developed using LC/ESI MS, one can imagine using MALDI to improve the throughput and make the analysis simpler and widely available. A different desalting procedure will be needed, and the conditions for quench (PLIMSTEX) and analysis will be controlled differently than when using LC/ESI MS. If successful, automated procedures for MALDI MS could be adopted for PLIMSTEX and FPOP. Nevertheless, we do not view these approaches for fast screening of libraries containing thousands or millions of compounds as there are simpler assay methods available, including direct MS measurements of complexes. Instead, we view these developments as refined approaches to determine affinity, solvent accessibility, and regions of interaction in the protein.

ABBREVIATIONS

H/D	Hydrogen/deuterium
PLIMSTEX	Protein – ligand interaction using mass spectrometry, titration, and H/D exchange
FPOP	Fast photochemical oxidation of proteins
ESI MS	Electrospray ionization-mass spectrometry
MALDI	Matrix-assisted laser desorption ionization
CaM	Calmodulin
IFABP	Intestinal fatty acid binding protein
GDP	Guanosine diphosphate
GTP	Guanosine triphosphate
HEPES	(N-[2-hydroxyethyl)piperazine-N'-[2-ethanesulfonic acid])
NMR	Nuclear magnetic resonance

ACKNOWLEDGMENTS

The research at Washington University was supported by the National Center for Research Resources of the National Institutes of Health, Grant P41RR00954, and by a supplemental grant from that resource. We acknowledge Don Rempel for his help in the development of modeling procedures for PLIMSTEX and to Drs Zhaohui Du and Raghu Chitta for some of the kinetics and titrations. We acknowledge Ilan Vidavsky, Jim Walters, and Henry Rohrs for assistance with data collection and analysis involving fast radical footprinting development. We acknowledge donations of protein from collaborators Dr. B. Pramanik (Schering-Plough Research Institute), Professor D. Cistola (Washington University), and Professor M. Shea (University of Iowa).

REFERENCES

Adams, M. J., Blundell, T. L., Dodson, E. J., Dodson, G. G., Vijayan, M., Baker, E. N., Harding, M. M., Hodgkin, D. C., Rimmer, B. and Sheat, S., 1969. Structure of rhombohedral 2 zinc insulin crystals. *Nature* 224 (5218), 491–495.

Bates, D. M. and Watts, D. G., 1988. *Nonlinear Regression Analysis and its Applications.* New York: Wiley.

Bogan, A. A. and Thorn, K. S., 1998. Anatomy of hot spots in protein interfaces. *J Mol Biol* 280 (1), 1–9.

Brange, J. and Volund, A., 1999. Insulin analogs with improved pharmacokinetic profiles. *Adv Drug Delivery Rev* 35 (2,3), 307–335.

Brange, J., Whittingham, J., Edwards, D., Zhang, Y.-S., Wollmer, A., Brandenburg, D., Dodson, G., and Finch, J., 1997. Insulin structure and diabetes treatment. *Curr Sci* 72 (7), 470–476.

Breuker, K., 2004. New mass spectrometric methods for the quantification of protein — ligand binding in solution. *Angew Chem Int Ed* 43 (1), 22 − 25.

Buxton, G. V., Greenstock, C. L., Helman, W. P., and Ross, A. B., 1988. Critical review of rate constants for reactions of hydrated electrons, hydrogen atoms and hydroxyl radicals (•OH/• O-) in aqueous solution. *J Phys Chem Ref Data* 17 (2), 513 − 886.

Chitta, R., Rempel, D. L., Grayson, M. A., and Gross, M. L., 2004. Evaluating self-association interactions using mass spectrometry, self-titration and amide exchange (SIMSTEX): studies on insulin. Proceedings of the 52nd Annual Conference on Mass Spectrometry and Allied Topics, Nashville, TN.

Chitta, R. K., Rempel, D. L., Grayson, M. A., Remsen, E. E., and Gross, M. L., 2006. Application of SIMSTEX to oligomerization of insulin analogs and mutants. *J Am Soc Mass Spectrom* 17 (11), 1526 − 1534.

Clark, S. M. and Konermann, L., 2004. Screening for noncovalent ligand-receptor interactions by electrospray ionization mass spectrometry-based diffusion measurements. *Anal Chem* 76 (23), 7077 − 7083.

Comte, M., Maulet, Y., and Cox, J. A., 1983. Ca^{2+}-dependent high-affinity complex formation between calmodulin and melittin. *Biochem J* 209 (1), 269 − 272.

Davies, M. J. and Dean, R. T., 1997. *Radical-Mediated Protein Oxidation: From Chemistry to Medicine*. Oxford: Oxford University Press.

de Vos, A. M., Tong, L., Milburn, M. V., Matias, P. M., Jancarik, J., Noguchi, S., Nishimura, S., Miura, K., Ohtsuka, E., and Kim, S. H., 1988. Three-dimensional structure of an oncogene protein: catalytic domain of human c-H-ras p21. *Science* 239 (4842), 888 − 893.

Dhavan, G. M., Chance, M. R., and Brenowitz, M., 2003. Kinetics analysis of DNA — protein interactions by time-resolved synchrotron X-ray footprinting. In: Johnson, K. and Williams, R., editors. *Kinetic Analysis of Macromolecules: A Practical Approach*. Oxford University Press: New York, NY: pp. 75 − 86.

Dyson, H. J. and Wright, P. E., 2004. Unfolded proteins and protein folding studied by NMR. *Chem Rev* 104 (8), 3607 − 3622.

Eliezer, D. and Wright, P. E., 1996. Is apomyoglobin a molten globule? Structural characterization by NMR. *J Mol Biol* 263 (4), 531 − 538.

Engen, J. R. and Smith, D. L., 2001. Investigating protein structure and dynamics by hydrogen exchange MS. *Anal Chem* 73, 256A − 265A.

Engen, J. R., Gmeiner, W. H., Smithgall, T. E., and Smith, D. L., 1999. Hydrogen exchange shows peptide binding stabilizes motions in Hck SH2. *Biochemistry* 38 (28), 8926 − 8935.

Evans, S. V. and Brayer, G. D., 1990. High-resolution study of the three-dimensional structure of horse heart metmyoglobin. *J Mol Biol* 213 (4), 885 − 897.

Fenton, J. H. and Jackson, H., 1899. The oxidation of polyhydric alcohols in presence of iron. *J Chem Soc Trans* 75, 1–11.

Ferguson, P. L. and Smith, R. D., 2003. Proteome analysis by mass spectrometry. *Annu Rev Biophys Biomol Struct* 32, 399 − 424.

Fraczkiewicz, R. and Braun, W., 1998. Exact and efficient analytical calculation of the accessible surface areas and their gradients for macromolecules. *J Comput Chem* 19 (3), 319 − 333.

Gilmanshin, R., Williams, S., Callender, R. H., Woodruff, W. H., and Dyer, R. B., 1997. Fast events in protein folding: relaxation dynamics of secondary and tertiary structure in native apomyoglobin. *Proc Natl Acad Sci USA* 94 (8), 3709 − 3713.

Greenfield, N. J., 2004. Circular dichroism analysis for protein — protein interactions. *Methods Mol Biol* 261, 55 — 78.

Guan, J.-Q., Almo, S. C., and Chance, M. R., 2004. Synchrotron radiolysis and mass spectrometry: a new approach to research on the actin cytoskeleton. *Acc Chem Res* 37 (4), 221 — 229.

Guan, J.-Q., Takamoto, K., Almo, S. C., Reisler, E., and Chance, M. R., 2005. Structure and dynamics of the actin filament. *Biochemistry* 44 (9), 3166 — 3175.

Gulotta, M., Gilmanshin, R., Buscher, T. C., Callender, R. H., and Dyer, R. B., 2001. Core formation in apomyoglobin: probing the upper reaches of the folding energy landscape. *Biochemistry* 40 (17), 5137 — 5143.

Haiech, J., Klee, C. B., Demaille, J. G., and Haiech, J., 1981. Effects of cations on affinity of calmodulin for calcium: ordered binding of calcium ions allows the specific activation of calmodulin-stimulated enzymes. Theoretical approach to study of multiple ligand binding to a macromolecule. *Biochemistry* 20 (13), 3890 — 3897.

Hambly, D. M. and Gross, M. L., 2005. Laser flash photolysis of hydrogen peroxide to oxidize protein solvent-accessible residues on the microsecond timescale. *J Am Soc Mass Spectrom* 16 (12), 2057 — 2063.

Hambly, D. and Gross, M., 2007. Laser flash photochemical oxidation to locate heme binding and conformational changes in myoglobin. *Int J Mass Spectrom* 259 (1 — 3), 124 — 129.

Hamuro, Y., Wong, L., Shaffer, J., Kim, J. S., Stranz, D. D., Jennings, P. A., Woods, V. L., and Adams, J. A., 2002. Phosphorylation driven motions in the COOH-terminal Src kinase, Csk, revealed through enhanced hydrogen — deuterium exchange and mass spectrometry (DXMS). *J Mol Biol* 323 (5), 871 — 881.

Hawkins, C. L. and Davies, M. J., 2001. Generation and propagation of radical reactions on proteins. *Biochem Biophys Acta* 1504 (2 — 3), 196 — 219.

Hillenkamp, F., 1998. Matrix-assisted laser desorption/ionization of non-covalent complexes. *NATO ASI Series, Series C: Mathematical and Physical Sciences* 510 *(New Methods for the Study of Biomolecular Complexes)*, 181 — 191.

John, J., Schlichting, I., Schiltz, E., Rosch, P., and Wittinghofer, A., 1989. C-terminal truncation of p21H preserves crucial kinetic and structural properties. *J Biol Chem* 264 (22), 13086 — 13092.

Johnson, B. M., Nikolic, D., and Breemen, R. B. v., 2002. Applications of pulsed ultrafiltration-mass spectrometry. *Mass Spectrom Rev* 21 (2), 76 — 86.

Kaltashov, I. A. and Eyles, S. J., 2002. Studies of biomolecular conformations and conformational dynamics by mass spectrometry. *Mass Spectrom Rev* 21 (1), 37 — 71.

Kiselar, J. G., Janmey, P. A., Almo, S. C., and Chance, M. R., 2003. Visualizing the Ca2+-dependent activation of gelsolin by using synchrotron footprinting. *Proc Natl Acad Sci USA* 100 (7), 3942 — 3947.

Klee, C. B., 1988. In: Cohen, P., and Klee, C. B., editors, *Calmodulin, Vol. 5: Molecular Aspects of Cellular Regulation* . Amsterdam, The Netherlands: Elsevier. p. 371.

Kurian, E., Kirk, W. R., and Prendergast, F. G., 1996. Affinity of fatty acid for rRat intestinal fatty acid binding protein: further examination. *Biochemistry* 35 (12), 3865 — 3874.

Liang, H. R., Foltz, R. L., Meng, M., and Bennett, P., 2003. Ionization enhancement in atmospheric pressure chemical ionization and suppression in electrospray ionization between target drugs and stable-isotope-labeled internal standards in quantitative liquid chromatography/tandem mass spectrometry. *Rapid Commun Mass Spectrom* 17 (24), 2815 — 2821.

Linse, S., Helmersson, A., and Forsén, S., 1991. Calcium binding to calmodulin and its globular domains. *J Biol Chem* 266 (13), 8050 – 8054.

Liu, R., Guan, J.-Q., Zak, O., Aisen, P., and Chance, M. R., 2003. Structural reorganization of the transferrin C-lobe and transferrin receptor upon complex formation: the C-lobe binds to the receptor helical domain. *Biochemistry* 42 (43), 12447 – 12454.

Loo, J. A., 1997. Studying noncovalent protein complexes by electrospray ionization mass spectrometry. *Mass Spectrom Rev* 16 (1), 1 – 23.

Loo, J. A., 2000. Electrospray ionization mass spectrometry: a technology for studying noncovalent macromolecular complexes. *Int J Mass Spectrom* 200 (1/3), 175 – 186.

Maleknia, S. D., Wong, J. W. H., and Downard, K. M., 2004. Photochemical and electrophysical production of radicals on millisecond timescales to probe the structure, dynamics and interactions of proteins. *Photochemical & Photobiological Sciences* 3 (8), 741 – 748.

McLaughlin, S. H. and Jackson, S. E., 2002. Folding and stability of the ligand-binding domain of the glucocorticoid receptor. *Protein Sci.* 11 (8), 1926 – 1936.

Muckenschnabel, I., Falchetto, R., Mayr, L. M., and Filipuzzi, I., 2004. SpeedScreen: label-free liquid chromatography – mass spectrometry-based high-throughput screening for the discovery of orphan protein ligands. *Anal Biochem* 324 (2), 241 – 249.

Nishimura, C., Dyson, H. J., and Wright, P. E., 2006. Identification of native and non-native structure in kinetic folding intermediates of apomyoglobin. *J Mol Biol* 355 (1), 139 – 156.

Nishimura, C., Lietzow, M. A., Dyson, H. J., and Wright, P. E., 2005. Sequence determinants of a protein folding pathway. *J Mol Biol* 351 (2), 383 – 392.

Pai, E. F., Kabsch, W., Krengel, U., Holmes, K. C., John, J., and Wittinghofer, A., 1989. Structure of the guanine-nucleotide-binding domain of the Ha-ras oncogene product p21 in the triphosphate conformation. *Nature* 341 (6239), 209 – 214.

Pocker, Y. and Biswas, S. B., 1981. Self-association of insulin and the role of hydrophobic bonding: a thermodynamic model of insulin dimerization. *Biochemistry* 20 (15), 4354 – 4361.

Powell, K. D. and Fitzgerald, M. C., 2003. Accuracy and precision of a new H/D exchange- and mass spectrometry-based technique for measuring the thermodynamic properties of protein – peptide complexes. *Biochemistry* 42 (17), 4962 – 4970.

Powell, K. D., Ghaemmaghami, S., Wang, M. Z., Ma, L., Oas, T. G., and Fitzgerald, M. C., 2002. A general mass spectrometry-based assay for the quantitation of protein – ligand binding interactions in solution. *J Am Chem Soc* 124 (35), 10256 – 10257.

Robinson, C. V., Chung, E. W., Kragelund, B. B., Knudsen, J., Aplin, R. T., Poulsen, F. M., and Dobson, C. M., 1996. Probing the nature of non-covalent interactions by mass spectrometry: a study of protein-CoA ligand binding and assembly. *J Am Chem Soc* 118, 8646 – 8653.

Scaloni, A., Miraglia, N., Orrù, S., Amodeo, P., Motta, A., Marino, G., and Pucci, P., 1998. Topology of the calmodulin-melittin complex. *J Mol Biol* 277 (4), 945 – 958.

Schermann, S. M., Simmons, D. A., and Konermann, L., 2005. Mass spectrometry-based approaches to protein ligand interactions. *Expert Rev Proteomics* 2 (4), 475 – 485.

Schiffman, A., Nelson, D. D., Jr., and Nesbitt, D. J., 1993. Quantum yields for hydroxyl production from 193 and 248 nm photolysis of nitric acid and hydrogen peroxide. *J Chem Phys* 98 (9), 6935 – 6946.

Scott, J. D. and Pawson, T., 2000. Cell communication: the inside story. *Sci Am* 282 (6), 72 – 79.

Sharp, J. S., Becker, J. M., and Hettich, R. L., 2003. Protein surface mapping by chemical oxidation: structural analysis by mass spectrometry. *Anal Biochem* 313 (2), 216 – 225.

Shcherbakova, I., Gupta, S., Chance, M. R., and Brenowitz, M., 2004. Monovalent ion-mediated folding of the tetrahymena thermophila ribozyme. *J Mol Biol* 342 (5), 1431 – 1442.

Slon-Usakiewicz, J. J., Ng, W., Dai, J.-R., Pasternak, A., and Redden, P. R., 2005. Frontal affinity chromatography with MS detection (FAC-MS) in drug discovery. *Drug Discovery Today* 10 (6), 409 – 416.

Smith, D. L. and Dharmasiri, K., 1998. Protein – ligand binding studied by amide hydrogen exchange and mass spectrometry. *NATO ASI Ser, Ser C.* 510, 45 – 58.

Sojo, L. E., Lum, G., and Chee, P., 2003. Internal standard signal suppression by co-eluting analyte in isotope dilution LC-ESI-MS. *Analyst* 128 (1), 51 – 54.

Sprang, S. R. and Coleman, D. E., 1998. Invasion of the nucleotide snatchers: structural insights into the mechanism of G protein GEFs. *Cell* 95 (2), 155 – 158.

Staub, O., Dho, S., Henry, P., Correa, J., Ishikawa, T., McGlade, J., and Rotin, D., 1996. WW domains of Nedd4 bind to the proline-rich PY motifs in the epithelial Na+ channel deleted in Liddle's syndrome. *EMBO J* 15 (10), 2371 – 2380.

Steiner, R. F., Albaugh, S., Fenselau, C., Murphy, C., and Vestling, M., 1991. A mass spectrometry method for mapping the interface topography of interacting proteins, illustrated by the melittin-calmodulin system. *Anal Biochem* 196 (1), 120 – 125.

Takamoto, K., Das, R., He, Q., Doniach, S., Brenowitz, M., Herschlag, D., and Chance, M. R., 2004. Principles of RNA compaction: insights from the equilibrium folding pathway of the P4 – P6 RNA domain in monovalent cations. *J Mol Biol* 343 (5), 1195 – 1206.

Urey, H. C., Dawsey, L. H., and Rice, F. O., 1929. Absorption spectrum and decomposition of hydrogen peroxide by light. *J Am Chem Soc* 51, 1371 – 1383.

van Mierlo, C. P. M. and Steensma, E., 2000. Protein folding and stability investigated by fluorescence, circular dichroism (CD), and nuclear magnetic resonance (NMR) spectroscopy: the flavodoxin story. *J Biotech* 79 (3), 281 – 298.

VanScyoc, W. S., Sorensen, B. R., Rusinova, E., Laws, W. R., Ross, J. B. A., and Shea, M. A., 2002. Calcium binding to calmodulin mutants monitored by domain-specific intrinsic phenylalanine and tyrosine fluorescence. *Biophys J* 83 (5), 2767 – 2780.

Vu, D. M., Myers, J. K., Oas, T. G., and Dyer, R. B., 2004. Probing the folding and unfolding dynamics of secondary and tertiary structures in a three-helix bundle protein. *Biochemistry* 43 (12), 3582 – 3589.

Wales, T. E. and Engen, J. R., 2006. Hydrogen exchange mass spectrometry for the analysis of protein dynamics. *Mass Spectrom Rev* 25 (1), 158 – 170.

Wang, F., Li, W., Emmett, M. R., Marshall, A. G., Corson, D., and Sykes, B. D., 1999. Fourier transform ion cyclotron resonance mass spectrometric detection of small Ca2+-induced conformational changes in the regulatory domain of human cardiac troponin C. *J Am Soc Mass Spectrom* 10 (8), 703 – 710.

Wang, W., Kitova, E. N., Sun, J., and Klassen, J. S., 2005. Blackbody infrared radiative dissociation of nonspecific protein – carbohydrate complexes produced by nanoelectrospray ionization: the nature of the noncovalent interactions. *J Am Soc Mass Spectrom* 16, 1583 – 1594.

Weinstein, H. and Mehler, E., 1994. Ca^{2+}-binding and structural dynamics in the function of calmodulin. *Annu Rev Physiol* 56, 213 – 236.

Wittinghofer, A. and Pai, E. F., 1991. The structure of Ras protein: a model for a universal molecular switch. *Trends Biochem Sci* 16 (10), 382 – 387.

Woods, V. L., Jr. and Hamuro, Y., 2001. High resolution, high-throughput amide deuterium exchange-mass spectrometry (DXMS) determination of protein binding site structure and dynamics: utility in pharmaceutical design. *Biochemistry* (Suppl. 37), 89 – 98.

Wu, Q., Bruce, J. E., Smith, R. D., Gao, J., Joseph-McCarthy, D., Sigal, G. B., and Whitesides, G. M., 1997. Carbonic anhydrase-inhibitor binding: from solution to the gas phase. *J Am Chem Soc* 119 (5), 1157 – 1158.

Xu, G. and Chance, M. R., 2007. Hydroxyl radical-mediated modification of proteins as probes for structural proteomics. *Chem Rev* 107 (8), 3514 – 3543.

Yao, Y. and Squier, T. C., 1996. Variable conformation and dynamics of calmodulin complexed with peptides derived from the autoinhibitory domains of target proteins. *Biochemistry* 35 (21), 6815 – 6827.

Yates, J. R., III, 1998. Mass spectrometry and the age of the proteome. *J Mass Spectrom* 33 (1), 1 – 19.

Zhang, J. and Matthews, C. R., 1998. Ligand binding is the principal determinant of stability for the p21H-Ras protein. *Biochemistry* 37 (42), 14881 – 14890.

Zhang, Z. and Smith, D. L., 1993. Determination of amide hydrogen exchange by mass spectrometry: a new tool for protein structure elucidation. *Protein Sci* 2 (4), 522 – 531.

Zhang, B., Zhang, Y., Wang, Z., and Zheng, Y., 2000. The role of Mg2+ cofactor in the guanine nucleotide exchange and GTP hydrolysis reactions of Rho family GTP-binding proteins. *J Biol Chem* 275 (33), 25299 – 25307.

Zhu, M. M., 2004. Determination of protein – ligand interactions using H/D exchange and mass spectrometry. (Dissertation). St. Louis (MO): Washington University.

Zhu, M. M., Rempel, D. L., Du, Z., and Gross, M. L., 2003. Quantification of protein – ligand interactions by mass spectrometry, titration, and H/D exchange: PLIMSTEX. *J Am Chem Soc* 125, 5252 – 5253.

Zhu, M. M., Chitta, R., and Gross, M. L., 2005. PLIMSTEX: a novel mass spectrometric method for the quantification of protein – ligand interactions in solution. *Int J Mass Spectrom* 240 (3), 213 – 220.

Zhu, M. M., Hambly, D., and Gross, M. L., 2007. Quantification of protein – ligand interactions in solution by hydrogen/deuterium exchange (PLIMSTEX). In: Wanner, K. and Höfner, G., editors. *Mass Spectrometry in Medicinal Chemistry: Applications in Drug Discovery. Methods and Principles in Medicinal Chemistry* . Wiley-VCH. pp. 341 – 376.

Note Added in Proof:

A paper published since the preparation of this chapter shows the application of PLIMSTEX at peptide resolution: Sperry, J.B., Shi, X., Rempel, D.L., Nishimura, Y., Akashi, S., Gross, M.L., 2008. A mass spectrometric approach to the study of DNA-binding proteins: Interaction of human TRF2 with telomeric DNA. *Biochem.* 47, 1797–1807.

Actin
    actin/cofilin binary complex, 191–210
    and ClusPro server, 192–198
    and cross-linking, 170–183
    F-, G-actin, 169
    free form oxidation rates, 198
    and synchrotron footprinting, 58, 170

Back-exchange, 3, 27–28, 126
Bottom-up proteomics, 158

Computational modeling, 157–168, 189–212
    CAPRI, 191
    ClusPro, 192, 210
        scheme, 193
    Rosetta structure prediction, 62
    summary of method, 208
Conformational dynamics
    and electrospray ionization mass spectrometry
        (ESI MS), 220–237
    and hydrogen/deuterium exchange, 25, 34–35,
        84–86, 129
Cross-linking
    and actin, 170–174
        with binding proteins, 176–183
    and computational modeling, 157
    and mass spectrometry
        advancements, 160–162
        analytical software, 165–167
        challenges, 158–160
        modeling, 165
        novel reagents, 162–165
    methodology, 158
    and radiolysis, 210
    strategies, 157
    and synchrotron footprinting, 61

Electron microscopy (EM)
    and actin binding proteins, 176, 178
    obstacles, 170

structure determination of protein
    complexes, 190
and viral capsids, 105–121
Electrospray fourier transform ion cyclotron
    resonance mass spectrometry
    (ES-FT-ICR-MS), 56–57, 160
Electrospray ionization mass
    spectrometry (ESI MS)
    dose-response curve, 54
    and hydrogen/deuterium exchange reactions
        advantages, 229
    and PLIMSTEX, 246, 264
    radiolysis, 57–58
    and radiolytic footprinting of actin, 198

Fenton reagent
    drawbacks, 56
    generation of hydroxyl radicals, 256
    and hydroxyl radical footprinting, 56
Footprinting
    quantitative, 52
        and solvent accessibility, 52
    synchrotron
        of α_1-antitrypsin, 76
        of actin/cofilin binary complex, 58, 170
        of Arp2/3 and WASp complex, 58
        and cross-linking, 61
        of cytochrome C, 55
        dose-response curve, 78
        generation of hydroxyl radicals, 58
        and mass spectrometry, 191
        radiolysis, 176, 198, 210
        scheme, 6, 71
        single hit approach, 257
Hydroxyl radical, 46–51, 258
    and actin/cofilin binary complex, 179
    chemistry of, 46
    generation of hydroxyl radicals, 47
        with Fenton reagent, 56, 256
        radiolysis, 45–46

Mass Spectrometry Analysis for Protein–Protein Interactions and Dynamics, Edited by Mark Chance
Copyright © 2008 John Wiley & Sons, Inc.

Footprinting (*continued*)
  versus hydrogen/deuterium exchange, 73–76,
    85–87
  scheme, 4, 70
  summary of method, 62–63
Fast photochemical oxidation of proteins (FPOP),
  256–264

High-performance liquid chromatography (HPLC)
  and electrospray ionization mass spectrometry
    (ESI MS), 210
  and mass spectrometry, 114
  reverse-phase fractionation, 171–182
  separation, 27, 128
Homology modeling, 189–192, 205, 211
Hydrogen/deuterium exchange
  and α_1-antitrypsin, 76
  and epitope mapping, 146
  and human growth hormone, 129–133
  versus hydroxyl radical footprinting, 73–76,
    85–87
  influence of gas phase ion chemistry, 231–234
  and mass spectrometry, 3, 4, 229–231
  and on-exchange, 98
  and PKA, 133
  versus radiolysis, 52–56
  theory, 124–126
  and viruses, 105–121
Hydroxyl radical(s), *see also* Footprinting,
    Hydroxyl radical
  and fast photochemical oxidation of proteins
    (FPOP), 256, 263
  in footprinting experiments, 80
  generation of, 55–58, 244
  and hydrogen/deuterium exchange, 124
Hydroxyl radical footprinting, *see* Footprinting,
    Hydroxyl radical

Liquid-chromatography mass spectrometry (LC-MS)
  and quantification of peptide oxidation, 52, 78
  and side-chain reactivity, 72
  and tandem mass spectrometry (MS/MS), 56–57

Matrix-assisted laser desorption/ionization-time
    of flight mass spectrometry
    (MALDI-TOF-MS), 161
  and cross-linking, 176, 180–182
  versus electrospray ionization mass
    spectrometry, 220, 245, 264
  to monitor hydrogen/deuterium exchange
    reactions, 229
  and PKA, 92–93

Off-exchange, 147
On-exchange, 93, 98, 126–128, 147

Protease(s)
  and hydrogen/deuterium exchange, 74–75,
    124, 127
  and LC/MS, 52
  in mass spectrometry, 210, 226
  proteolysis before mass spectrometry, 28–29
  and serpins, 87
  and top-down proteomics, 158, 229
  viral, 109–112
Protein ligand interaction using mass spectrometry
    titration and H/D exchange
    (PLIMSTEX), 243–265
  to determine binding stoichiometry, 246–252
  and ESI MS, 245, 264

Quantitative footprinting, *see* Footprinting,
    Quantitative

Radiolytic footprinting, *see* Footprinting,
    Synchrotron

Solvent accessibility
  and amide hydrogen exchange, 14, 18–19
  of Arp2/3 and WASp complex, 61
  and hydrogen/deuterium exchange, 3,
    35, 94
  and hydroxyl radical footprinting, 45–46, 51,
    72, 79, 83, 85
  introduction, 69–71
  and PLIMSTEX, 244, 248, 250, 259
Structural biology, 218
  challenges of, 2
  and hydrogen/deuterium exchange mass
    spectrometry, 31
Structural genomics
  computational modeling techniques,
    190–192
  and interactome, 189
Synchrotron footprinting, *see* Footprinting,
    Synchrotron

Tandem mass spectrometry (MS/MS), 54, 75–76,
    210, 231
Thermodynamics
  and hydrogen/deuterium exchange mass
    spectrometry, 124–126, 130
Top-down proteomics, 158, 229

X-ray crystallography
  and FPOP, 263
  and hydrogen/deuterium exchange mass
    spectrometry, 147
  obstacles, 33, 34 146, 191
  and protein-protein interactions, 123
  and viral capsids, 105–121

Printed and bound by CPI Group (UK) Ltd, Croydon, CR0 4YY